Phased Array Antennas

WILEY SERIES IN MICROWAVE AND OPTICAL ENGINEERING

KAI CHANG, Editor
Texas A&M University

A complete list of the titles in this series appears at the end of this volume.

Phased Array Antennas

R. C. HANSEN

A WILEY-INTERSCIENCE PUBLICATION
JOHN WILEY & SONS, INC.
NEW YORK/CHICHESTER/WEINHEIM/BRISBANE/SINGAPORE/TORONTO

Cover illustration: Canadian Radarsat SAR Antenna under test. Courtesy of Canadian Space Agency and Spar Aerospace.

This book is printed on acid-free paper. ∞

Copyright © 1998 by John Wiley & Sons, Inc. All rights reserved.

Published simultaneously in Canada.

No part of this publication may be reproduced, stored in a retrieval system or transmitted in any form or by any means, electronic, mechanical, photocopying, recording, scanning or otherwise, except as permitted under Sections 107 or 108 of the 1976 United States Copyright Act, without either the prior written permission of the Publisher, or authorization through payment of the appropriate per-copy fee to the Copyright Clearance Center, 222 Rosewood Drive, Danvers, MA 01923, (508) 750-8400, fax (508) 750-4744. Requests to the Publisher for permission should be addressed to the Permissions Department, John Wiley & Sons, Inc., 605 Third Avenue, New York, NY 10158-0012, (212) 850-6011, fax (212) 850-6008, E-Mail: PERMREQ @ WILEY.COM.

Library of Congress Cataloging-in-Publication Data:
Hansen, Robert C.
    Phased array antennas / R.C. Hansen.
    p. cm. — (Wiley series in microwave and optical engineering)
    "A Wiley–Interscience publication."
    Includes bibliographical references and index.
    ISBN 0-471-53076-X (alk. paper)
    1. Microwave antennas. 2. Phased array antennas. I. Title.
II. Series.
TK7871.67.M53H36 1997
621.382′4—dc21   97-23708
                                                                                                           CIP

Printed in the United States of America.

10 9 8 7 6 5 4 3 2 1

This book is dedicated to those who made Microwave Scanning Antennas possible:

Nicolas A. Begovich
Robert W. Bickmore
Jesse L. Butler
Lorne K. De Size
Robert S. Elliott
Richard C. Johnson
H. C. Ko
Wolfgang H. Kummer
Robert G. Malech
Donald L. Margerum
Arthur A. Oliner
Jack F. Ramsay
Joseph A. Vitale

Contents

Preface		xv
1	**Introduction**	**1**
1.1	Array Background	1
1.2	Systems Factors	2
1.3	Annotated Reference Sources	4
2	**Basic Array Characteristics**	**7**
2.1	Uniformly Excited Linear Arrays	7
	2.1.1 Patterns	7
	2.1.2 Beamwidth	9
	2.1.3 Sidelobes	11
	2.1.4 Grating Lobes	11
	2.1.5 Bandwidth	15
2.2	Planar Arrays	17
	2.2.1 Array Coordinates	17
	2.2.2 Beamwidth	18
	2.2.3 Grating Lobes: Rectangular Lattice	20
	2.2.4 Grating Lobes: Hexagonal Lattice	23
2.3	Beam Steering and Quantization Lobes	25
	2.3.1 Steering Increment	25
	2.3.2 Steering Bandwidth	26
	2.3.3 Phaser Quantization Lobes	26
	2.3.4 Subarray Quantization Lobes	31
	2.3.5 QL Decollimation; Overlapped Subarrays	33
2.4	Directivity	34
	2.4.1 Linear Array Directivity	34
	2.4.2 Directivity of Arrays of Short Dipoles	38
	2.4.3 Directivity of Arrays of Resonant Elements	39
	2.4.4 Planar Array Directivity	41
	References	45

3 Linear Array Pattern Synthesis   47

- 3.1 Introduction   47
  - 3.1.1 Pattern Formulations   47
  - 3.1.2 Physics Versus Mathematics   49
  - 3.1.3 Taylor Narrow-Beam Design Principles   50
- 3.2 Dolph–Chebyshev Arrays   51
  - 3.2.1 Half-Wave Spacing   51
  - 3.2.2 Spacing Less Than Half-Wave   57
- 3.3 Taylor One-Parameter Distribution   59
  - 3.3.1 One-Parameter Design   59
  - 3.3.2 Bickmore–Spellmire Two-Parameter Distribution   63
- 3.4 Taylor N-Bar Aperture Distribution   64
- 3.5 Low-Sidelobe Distributions   71
  - 3.5.1 Comparison of Distributions   71
  - 3.5.2 Average Sidelobe Level   73
- 3.6 Villeneuve N-Bar Array Distribution   74
- 3.7 Difference Patterns   77
  - 3.7.1 Canonical Patterns   77
  - 3.7.2 Bayliss Patterns   78
  - 3.7.3 Sum and Difference Optimization   83
  - 3.7.4 Discrete Zolotarev Distributions   84
- 3.8 Sidelobe Envelope Shaping   85
- 3.9 Shaped Beam Synthesis   90
  - 3.9.1 Woodward–Lawson Synthesis   90
  - 3.9.2 Elliott Synthesis   93
- 3.10 Thinned Arrays   96
  - 3.10.1 Probabilistic Design   96
  - 3.10.2 Space Tapering   101
  - 3.10.3 Minimum Redundancy Arrays   102
  - References   102

4 Planar and Circular Array Pattern Synthesis   106

- 4.1 Circular Planar Arrays   106
  - 4.1.1 Flat Plane Slot Arrays   106
  - 4.1.2 Hansen One-Parameter Pattern   107
  - 4.1.3 Taylor Circular $\bar{n}$ Pattern   112
  - 4.1.4 Circular Bayliss Difference Pattern   116
  - 4.1.5 Difference Pattern Optimization   121
- 4.2 Noncircular Apertures   121
  - 4.2.1 Two-Dimensional Optimization   121
  - 4.2.2 Ring Sidelobe Synthesis   123
  - References   125

5 Array Elements — 127

- 5.1 Dipoles — 127
  - 5.1.1 Thin Dipoles — 127
  - 5.1.2 Bow-Tie and Open Sleeve Dipoles — 133
- 5.2 Waveguide Slots — 137
  - 5.2.1 Broad Wall Longitudinal Slots — 138
  - 5.2.2 Edge Slots — 143
  - 5.2.3 Stripline Slots — 145
  - 5.2.4 Open End Waveguides — 145
- 5.3 TEM Horns — 146
  - 5.3.1 Development of TEM Horns — 146
  - 5.3.2 Analysis and Design of Horns — 148
  - 5.3.3 TEM Horn Arrays — 149
  - 5.3.4 Millimeter Wave Antennas — 150
- 5.4 Microstrip Patches and Dipoles — 150
  - 5.4.1 Transmission Line Model — 153
  - 5.4.2 Cavity and Other Models — 155
  - 5.4.3 Parasitic Patch Antennas — 156
  - References — 158

6 Array Feeds — 164

- 6.1 Series Feeds — 164
  - 6.1.1 Resonant Arrays — 164
    - 6.1.1.1 *Impedance and Bandwidth* — 164
    - 6.1.1.2 *Resonant Slot Array Design* — 168
  - 6.1.2 Travelling Wave Arrays — 170
    - 6.1.2.1 *Frequency Squint and Single Beam Condition* — 171
    - 6.1.2.2 *Calculation of Element Conductance* — 175
    - 6.1.2.3 *TW Slot Array Design* — 177
  - 6.1.3 Frequency Scanning — 181
  - 6.1.4 Phaser Scanning — 184
- 6.2 Shunt (Parallel) Feeds — 188
  - 6.2.1 Corporate Feeds — 188
  - 6.2.2 Distributed Arrays — 189
- 6.3 Two-Dimensional Feeds — 191
  - 6.3.1 Fixed Beam Arrays — 191
  - 6.3.2 Sequential Excitation Arrays — 193
  - 6.3.3 Electronic Scan in One Plane — 193
  - 6.3.4 Electronic Scan in Two Planes — 195
- 6.4 Photonic Feed Systems — 200
  - 6.4.1 Fiber Optic Delay Feeds — 202
    - 6.4.1.1 *Binary Delay Lines* — 202
    - 6.4.1.2 *Acousto-Optical Switched Delay* — 204

x CONTENTS

6.4.1.3 Modulators and Photodetectors	204
6.4.2 Wavelength Division Fiber Delay	205
6.4.2.1 Dispersive Fiber Delay	205
6.4.2.2 Bragg Fiber Grating Delay	206
6.4.2.3 Travelling Wave Fiber Delay	207
6.4.3 Optical Delay	207
6.4.4 Optical Fourier Transform	207
6.5 Systematic Errors	208
6.5.1 Parallel Phasers	208
6.5.2 Series Phasers	210
6.5.3 Systematic Error Compensation	210
References	211

7 Mutual Coupling   215

7.1 Introduction	215
7.2 Fundamentals of Scanning Arrays	215
7.2.1 Current Sheet Model	215
7.2.2 Free and Forced Excitations	217
7.2.3 Scan Impedance and Scan Element Pattern	219
7.2.4 Minimum Scattering Antennas	222
7.3 Spatial Domain Approaches to Mutual Coupling	224
7.3.1 Canonical Couplings	224
7.3.1.1 Dipole and Slot Mutual Impedance	224
7.3.1.2 Microstrip Patch Mutual Impedance	228
7.3.1.3 Horn Mutual Impedance	229
7.3.2 Impedance Matrix Solution	231
7.3.3 The Grating Lobe Series	232
7.4 Spectral Domain Approaches	235
7.4.1 Dipoles and Slots	235
7.4.2 Microstrip Patches	246
7.4.3 Printed Dipoles	250
7.4.4 Printed TEM Horns	252
7.4.5 Unit Cell Simulators	254
7.5 Scan Compensation and Blind Angles	254
7.5.1 Blind Angles	254
7.5.2 Scan Compensation	257
7.5.2.1 Coupling Reduction	257
7.5.2.2 Compensation Feed Networks	258
7.5.2.3 Multimode Elements	261
7.5.2.4 External Wave Filter	264
References	265

8 Finite Arrays   273

8.1 Methods of Analysis	273

	8.1.1 Overview	273
	8.1.2 Finite-by-Infinite Arrays	276
8.2	Scan Performance of Small Arrays	281
8.3	Finite-by-Infinite Array Gibbsian Model	287
	8.3.1 Salient Scan Impedance Characteristics	287
	8.3.2 A Gibbsian Model for Finite Arrays	297
	References	301

9 Superdirective Arrays 304

9.1	Historical Notes	304
9.2	Maximum Array Directivity	305
	9.2.1 Broadside Directivity for Fixed Spacing	305
	9.2.2 Directivity as Spacing Approaches Zero	307
	9.2.3 Endfire Directivity	308
	9.2.4 Bandwidth, Efficiency, and Tolerances	309
9.3	Constrained Optimization	317
	9.3.1 Dolph–Chebyshev Superdirectivity	317
	9.3.2 Constraint on Q or Tolerances	323
9.4	Matching of Superdirective Arrays	325
	9.4.1 Network Loss Magnification	326
	9.4.2 HTS Arrays	326
	References	327

10 Multiple-Beam Antennas 330

10.1	Introduction	330
10.2	Beamformers	330
	10.2.1 Networks	331
	10.2.1.1 Power Divider BFN	331
	10.2.1.2 Butler Matrix	331
	10.2.1.3 Blass and Nolen Matrices	335
	10.2.1.4 2-D BFN	337
	10.2.1.5 McFarland 2-D Matrix	337
	10.2.2 Lenses	341
	10.2.2.1 Rotman Lens BFN	341
	10.2.2.2 Bootlace Lenses	356
	10.2.2.3 Dome Lenses	361
	10.2.2.4 Other Lenses	362
	10.2.3 Digital Beamforming	364
10.3	Low Sidelobes and Beam Interpolation	365
	10.3.1 Low-Sidelobe Techniques	365
	10.3.1.1 Interlaced Beams	365
	10.3.1.2 Resistive Tapering	366

xii CONTENTS

10.3.1.3 Lower Sidelobes via Lossy Networks	366
10.3.1.4 Beam Superposition	369
10.3.2 Beam Interpolation Circuits	370
10.4 Beam Orthogonality	372
10.4.1 Orthogonal Beams	372
10.4.1.1 Meaning of Orthogonality	372
10.4.1.2 Orthogonality of Distributions	373
10.4.1.3 Orthogonality of Arrays	375
10.4.2 Effects of Nonorthogonality	376
10.4.2.1 Efficiency Loss	376
10.4.2.2 Sidelobe Changes	377
References	380
11 Conformal Arrays	**384**
11.1 Scope	384
11.2 Ring Arrays	385
11.2.1 Continuous Ring Antenna	385
11.2.2 Discrete Ring Array	389
11.2.3 Beam Cophasal Excitation	393
11.3 Arrays on Cylinders	395
11.3.1 Slot Patterns	397
11.3.2 Array Pattern	397
11.3.2.1 Grating Lobes	401
11.3.2.2 Principal Sidelobes	406
11.3.2.3 Cylindrical Depolarization	407
11.3.3 Slot Mutual Admittance	411
11.3.3.1 Modal Series	413
11.3.3.2 Admittance Data	417
11.3.4 Scan Element Pattern	420
11.4 Sector Arrays on Cylinders	422
11.4.1 Patterns and Directivity	422
11.4.2 Comparison of Planar and Sector Arrays	424
11.4.3 Ring and Cylindrical Array Hardware	426
11.5 Arrays on Cones and Spheres	429
11.5.1 Conical Arrays	430
11.5.1.1 Lattices on a Cone	432
11.5.1.2 Conical Depolarization and Coordinate Systems	435
11.5.1.3 Projective Synthesis	443
11.5.1.4 Patterns and Mutual Coupling	443
11.5.1.5 Conical Array Experiments	443
11.5.2 Spherical Arrays	445
References	446

12 Measurements and Tolerances   **453**

12.1 Measurement of Low-Sidelobe Patterns   453
12.2 Array Diagnostics   455
12.3 Waveguide Simulators   458
12.4 Array Tolerances   464
    12.4.1 Directivity Reduction and Average Sidelobe Level   465
    12.4.2 Beam Pointing Error   466
    12.4.3 Peak Sidelobes   468
    References   470

Author Index   **473**

Subject Index   **482**

Preface

Although array antennas have many decades of history, the last two decades have experienced a maturation, both in the understanding and design of arrays, and in the use of large sophisticated arrays. Radars utilizing electronic scanning arrays are in common use, from airport surveillance to missile detection and tracking; names of U.S. military systems such as Aegis, Patriot, and Pave Paws are well known. This book is a comprehensive treatment of all aspects of phased arrays; much has changed since the only other such work, *Microwave Scanning Antennas*, appeared in 1966. Most noteworthy has been the parallel development of inexpensive computer power and the theoretical understanding of nearly all aspects of phased array design. Design algorithms suitable for computers are emphasized here, with numerical tips and short algorithms sprinkled throughout the chapters. The work is prepared from the dual viewpoint of a design engineer and an antenna array analyst.

Following an introductory chapter is Chapter 2 on basic array characteristics, covering grating lobes, quantization lobes, bandwidth, and directivity. Highly efficient linear aperture and array synthesis techniques are covered next, including sum and difference patterns. Chapter 4 treats synthesis of planar arrays. Array elements are covered next, including not only the classic dipoles and slots, but TEM horns and patches. In Chapter 6, feeds for linear and planar arrays, both fixed beam and scanning, are examined; photonic time delay and feeders are included. Array performance is strongly affected by mutual impedance; Chapter 7 investigates ways of calculating this for various arrays elements, including an extensive treatment of ways of calculating array performance with mutual effects included. Among these are unit cell, spectral moment method, finite impedance matrix, and scattering techniques. Finite arrays are examined in Chapter 8, including the recently developed Gibbsian models. Next is an extensive view of superdirective arrays; the implications of high-temperature superconductors for antennas is an important feature. Multiple-beam arrays, as opposed to multiple-beam reflector feeds, are treated in Chapter 10. Included are one-dimensional and two-dimensional Butler and Rotman lenses, and the practical meaning of beam orthogonality. Conformal arrays, ranging from ring arrays to arrays on cones, are covered next; much previously unpublished material is included in this chapter. Finally, Chapter 12 discusses array diagnostics, waveguide simulators in depth, and array tolerances. Extensive references to the archival literature are used in each chapter to offer additional sources of data.

ROBERT C. HANSEN
Tarzana, CA

CHAPTER ONE

Introduction

1.1 ARRAY BACKGROUND

Discovery of the first works on array antennas in a task best left to historians, but the two decades before 1940 contained much activity on array theory and experimentation. Some of the researchers were G. H. Brown, E. Bruce, P. S. Carter, C. W. Hansell, A. W. Ladner, N. E. Lindenblad, A. A. Pistolkors, S. A. Schelkunoff, G. C. Southworth, E. J. Sterba, and T. Walmsley. Primary journals were *Proc. IRE, Proc. IEE, BSTJ, RCA Review*, and *Marconi Review*. During World War II, much array work was performed in the United States and Britain. Interest in arrays returned in the early 1960s, with research projects at Lincoln Laboratories, General Electric, RCA, Hughes and others. Some of the array conferences are mentioned in the annotated reference list in Section 1.3.

A salient event was the publication by Academic Press of the three-volume book *Microwave Scanning Antennas* (*MSA*), with volume 1 appearing in 1964, and volumes 2 and 3 in 1966. This work was the first extensive coverage of phased arrays, with emphasis on mutual coupling theory, which is the basis of all array characteristics. After 30 years, *MSA* is still in print, through Peninsula Publishing.

It is the purpose of this book to present a thorough and extensive treatment of phased arrays, adding to and updating the array portions of *MSA*. The scope of the book is all types of arrays except adaptive, for which several excellent books exist; see references at the end of the chapter. Multiple-beam arrays are included. Because most arrays operate at frequencies that allow spacing above ground to be sufficiently large to preclude ground affecting the array internal parameters, all arrays are presumed to be in free space. Active arrays, that is, those containing active devices, are not treated, nor are array-related circuit components, except for phasers, which are discussed briefly. It is also assumed that all array elements are identical, although the impedance matching may vary with the element position. A semantic difficulty arises with the phrase "phased array". For some people this implies beam steering or scanning. But for others all arrays are phased; fixed beam broadside arrays

2 INTRODUCTION

are also phased. There are more important questions of terminology; these are addressed next.

1.2 SYSTEMS FACTORS

Important array factors for the systems designer are broadside pattern, gain versus angles, element input impedance, and efficiency. For all regular arrays, the pattern is given by the product of the element pattern and the pattern of the isotropic array, where the array elements are replaced by isotropes. However, the element excitations must be those of the real array; as discussed later, these are found by solving equations associated with a self-impedance and mutual-impedance or admittance matrix. In general, each element of an array will have a different input impedance. For a fixed beam array these are called "embedded impedances"; the obsolete and misleading term "active impedance" is deprecated. A scanning array not only has different element impedances, but each of them varies with scan angle. These element input impedances are called *scan impedances*.

The pattern of array gain versus angles is called *scan element pattern*; this term replaces *active element pattern*. The *scan element pattern* (SEP) is an extremely useful design factor. The element pattern and mutual coupling effects are subsumed into the *scan element pattern*; the overall radiated pattern is the product of the *scan element pattern* and the pattern of an isotropic array of elements scanned to the proper angle. The isotropic array factor incorporates the effects of array size and array lattice, while the *scan element pattern*, as mentioned, incorporates element pattern, backscreen if used, and mutual coupling. Since the *scan element pattern* is an envelope of array gain versus scan angles, it tells the communications system or radar designer exactly how the array performs with scan, whether blind angles exist, and whether matching at a particular scan angle is advantageous. *Scan element pattern* is used for antenna gain in the conventional range equations. For an infinite array the SEP is the same for all elements, but for a finite array each element sees a different environment, so that the SEP is an overall array factor. Use of infinite array *scan element patterns* allows array performance to be separated into this SEP and edge effects. Formulas for both finite array and infinite array *scan element pattern* are derived later; edge effects are also discussed later.

A similar parameter, appropriate for backscattering from antenna arrays, is the *scattering scan element pattern* (SSEP). This parameter gives the backscattered field intensity from an array element, when the array is excited by an incident plane wave. This then is different from the SEP, which relates radiated field intensity to total radiated power. The radar cross section (RCS) relates re-radiated field intensity to incident field intensity, with a $4\pi R^2$ factor. SSEP is this ratio of re-radiated to incident intensity; a convenient normalization is to the broadside value. Just as in the case of a radiating array, the scattering array finite size and edge effects have been separated, so that the SSEP relates the effects of element design and array lattice. It can then be used to make design trades for type of element and lattice; the features due to the array size are included simply by multiplying by the isotropic array factor. SSEP is, of course, related to the RCS pattern. It can be considered as the RCS pattern of one unit cell of the array.

System factors also arise in arrays used for wideband baseband (no carrier) applications. The one-way (communications) range equation, written without explicit wavelength dependence, is

$$P_r = \frac{P_t G A_e}{4\pi R^2} \qquad (1.1)$$

where as usual P_r and P_t are received and transmitted powers, R is the range, and G and A_e are the gain of one antenna and the effective area of the other antenna. Both gain and effective area include an impedance mismatch factor:[1] $(1 - |\Gamma|^2)$. It is assumed that P_t is fixed, independent of frequency. If the GA_e product is relatively constant over the frequency band of interest, then the signal is transferred without significant dispersion, providing that the antenna and matching unit phase are well-behaved also (Hansen and Libelo, 1995). Otherwise significant dispersion can occur.

From a casual look at array antennas one might assume a planar array to be a constant effective area antenna. However, for a regularly spaced array of low-gain elements, as the frequency increases from nominal half-wave spacing, the gain increases until the first grating lobe appears, with the gain then dropping back to the original level. Further increases in frequency produce additional rises in gain followed by drops as grating lobes appear. The net result is that over a wide bandwidth the gain of an array is at best roughly constant and equal to the half-wave spaced value (Hansen, 1972). This does not include effects of embedded element impedance mismatch with frequency, a phenomenon which further greatly reduces gain. Thus the regularly spaced array is not a candidate for compensation of dispersion. An array with pseudorandom spacing does not experience the appearance of regular grating lobes as frequency is increased. The fraction of power in the sidelobes is roughly constant in a well-designed nonuniformly spaced array, and thus the gain is roughly constant with frequency. Of more importance, however, is the fact that very large numbers of elements are needed to achieve even moderately low sidelobe levels. Thus these types of arrays are not suitable for dispersion compensation either. Arrays of higher-gain elements experience, in addition, the dispersion introduced by the elements themselves and are even less suitable.

---

[1] Note that "effective length", which is defined as open circuit voltage divided by incident electric field, does not include impedance mismatch, and is therefore useless by itself.

1.3 ANNOTATED REFERENCE SOURCES

Many textbooks discuss arrays, but the books and digests listed here provide in-depth resources on phased arrays.

Microwave Scanning Antennas, R. C. Hansen, Ed., 3 vols., Academic Press, 1964, 1966 [Peninsula Publishing, 1985, 442 pp., 400 pp., 422 pp. (Peninsula combined volumes)].

This, the first extensive work on phased arrays, is still quite useful. Volume 1 has a chapter on aperture distributions. Volume 2 includes array theory, and infinite and finite array analysis; probably the first development of the spectral domain analysis technique for arrays. Feeds, frequency scanning, and multiple beams are covered in vol. 3; multiple beams by Butler of matrix fame.

Proceedings of the 1964 RADC Symposium on Electronically Scanned Array Techniques and Applications, report RADC-TDR-64-225, AD-448 481.

Contained here are early papers on phase quantization errors, ferrite and semiconductor phasers, and beam forming matrices.

The Theory and Design of Circular Antenna Arrays, James D. Tillman, University of Tennessee Engineering Experiment Station, 1966, 235 pp.

This treatise on ring arrays and concentric ring arrays applies sequence theory of azimuthal modes, called symmetrical components in electric power work, to the analysis of impedance and pattern. Array scanning is also discussed.

Proceedings of the 1970 NELC Conformal Array Conference, TD-95, Naval Electronics Lab. Center, AD-875 378.

Both ring arrays and cylindrical arrays are treated in papers, both theoretically and for applications.

Phased Array Antennas, A. A. Oliner and G. H. Knittel, Artech, 1972, 381 pp.

This book is a record of the 1970 Phased Array Antenna Symposium held at Polytechnic Institute of Brooklyn. Included are many papers on impedance calculations, blind angles, etc., and also on practical aspects such as scan compensation and feeding and phasing.

Theory and Analysis of Phased Array Antennas, N. Amitay, V. Galindo, and C. P. Wu, Wiley-Interscience, 1972, 443 pp.

Arrays of waveguide radiators are the subject here. The spectral domain method is used extensively. Small finite arrays are solved via equations over the modes and elements. This work is one of the first using multimode spectral analysis.

Proceedings of the 1972 NELC Array Antenna Conference, TD-155, 2 Parts, Naval Electronics Lab. Center, AD-744 629, AD-744 630.

Many papers cover array techniques and components; adaptive arrays, and conformal arrays.

Theory and Application of Antenna Arrays, M. T. Ma, Wiley-Interscience, 1974, 413 pp.

Transform analysis and synthesis of fixed beam arrays is covered, along with many general array examples. The effect of ground on arrays represents a significant part of this book.

Conformal Antenna Array Design Handbook, R. C. Hansen, Ed., Naval Air Systems Command, 1982, AD-A110 091.

This report summarizes a decade of Navair-supported work on cylindrical and conical slot arrays, including mutual impedance algorithms.

Antenna Theory and Design, R. S. Elliott, Prentice-Hall, 1981, 594 pp.

This text is an excellent source for waveguide slot array analysis and synthesis. Sidelobe envelope shaping is treated in detail.

The Handbook of Antenna Design, A. W. Rudge, K. Milne, A. D. Olver, and P. Knight, Eds., IEE/Peregrinus, 1983, vol. 2, 945 pp.

This handbook contains chapters on linear arrays, planar arrays, conformal arrays, ring arrays, and array signal processing. Extensive data are included on array analysis and synthesis, including mutual coupling effects.

Proceedings of the 1985 RADC Phased Array Symposium, H. Steyskal, Ed., report RADC-TR-85-171, AD-A169 316.

This symposium record contains papers on microstrip arrays, adaptive arrays, and scan impedance, among others. A second volume has restricted distribution.

Antenna Handbook, Y. T. Lo and S. W. Lee, Van Nostrand Reinhold, 1988.

This handbook contains chapters on array theory, slot arrays, periodic and aperiodic arrays, practical aspects, and multiple-beam arrays.

Antenna Engineering Handbook, R. C. Johnson and H. Jasik, McGraw-Hill, 1993.

This updated edition of an old classic contains chapters on array theory, slot arrays, frequency scan and phased arrays, and conformal arrays.

Phased Array Antenna Handbook, R. J. Mailloux, Artech, 1994. 534 pp.

This specialized handbook covers most array topics, with emphasis on analysis and synthesis. A chapter covers limited scan arrays and time delayed arrays.

Phased Array-Based Systems and Applications, N. Fourikis, Wiley-Interscience, 1997.

This book emphasizes systems aspects of arrays.

Adaptive Antenna Reference Books

Compton, R. T., Jr., *Adaptive Antennas*, Prentice-Hall, 1988.
Hudson, J. E., *Adaptive Array Principles*, IEE/Peregrinus, 1981.
Monzingo, R. A. and Miller, T. W., *Introduction to Adaptive Arrays*, Wiley, 1980.
Widrow, B. and Stearns, S. D., *Adaptive Signal Processing*, Prentice-Hall, 1985.

REFERENCES

Hansen, R. C., "Comparison of Square Array Directivity Formulas," *Trans. IEEE*, Vol. AP-20, Jan. 1972, pp. 100–102.

Hansen, R. C. and Libelo, L. F., "Wideband Dispersion in Baseband Systems," *Trans. IEEE*, Vol. AES-31, July 1995, pp. 881–890.

CHAPTER TWO

Basic Array Characteristics

This chapter is concerned with basic characteristics of linear and planar arrays, primarily with uniform excitation. The theory of, and procedures for, the design of array distributions to produce narrow-beam, low-sidelobe patterns, or shaped beams, are covered in detail in Chapter 3. Impedance effects due to mutual coupling are treated in Chapter 7. Covered here are such parameters as pattern, beamwidth, bandwidth, sidelobes, grating lobes, quantization lobes, and directivity.

2.1 UNIFORMLY EXCITED LINEAR ARRAYS

2.1.1 Patterns

In general, the excitation of an array consists of an amplitude and a phase at each element. This discrete distribution is often called an aperture distribution, where the discrete array is the aperture. The far-field radiation pattern is just the discreet Fourier transform of the array excitation. The array pattern is the product of the isolated element pattern and the isotropic array factor; this is the "forced excitation" problem. To achieve this, the element drives are individually adjusted so that the excitation of each element is exactly as desired. More common is the "free excitation" situation, where the element drives are all fixed, and the element excitations are those allowed by the *scan impedance*. The latter is discussed in detail in Chapter 7. Here the concern will be only with the forced excitation array, where the excitations are constant in amplitude, but may have a scan phase.

A common notation in the antenna literature is used here, where λ is wavelength, d is element spacing, $k = 2\pi/\lambda$, and the angular variable is u. The latter is $u = (\sin\theta - \sin\theta_0)$ where θ_0 is the scan angle. Uniform (equal spacing) is assumed in this chapter; unequally spaced arrays are discussed in Chapter 3. Although it is simpler to have a coordinate system axis in the center of a linear array, complications ensue for even and odd numbers of elements. A more

8 BASIC ARRAY CHARACTERISTICS

general case starts the coordinate system at one end of the array, as shown in Fig. 2.1. The pattern, sometimes called a space factor, is

$$F(u) = \sum A_n \exp[jkd(n-1)u]. \tag{2.1}$$

A_n is the complex excitation, which for much of this section will be assumed constant.

For uniform excitation, the array pattern becomes a simple result, where the exponential in Eqn. (2.2) can be discarded, leaving a real pattern:

$$F(u) = \exp[j\pi(N-1)u] \frac{\sin \tfrac{1}{2} Nkdu}{N \sin \tfrac{1}{2} kdu} \tag{2.2}$$

This interelement phase shift is $kd \sin \theta_0$. By varying this phase shift, the beam position can be scanned. Figure 2.2 shows patterns produced by the various spacings and phases (Southworth, 1930).

Many linear arrays are designed to produce a narrow beam. Figure 2.3 depicts how the beam changes with scan. With no scan the narrow beam is omnidirectional around the array axis. As the beam is scanned this "disk" beam forms into a conical beam as shown in the center sketch. When the 3 dB point gets to 90 deg, a singular situation occurs. Beyond this scan angle the beam has two peaks, and the "beamwidth" will double as the outside 3 dB points are used. Finally, at endfire a pencil beam results; thus a linear array at broadside yields directivity in one dimension while at endfire it yields directivity in two dimensions. It might be expected as a result that the endfire beamwidth is broader; this will be shown next.

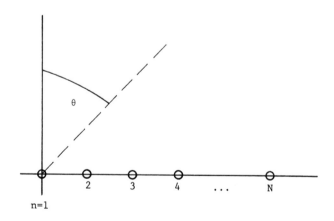

Figure 2.1 Linear array geometry.

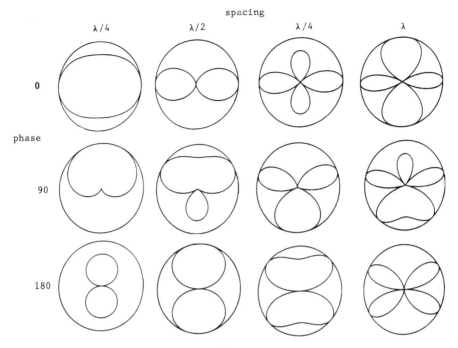

Figure 2.2 Two-element array patterns.

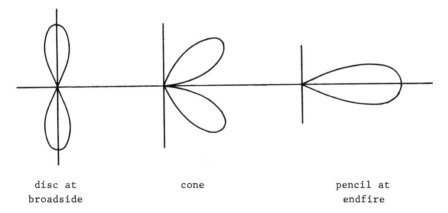

disc at broadside   cone   pencil at endfire

Figure 2.3 Linear array beams.

2.1.2 Beamwidth

The half-power points on a uniform array pattern are found by putting $\sin \frac{1}{2} Nkdu_3 / (N \sin \frac{1}{2} kdu_3) = \sqrt{0.5}$. Figure 2.4 gives the solution of this as a function of the number of elements in the array. For $N \geq 7$, the variation in normalized beamwidth Nu_3 is less than 1%, and the error is only 5% for $N = 3$. Thus for large arrays, the half-power points are given simply by $\frac{1}{2} Nkdu_3 = \pm 0.4429$. For a beam scanned at angle θ_0, this gives the 3 dB beamwidth θ_3 as

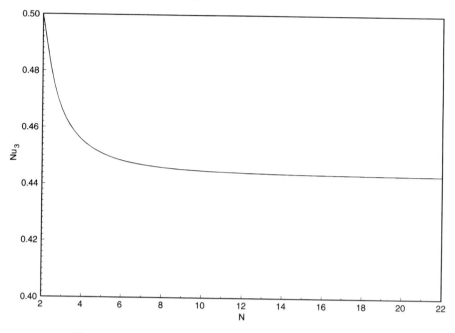

Figure 2.4 Normalized beamwidth vs. number of elements.

$$\theta_3 = \arcsin\left(\sin\theta_0 + 0.4429\frac{\lambda}{Nd}\right)$$
$$- \arcsin\left(\sin\theta_0 - 0.4429\frac{\lambda}{Nd}\right). \quad (2.3)$$

For large N, this reduces to

$$\theta_3 \simeq \frac{0.8858\lambda}{Nd\cos\theta_0}. \quad (2.4)$$

The beam collapse near endfire, where the 3 dB point is at 90 deg, occurs for a scan angle of

$$\theta_0 = \arcsin\left(1 - 0.4429\frac{\lambda}{Nd}\right). \quad (2.5)$$

The beamwidth broadening near endfire is shown in Fig. 2.5 for several arrays. For large N, the endfire beamwidth is

$$\theta_3 \simeq 2\sqrt{\frac{0.8858\lambda}{Nd}}. \quad (2.6)$$

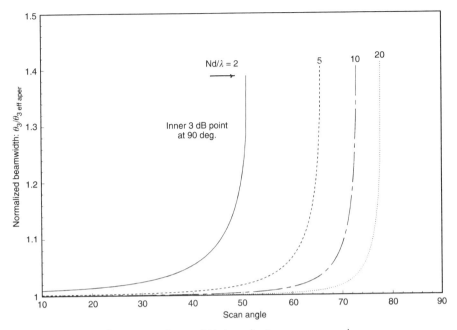

Figure 2.5 Beamwidth broadening vs. scan angle.

The accuracy of this is better than 1% for $Nd/\lambda > 4$. The endfire beamwidth is larger than the broadside value by $2.14\sqrt{Nd/\lambda}$. Thus the endfire pencil beam is broader than the broadside pancake beam.

2.1.3 Sidelobes

Uniform array nulls and sidelobes are well behaved and equally spaced. The nulls occur at $u = n/N$, with $n = 1$ to $N - 1$. The peaks of $F(u)$ occur for u that are solutions of $N \tan \pi u = \tan N\pi u$. For large N, this reduces to $\tan N\pi u = N\pi u$; the first solution is $Nu = 1.4303$. A convenient term is sidelobe ratio, which is the ratio of the main beam amplitude to that of the first sidelobe. For large arrays, the sidelobe ratio (SLR) is the same as that for uniform line sources, and is independent of the main beam angle. It is 13.26 dB. For smaller arrays the value of u for the first sidelobe is shown in Fig. 2.6. Figure 2.7 gives SLR versus number of elements. Arrays of less than 8 elements are shown to experience a significant sidelobe ratio degradation. The uniform linear array has a sidelobe envelope that decays as $1/\pi u$, and, as will be discussed in Chapter 3, this decay allows a low-Q and tolerance-insensitive array design.

2.1.4 Grating Lobes

The array pattern equation (Eqn. 2.2) allows the inference that a maximum pattern value of unity occurs whenever $u = N$. If d/λ and θ_0 are chosen properly, only one main beam will exist in "visible" space, which is for

12 BASIC ARRAY CHARACTERISTICS

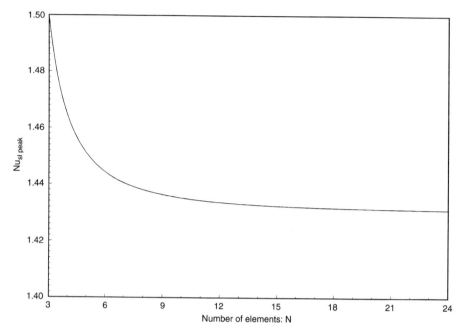

Figure 2.6 First sidelobe position vs. number of elements.

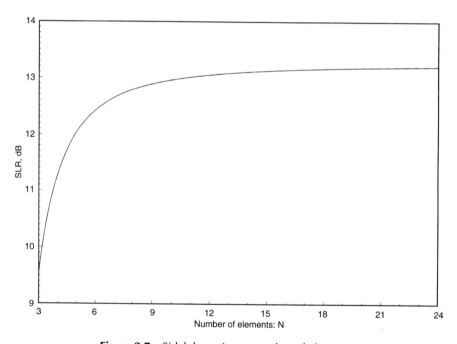

Figure 2.7 Sidelobe ratio vs. number of elements.

$-90 \leq \theta \leq 90$ deg. Large spacings will produce additional main beams called grating lobes (GL); this is because the larger spacing allows the waves from each element to add in phase at the GL angle as well as at the main beam angle. The equation for grating lobes is easily determined:

$$\frac{d}{\lambda} = \frac{n}{\sin\theta_0 - \sin\theta_{gl}}. \tag{2.7}$$

For half-wave spacing, a grating lobe appears at −90 deg for a beam scanned to +90 deg. A one-wavelength spacing allows grating lobes at ±90 deg when the main beam is broadside. Figure 2.8 shows a grating lobe at −45 deg when the beam is scanned to ±45 deg for a spacing of 0.707λ. The onset of grating lobes versus scan angle is shown in Fig. 2.9. The common rule that half-wave spacing precludes grating lobes is not quite accurate, as part of the grating lobe may be visible for extreme scan angles.

For any scan angle it is desirable to keep all of the grating lobes out of visible space. In principle one could adjust the spacing so that the grating lobe amplitude at the edge of visible space is just equal to the sidelobe level. However, the sides of the grating lobe are steep, which means that tight tolerances would be required to avoid an excessive amount of grating lobe. A better scheme puts the pattern null adjacent to the grating lobe at −90 deg; this comfortably excludes the entire grating lobe. Figure 2.10 depicts part of a pattern, where the main beam is at θ_0 and the grating lobe peak at θ_{gl}. When

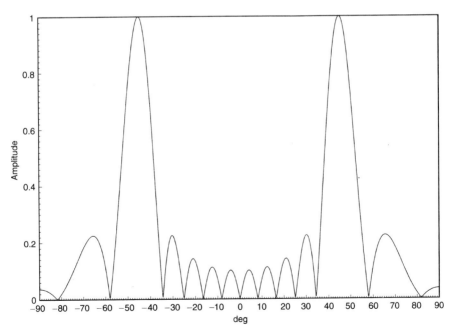

Figure 2.8 Array pattern scanned to 45 deg, $d/\lambda = 0.7071$, $N = 10$.

14 BASIC ARRAY CHARACTERISTICS

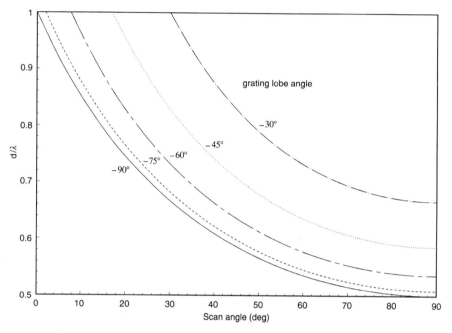

Figure 2.9 Grating lobe appearance vs. element spacing and scan.

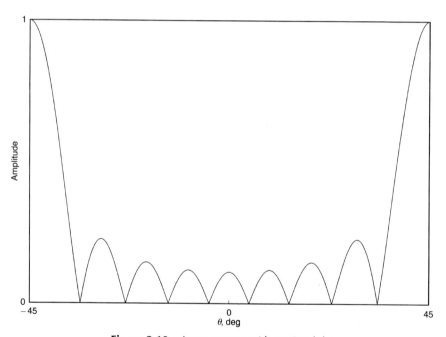

Figure 2.10 Array pattern with grating lobe.

the $\theta_{gl} - \theta_1$ null is placed at 90 deg, the grating lobe is in invisible space. The array spacing reduction required to accomplish this is given by:[1]

$$\frac{d}{\lambda} = \frac{N - \sqrt{1 + B^2}}{N(1 + \sin\theta_0)}. \quad (2.8)$$

This is a general formula that applies to amplitude tapered distributions as well. B is a taper constant that is discussed in Chapter 3. This spacing reduction for grating lobe null placement is shown in Fig. 2.11 where SLR is a parameter. Recall that for the uniform array the SLR = 13.26 dB curve applies.

2.1.5 Bandwidth

Bandwidth of an array is affected by many factors, including change of element input impedances with frequency, change of array spacing in wavelengths that may allow grating lobes, change in element beamwidth, etc. When an array is scanned with fixed units of phase shift, provided by phasers, there is also a bandwidth limitation as the position of the main beam will change with frequency. When the array is scanned with true time delay, the beam position is independent of frequency to first order. But with fixed phase shift, the beam

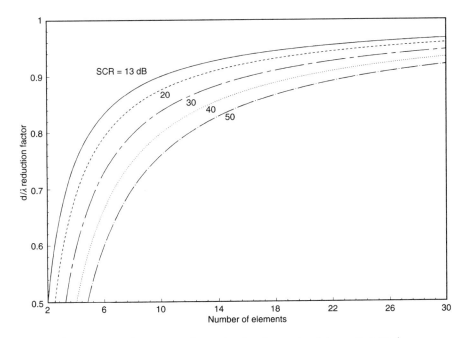

Figure 2.11 Element spacing reduction for grating lobe null at 90 deg.

---

[1]This analysis is due to David Munger and Richard Phelan.

movement is easily calculated. Beam angle θ is simply related to scan angle θ_0 by $\sin\theta = (f_0/f)\sin\theta_0$; Figure 2.12 shows this behavior. To calculate steering bandwidth, assume that the main beam has moved from scan angle θ_0 to the 3 dB points for frequencies above and below nominal. Let subscripts 1 and 2 represent the lower and upper frequencies, respectively. Fractional bandwidth is then given by

$$\text{BW} = \frac{f_2 - f_1}{f_0} = \frac{(\sin\theta_2 - \sin\theta_1)\sin\theta_0}{\sin\theta_1 \sin\theta_2}. \tag{2.9}$$

For large arrays

$$\text{BW} \simeq \frac{\theta_3}{\sin\theta_0}. \tag{2.10}$$

The bandwidth is then given by

$$\text{BW} \simeq \frac{0.866\lambda}{L\sin\theta_0} \tag{2.11}$$

for a uniform array, and

$$\text{BW} \simeq \frac{\lambda}{L\sin\theta_0}. \tag{2.12}$$

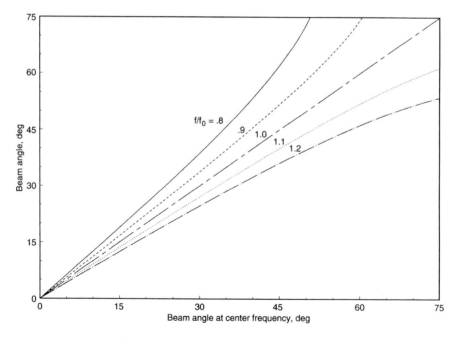

Figure 2.12 Beam angle shift with frequency.

for tapered arrays. When the beam angle is 30 deg, the commonly used formula for fractional bandwidth results:

$$\text{BW} \simeq \frac{2\lambda}{L}. \tag{2.13}$$

As a result, long arrays have smaller bandwidth in terms of beam shift at the band edges. See also Section 2.3.2.

2.2 PLANAR ARRAYS

2.2.1 Array Coordinates

In this section the characteristics of linear arrays are extended to planar arrays. Spherical coordinates with the polar axis normal to the plane of the array are convenient; see Fig. 2.13. This figure shows an array of elements on a rectangular lattice with even number of elements along the x-axis and also along the y-axis. For most pencil beam applications, the array excitations are symmetric, so that the pattern is given by summing over only half the elements along each axis:

$$F(u, v) = \sum_{n=1}^{N/2} \sum_{m=1}^{M/2} A_{nm} \cos\left[(n - \tfrac{1}{2})kd_x u\right] \cos\left[(m - \tfrac{1}{2})kd_y v\right]. \tag{2.14}$$

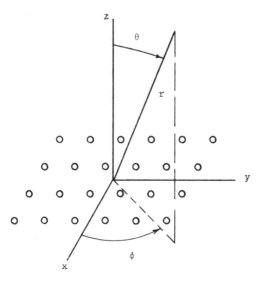

Figure 2.13 Spherical coordinate system and rectangular lattice.

18 BASIC ARRAY CHARACTERISTICS

The direction cosine plane variables are

$$u = \sin\theta \cos\phi - \sin\theta_0 \cos\phi_0,$$
$$v = \sin\theta \sin\phi - \sin\theta_0 \sin\phi_0. \quad (2.15)$$

with θ_0, ϕ_0 the beam pointing angles. The element spacings are d_x and d_y. The interelement phase shifts needed to scan the beam are

$$\Phi_u = kd_x u_0 = kd_x \sin\theta_0 \cos\phi_0,$$
$$\Phi_v = kd_y v_0 = kd_y \sin\theta_0 \sin\phi_0. \quad (2.16)$$

A triangular lattice, as shown in Fig. 2.14, is often used, as it allows slightly larger element spacing without appearance of grating lobes.

2.2.2 Beamwidth

For scanning in a principal plane, the beamwidth is just that given before in Eqn. (2.3). Cross-plane beamwidth is more complex, as both the array length in the principal plane, L, and the array width, W, affect it. Taking principal plane scan, with $\phi_0 = 0$ and $\theta = \theta_0$, the direction cosine variables are

$$u = \frac{L}{\lambda}\sin\theta_0(1 - \cos\phi), \qquad v = \frac{W}{\lambda}\sin\theta_0 \sin\phi. \quad (2.17)$$

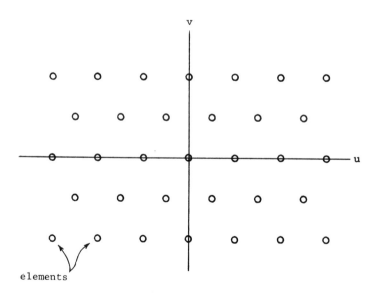

Figure 2.14 Triangular lattice.

The half-power values u_3 and v_3 are solutions of

$$\operatorname{sinc} \pi u_3 \operatorname{sinc} \pi v_3 = \frac{\pi^2 u_3 v_3}{\sqrt{2}}. \tag{2.18}$$

The cross-plane beamwidth is found from

$$\phi_{3y} = 2 \arcsin(\sin \theta_0 \sin \phi_3), \tag{2.19}$$

where the $\sin \theta_0$ factor represents a projected aperture. For large arrays, an excellent approximation is

$$\phi_{3y} \simeq 2 \sin \theta_0 \sin \phi_3. \tag{2.20}$$

The array aspect ratio W/L is a convenient parameter, as the root ϕ_{3y} can be determined as a function of $(W\lambda) \sin \theta_0$. The cross-plane beamwidth can also be written in terms of v_3:

$$\frac{W}{\lambda} \phi_{3y} \simeq 2 v_3. \tag{2.21}$$

Figure 2.15 shows beamwidth in the scan plane as a function of array length and scan angle; Fig. 2.16 gives cross-plane beamwidth as a function of array aspect ratio and normalized scan parameter. It can be seen that for a normal-

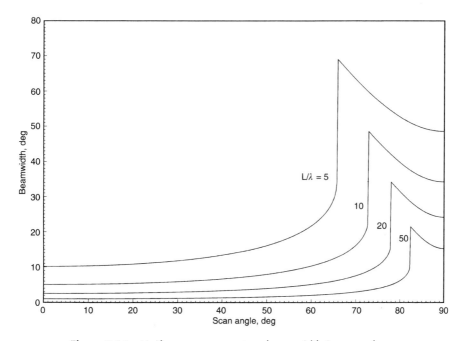

Figure 2.15 Uniform square aperture beamwidth in scan plane.

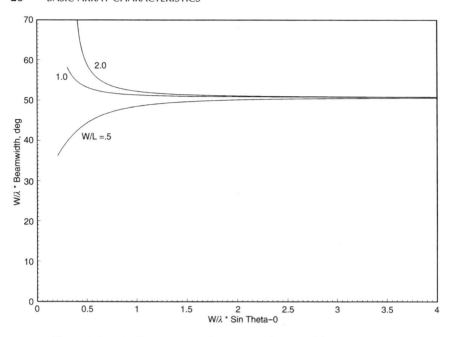

Figure 2.16 Uniform rectangular aperture beamwidth in cross-plane.

ized scan parameter of 1 or greater, the cross-plane beamwidth is close to the nominal value.

As the beam is scanned, its shape changes as the projected aperture of the array changes. For rectangular arrays the 3 dB beam contour is approximately elliptical. However, for scans not in the principal planes, the combination of projected aperture width and length and the scan angles results in an elliptical beam whose major diameter is generally not oriented along the scan plane or principal planes (Elliott, 1966). Since the area of an ellipse is proportional to the product of major and minor diameters, the 3 dB beam area, called areal beamwidth, is proportional to the product of the major-axis and minor-axis beamwidths. The areal beamwidth is to first order independent of azimuth angle ϕ, although the beam shape and orientation of major axis may change with θ. Figure 2.17 shows several beams as they change with scan.

2.2.3 Grating Lobes: Rectangular Lattice

A rectangular lattice with scanning in either principal plane behaves exactly like a linear array, as described in Section 2.2. For other scan angles the situation is less simple. The u,v-plane, sometimes called the direction cosine plane, was developed by Von Aulock (1960) and is extremely useful for understanding grating lobe behavior. The grating lobe positions can be plotted in the u,v-plane; they occur at the points of an inverse lattice, that is, the lattice spacing is λ/d_x and λ/d_y; see Fig. 2.18. All real angles, representing visible space, are inside or on the unit circle. The latter represents $\theta = 90$ deg.

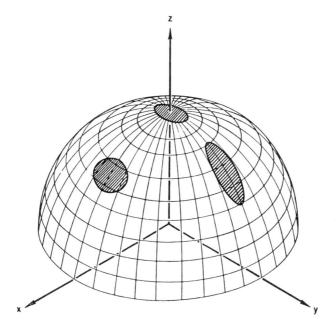

Figure 2.17 Beam shape vs. scan position for a pencil beam.

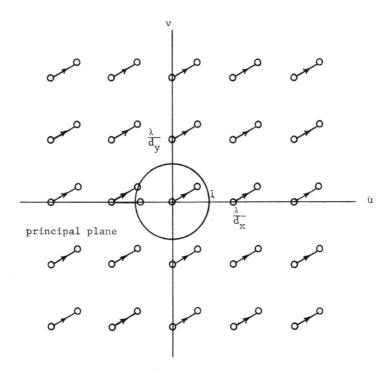

Figure 2.18 Direction cosine plane.

22 BASIC ARRAY CHARACTERISTICS

Angles outside the unit circle are "imaginary," or in invisible space. When the main beam is scanned, the origin of the u,v plot moves to a new value, and all grating lobes move correspondingly. However the unit circle remains fixed. Thus, for scan in the u-plane ($\phi = 0$) the main beam point moves towards $+1$ for $\theta > 0$, and all GL points move the same amount. When the GL just outside the unit circle moves enough to intersect the unit circle, that GL becomes visible. Thus the distance between the broadside GL and the unit circle must be no larger than unity, or the GL will not intersect the unit circle before the main beam stops at the right side of the unit circle. The result is

$$u_{gl} \leq \frac{\lambda}{d_x} - 1, \quad \text{or} \quad \frac{d_x}{\lambda} \geq \frac{1}{1 + \sin\theta}. \tag{2.22}$$

The inequality is such that exceeding the equality allows a grating lobe to appear. Scan along the v-axis ($\phi = 90\,\text{deg}$) is analogous:

$$\frac{d_y}{\lambda} \geq \frac{1}{1 + \sin\theta}. \tag{2.23}$$

If the array spacings are sufficiently large, scan to any pair of angles may produce one or more grating lobes. The limiting cases are of interest, and these are of two types: diagonal GL, and tangential GL. The diagonal plane scan sketched in Fig. 2.19 produces a GL when the diagonal point intersects the unit circle normally. The u,v values that allow this are

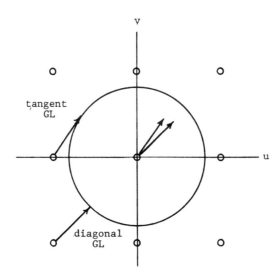

Figure 2.19 Grating lobe limiting cases.

$$u_{gl} = \frac{SQ-1}{SQd_x/\lambda}, \qquad v_{gl} = \frac{SQ-1}{SQd_y/\lambda};$$
$$\sin\theta_{gl} = SQ - 1, \qquad \tan\phi_{gl} = \frac{d_x}{d_y}. \qquad (2.24)$$

where $SQ = \sqrt{\lambda^2/d_x^2 + \lambda^2/d_y^2}$. This diagonal plane GL occurs only for $SQ \geq 2$. In general, a grating lobe appears whenever the propagation constant is real; at the transition it is zero:

$$\frac{\beta}{k} = \sqrt{1 - (u - n\lambda/d_x)^2 - (v - m\lambda/d_y)^2}. \qquad (2.25)$$

Scan in principal planes corresponds to $n = 0$ or $m = 0$. The diagonal case is for $n = 1$, $m = 1$. Sufficiently large spacings allow several GL to appear, with the result that the propagation constant is real for larger n and/or larger m. The tangential limiting case occurs when the GL is located on the u-axis just left of the unit circle. A diagonal scan can place the GL just tangent to the circle (see Fig. 2.19). Only d_x is involved here, and the limiting values are

$$u_{gl} = \frac{1 - d_x^2/\lambda^2}{d_x/\lambda}, \qquad v_{gl} = \sqrt{1 - d_x^2/\lambda^2};$$
$$\sin\theta_{gl} = \frac{\sqrt{1 - d_x^2/\lambda^2}}{d_x/\lambda}, \qquad \sin\phi_{gl} = \frac{d_x}{\lambda}. \qquad (2.26)$$

The minimum value of d_x/λ is $1/\sqrt{2}$. The v tangent case is analogous, as might be expected, with u and v interchanged and d_x and d_y interchanged. For unrestricted scan angles, the principal plane GL appears first, controlled by the smaller of λ/d_x and λ/d_y. The u,v-plane not only gives an excellent physical picture of GL occurrence, but also allows the formulas to be derived easily. In summary, if $d_x/\lambda < 1/(1 + \sin\theta)$ and $d_y/\lambda < 1/(1 + \sin\theta)$, no grating lobes will exist for the rectangular lattice.

2.2.4 Grating Lobes: Hexagonal Lattice

Isosceles triangular lattices are sometimes used; see the brick waveguide array in Chapter 7. A commonly used special case is the equilateral triangular lattice, or regular hexagonal lattice. Let the elements be all equally spaced by $2d$, and let the x-axis go through a row of elements spaced this distance apart. Then $d_x = d$ and $d_y = \sqrt{3}d/2$, where d_x and d_y are half the distance to the next element along the x- or y-axis. The hexagonal array is simply analyzed by breaking it into two interlaced rectangular lattice arrays. The pattern is then the sum of the two array patterns. The grating lobe locations for no beam scan are simply

$$u_{gl} = \pm\frac{n\lambda}{2d}, \qquad v_{gl} = \pm\frac{m\pi}{\sqrt{3}d} = \pm\frac{m\lambda}{2d_y}. \qquad (2.27)$$

24 BASIC ARRAY CHARACTERISTICS

The inverse lattice in the u,v-plane is shown in Fig. 2.20; the GL points are equidistant by λ/d. Scan in the u-plane gives the same results as for the rectangular lattice. Diagonal plane scan, approaching the unit circle normally, yields a GL for

$$u_{gl} = \frac{1 - d/\lambda}{2d/\lambda}, \qquad v_{gl} = \frac{\sqrt{3}(1 - d/\lambda)}{2d/\lambda};$$
$$\sin\theta_{gl} = \frac{1 - d/\lambda}{d/\lambda}, \qquad \phi_{gl} = \frac{\pi}{3}. \tag{2.28}$$

The minimum value of d/λ for this lobe to appear is $1/\sqrt{2}$. Now there are two tangent cases, where the GL normally on the u-axis becomes tangent to the unit circle, and where the normally diagonal GL becomes tangent to the unit circle. The first of these two cases occurs for

$$u_{gl} = \frac{1 - d^2\lambda^2}{d/\lambda}, \qquad v_{gl} = \sqrt{1 - d^2/\lambda^2};$$
$$\sin\theta_{gl} = \frac{\sqrt{1 - d^2/\lambda^2}}{d/\lambda}, \qquad \sin\phi_{gl} = \frac{d}{\lambda}. \tag{2.29}$$

The minimum value of d/λ for this GL to appear is $1/\sqrt{2}$.

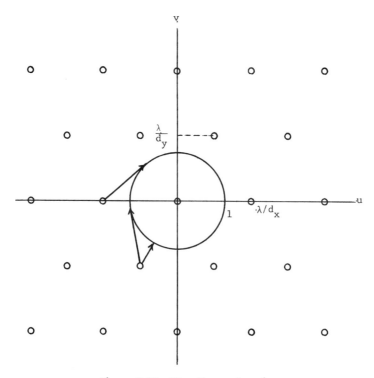

Figure 2.20 Direction cosine plane.

Tangency from the diagonal point is somewhat more complicated, and occurs for

$$u_{gl} = \frac{\sqrt{1-d^2/\lambda^2}}{2}\left(\frac{\sqrt{1-d^2/\lambda^2}}{d/\lambda}+\sqrt{3}\right),$$

$$v_{gl} = \frac{\sqrt{1-d^2/\lambda^2}}{2}\left(\frac{\sqrt{3(1-d^2/\lambda^2)}}{d/\lambda}-1\right); \quad (2.30)$$

$$\sin\theta_{gl} = \sqrt{\frac{1-d^2/\lambda^2}{d/\lambda}}, \quad \phi_{gl} = \frac{\pi}{3}-\arcsin\frac{d}{\lambda}.$$

Comparing the square lattice and the hexagonal lattice, both have grating lobe appearance when $\sin\theta = \lambda/d - 1$. The square lattice element area is d^2, while that for the hexagonal lattice is $2d^2/\sqrt{3}$. Thus the hexagonal lattice requires only 0.866 as many elements to give the same grating lobe free area. This results in a saving of 15% in number of elements (Sharp, 1961; Lo and Lee, 1965).

2.3 BEAM STEERING AND QUANTIZATION LOBES

Interelement phase shift is necessary to provide beam scan, the devices that produce this phase shift are called *phasers*. Row and column phasing is the simplest, even for circular arrays. Thus each phaser is driven by a command to produce a specified x- and a specified y-axis phase. The steering bits affect the precision of the beam steering, control bandwidth, and produce phase quantization lobes. Each of these will be discussed.

2.3.1 Steering Increment

The smallest steering increment is related to the smallest phase bit (Hansen, 1983).

$$\frac{\theta_{err}}{\theta_3} = \frac{1}{2^M 1.029} \quad (2.31)$$

The factor 1.029 is the circular distribution normalized beamwidth mentioned earlier, and M is the number of bits. Table 2.1 gives the steering increment divided by the half-power beamwidth θ_3. Note that in this case the largest bit is π. Thus 4 bits gives a steering least count of 0.061 beamwidth, or roughly 1/16 of a beamwidth. Adding a bit, of course, decreases the steering increment by a factor of 2. Adequate steering precision will often be provided by 4 bits. This phaser would be located at each element; the phases would be 180, 90, 45, and 22.5 deg.

TABLE 2.1 Steering Precision Phase Bits

Phaser Bits	$\theta_{\text{err}}/\theta_3$
3	$0.121 \sim 1/8$
4	$0.0607 \sim 1/16$
5	$0.0304 \sim 1/32$

2.3.2 Steering Bandwidth

For an electronic scanning array, bandwidth is usually defined such that at the band edges the main beam has moved through half the beamwidth, that is, the gain has dropped 3 dB. For an array scanned by phase alone, the bandwidth is given approximately by $\text{BW} = \theta_3 \sin\theta_0$. Using the rough value of $\theta_3 \simeq \lambda/L$, for a 30 deg scan the bandwidth is given by $\text{BW} \simeq 2\lambda/L$, an oft-quoted but imprecise result.

Use of time delay for some bits will increase the bandwidth; the amount of the increase will be determined here. Let the center frequency be represented by f_0. Let the upper and lower band edges be represented by f_2 and f_1. Similarly, let the sine of the center frequency main beam angle at the scan limit be S_0 and at the band edges be S_2 and S_1. Further, let N be the number of time delay bits. The frequency excursions that define the bandwidth are now given by

$$(2^N - 1)f_1 S_0 + f_0 S_0 = 2^N f_1 S_1,$$
$$(2^N - 1)f_2 S_0 + f_0 S_0 = 2^N f_2 S_2. \quad (2.32)$$

For small bandwidths the band edges may be approximated by $f_2 = f_0 + \Delta_2$ and $f_1 = f_0 - \Delta_1$. The normalized bandwidth is then $(\Delta_1 + \Delta_2)$. For large scan angles the expressions for S_2 and S_1 can also be simplified. However, for the small scan angle used here, this simplification becomes inaccurate as A approaches unity. Thus the exact equations were solved.

It is convenient to normalize the bandwidth by a factor $\sin\theta_0/\theta_3$. When all bits are phase, the normalized bandwidth is 1. Changing each bit from phase to time delay roughly doubles the bandwidth factor, and the bandwidth. Note that the bandwidth increases faster than 2^M, as it must since when all bits are time delay the steering bandwidth is infinite (see Table 2.2). However, the increase is less for larger arrays.

2.3.3 Phaser Quantization Lobes

Most phasers are now digitally controlled, whether the intrinsic phase shift is analog or digital. Such phasers have a least phase, corresponding to one bit. An M-bit phaser has phase bits of $2\pi/2^M$, $2\pi/2^{M-1}$, ..., π. The ideal linear phase curve for electronic scanning is approximated by stair-step phase, producing a sawtooth error curve; see Fig. 2.21. Since the array is itself discrete, the positions of the elements on the sawtooth are important. There are two well-defined cases. The first case is when the number of elements is less than the

BEAM STEERING AND QUANTIZATION LOBES 27

TABLE 2.2 Time Delay Bits Versus Bandwidth, $\theta_0 = 30\,\text{deg}$, $\theta_3 = 1\,\text{deg}$

Bits of Time Delay	Normalized Bandwidth Factor
0	1
1	2.002
2	4.016
3	8.124
4	17.016

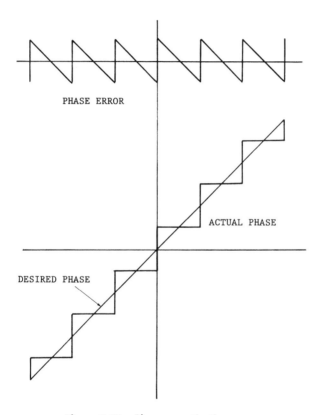

Figure 2.21 Phase quantization errors.

number of steps. In this case the phase errors assume a random nature. This can be evaluated by approximating the phase variance by 1/3 of the peak sawtooth error of $\pi/2^M$. The variance is

$$\sigma^2 = \frac{\pi^2}{3 \cdot 4^M}. \tag{2.33}$$

and the RMS sidelobe level is σ^2/G. For uniform excitation of the array the RMS sidelobe level is σ^2/N. Figure 2.22 gives RMS sidelobe level due to

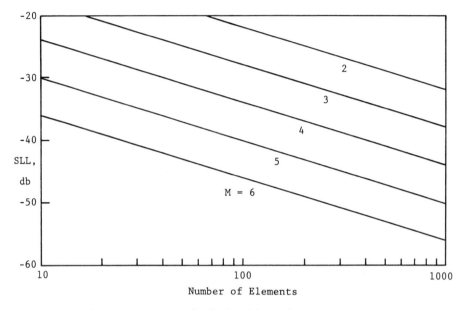

Figure 2.22 RMS sidelobe level from phaser quantization.

random phaser quantization error, for broadside radiation. At scan angle θ_0 the main beam is reduced by approximately $\cos^2 \theta_0$, so that the RMS sidelobe level with respect to the main beam is reduced. A modest gain decrease accompanies the increased sidelobes:

$$\frac{G}{G_0} \simeq 1 - \sigma^2. \tag{2.34}$$

This random gain decrease is given in Table 2.3.

The second case has two or more elements per phase step, and the discrete (array) case is approximated by a continuous case. This quantization produces a set of lobes called quantization lobes (QL), that have predictable amplitudes and positions. An N-element array has an end-to-end phase of $(N-1)kd \sin \theta_0$. The number of elements per phase step, with M bits of phase, is J:

TABLE 2.3 Phaser Gain Decrease for Small Arrays

M	Gain Decrease (dB)
2	1.000
3	0.229
4	0.056
5	0.014

$$J = \frac{N}{(N-1)2^M(d/\lambda)\sin\theta_0}. \tag{2.35}$$

This allows the largest scan angle for which $J \geq 2$ to be found. For large arrays, N approximately cancels, leaving $\sin\theta_0 \simeq 1/[(d/\lambda)2^{M+1}]$. For half-wave element spacing the maximum angles are given in Table 2.4. Larger angles produce pattern errors that are more complex. For J between 1 and 2 there is a transition region between the random sidelobes regime ($J < 1$) discussed and the QL regime.

The QL angle θ_q is governed by the step width W:

$$\frac{W}{\lambda} = \frac{1}{2^M \sin\theta_0}. \tag{2.36}$$

This gives a peak sawtooth error of $\beta = \pi/2^M$, as mentioned above. The QL amplitudes are given by $\mathrm{sinc}\,(\beta \pm i\pi)$, where $i = 0$ gives the main beam. The first two QL and the main beam are given by Miller (1964):

$$\begin{aligned}
\text{MAIN BEAM:} \quad & \mathrm{sinc}\,\beta. \\
\text{FIRST QL:} \quad & \frac{\sin\beta}{\pi \pm \beta}. \\
\text{SECOND QL:} \quad & \frac{\sin\beta}{2\pi \pm \beta}.
\end{aligned} \tag{2.37}$$

When $\beta = \pi/2$, the main beam and first QL are equal, clearly a bad case. But this only occurs for one-bit phasers, an extreme case. Table 2.5 gives lobe amplitudes versus number of phaser bits.

Gain decrease is given approximately by the main beam decrease (Allen et al., 1960):

$$\frac{G}{G_0} \simeq \mathrm{sinc}^2\beta. \tag{2.38}$$

These data allow the array designer to make an intelligent trade on phaser bits.

For rectangular lattices, the grating lobe appearances derived for linear arrays are still valid along the u- and along the v-axes. These intersect to

TABLE 2.4 Maximum Scan Angle for Well Defined QL

M bits	$\theta_{\max}$ (deg)
1	30
2	14.48
3	7.18
4	3.58
5	1.79

TABLE 2.5 Phaser Quantization Lobe Amplitudes

M	Main Lobe (dB)	QL$_1$ (dB)	QL$_2$ (dB)
1	−3.92	−3.92	−13.46
2	−0.912	−10.45	−14.89
3	−0.224	−17.13	−19.31
4	−0.056	−23.58	−24.67
5	−0.014	−29.84	−30.38
6	0	−35.99	−36.26

form bands where the grating lobes tend to appear, as shown in Fig. 2.23. The discrete spacing of the grating lobes depends, of course, upon the scan angles. As always, only grating lobes on bands within the unit circle represent visible QL.

When a hexagonal lattice is used with a hexagonal phaser network, which is a two-dimensional structure, the QL bands follow the principal diagonals. In most cases only two bands would cross the unit circle. However, hexagonal networks require many circuit board crossovers, and hence are seldom used. A more common scheme uses row and column phasers: these produce QL bands parallel to the u- and v-axes (Nelson, 1969). Figure 2.24 shows a hexagonal lattice with row and column phasers. Additional QL bands can occur in this situation. In summary, the hexagonal lattice offers grating lobe advantages, but no quantization lobe advantages when row-column (u,v) phasers are used.

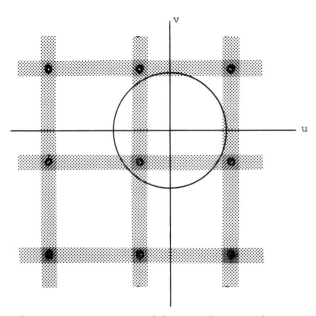

Figure 2.23 Quantization lobe strips for square lattice.

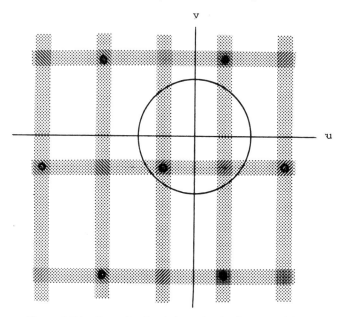

Figure 2.24 Quantization lobe strips for hexagonal lattice.

2.3.4 Subarray Quantization Lobes

An array composed of contiguous subarrays, when scanned, has a phase consisting of stair-steps, with one step over each subarray. This, just as in the case of digital phasers, produces quantization lobes, except that a subarray has at least two elements. So the QL are always deterministic here. Call the sawtooth width (subarray width) W, and the peak sawtooth error β, as before. For large arrays the discrete subarray may be replaced by a continuous aperture, but with the sawtooth phase. Consider N equal subarrays, and with uniform excitation. The pattern is

$$F(u) = \tfrac{1}{2} \sum_{n=1}^{N} \int_{N_1}^{N_2} \exp\{j[(\tfrac{1}{2}kdu - N\beta)p + \beta(2n - 1 - N)]\}dp, \qquad (2.39)$$

where the integration is over each step, and

$$u = (\sin\theta_0 - \sin\theta), \qquad N_1 = \frac{2n - N}{N} \qquad \text{and} \qquad N_2 = \frac{2n - 2 - N}{N}.$$

Convenient variables are

$$\begin{aligned} w &= \frac{W}{\lambda}(\sin\theta_0 - \sin\theta), & \beta &= \pi v; \\ v &= \frac{W}{\lambda}\sin\theta, & v_0 &= \frac{W}{\lambda}\sin\theta_0. \end{aligned} \qquad (2.40)$$

The integration and summation can both be performed in closed form, with the result

$$F(u) = \frac{\sin N\pi w}{N \sin \pi w} \operatorname{sinc} \pi v. \qquad (2.41)$$

This is immediately recognized as the pattern of a uniform array of N isotropic elements spaced W wavelengths apart, times the pattern of a uniform line source of length W wavelengths, with main beam at $\theta = v = 0$. For W larger than one wavelength, grating lobes will exist from the first factor. These GL have unit amplitudes and positions for $w = 0, 1, 2, \ldots$. Of these the first is the main beam. These lobes are weighted by the sinc beam. With the GL spaced by unity in w, and with the line beamwidth of 2 (in v), the first GL is always at the edge of, or in, the sinc main beam. For example, for no scan the first GL is at the sinc factor first null. As the array beam scans, it moves away from the sinc peak, thus decreasing. Concomitantly the GL moves from the sinc null toward the peak. Thus the level of the first quantization is closely that of the sinc factor at the first GL position: $v = v_0 - 1$. The next QL will occur at the second GL: $v = v_0 - 2$, with the sinc pattern sidelobes there given the QL amplitude. An exact determination of QL peaks would require a differentiation of Eqn. (2.31), with the result set to zero. As long as N is large, the GL beamwidth will be small compared with the sinc beamwidth, and the results given here will be accurate. When the N-element array pattern beam occurs at a null in the subarray sinc pattern, the QL bifurcates, with a small lobe each side of the null (Mailloux, 1994). These are of little interest owing to their low amplitude. The QL amplitudes, and that of the main beam, are formed from Eqn. (2.37), using the subarray β of course. The QL locations are approximately at the GL angles:

$$\sin \theta_{gl} = \sin \theta_0 \pm \frac{m\lambda}{W}, \quad m = 1, 2, 3, \ldots. \qquad (2.42)$$

Approximate quantization lobe amplitudes are given by $\operatorname{sinc} \pi v_q$; these versus $v_0 = (W/\lambda) \sin \theta_0$ are given in Table 2.6 for various levels. The table allows

TABLE 2.6 Subarray Quantization Lobes

β/π	Main Lobe (dB)	QL$_1$ (dB)	QL$_2$ (dB)
0.025	−0.01	−31.83	−32.26
0.05	−0.04	−25.61	−26.48
0.1	−0.14	−19.23	−20.97
0.15	−0.32	−15.39	−18.02
0.2	−0.58	−12.62	−16.14
0.3	−1.33	−8.69	−14.06
0.4	−2.42	−5.94	−13.30
0.5	−3.92	−3.92	−13.46

various combinations of subarray size and scan angle to be traded. For example, QL no higher than uniform sidelobe level obtains for $W/\lambda = 2$ at 5.37 deg scan, or for $W/\lambda = 1$ at 10.79 deg scan. These amplitudes are relative to the main beam at broadside. Figure 2.25 shows QL for an array with half-wave spacing and 75 elements. Each subarray contains 5 elements; scan angle is 3 deg, with $\beta/\pi = 0.1308$. The closest QL are -16.69 dB at -20.34 deg, and -18.98 dB at 26.89 deg.

A more precise determination of QL behavior would use a discrete subarray instead of a continuous subarray. However, the accuracy of the model used here is adequate, and it is simpler.

These results point up the reasons why these lobes are called quantization lobes rather than grating lobes. Although they occur at the grating lobe angles, their appearance and amplitude are functions of scan angle. While grating lobes appear even for zero scan, phaser and subarray quantization lobes do not exist for zero scan. Their amplitude increases with scan. Further grating lobes tend to have high amplitude, diminished from main beam amplitude only by a slowly varying element pattern. Quantization lobes, on the other hand, are typically much lower than the main beam, and each additional QL is significantly lower than the previous one.

2.3.5 QL Decollimation; Overlapped Subarrays

There are several schemes for decollimation of phaser quantization lobes, of which the best is the phase added technique (Smith and Guo, 1983). In this

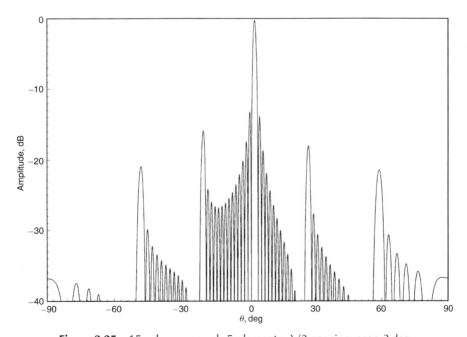

Figure 2.25 15 subarrays, each 5 elements, $\lambda/2$ spacing, scan 3 deg.

34 BASIC ARRAY CHARACTERISTICS

technique the element phases are calculated to a number of bits that would provide sufficiently low quantization lobes if used. Call this number of bits N. The phasers are driven by a smaller number of bits, M. A set of random numbers, one for each phaser, with each number N bits long, is generated and stored. The random number table is then truncated to M bits and stored separately. When the set of phaser drive bits is calculated for a given scan direction, the N-bit random numbers are added, one to each phaser drive word. The phaser drive words are then truncated to M bits, and the M-bit random numbers are then subtracted, one for each phaser drive word. The resulting words are then used to drive the phasers. The result of this is a decollimation of the quantization lobe, and a small rise in the average sidelobe envelope. It is important that the results of this method, unlike those of some others, have zero mean value. The RMS sidelobe level is given by

$$SLL_{rms} = \frac{\alpha^2}{(1-\alpha^2)N_{elem}}, \qquad (2.43)$$

where the parameter $\alpha = \pi/(\sqrt{3}\, 2^M)$, and the array has N_{elem} elements. The number of phaser bits is M, where again the largest is π. See Table 2.7.

How decollimation works on large elements or subarrays needs to be investigated, but the gain drop cannot be recovered. The only scheme for significantly reducing these effects, and obviating the gain loss, uses overlapped subarrays (Mailloux, 1994). As shown in Fig. 2.26, the output of each element is divided into two or more outputs. This allows two or more configurations of subarrays to be connected to the same element. These subarray outputs can then be combined to produce a virtual amplitude taper. The result is reduced QL and restoration of gain. Unfortunately, the feed structure is very complex even for a linear array. The element pattern significantly reduces the effect of these quantization lobes, as it has the lowest value at the extreme scan angle, for which the QL are the largest.

2.4 DIRECTIVITY

2.4.1 Linear Array Directivity

The directivity of a linear array, as usual, is the integrated power radiation pattern over a sphere divided by the power density at the angle of interest. With θ the angle from the normal to the array, the directivity G becomes

TABLE 2.7 QL Decollimation

Phaser Bits	α	RMS SLL (dB)
3	0.2267	−44.1
4	0.1134	−50.1
5	0.0567	−56.1

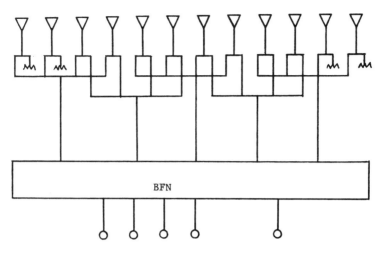

Figure 2.26 Overlapped subarrays.

$$G = \frac{4\pi |F|^2}{\int_0^{2\pi} \int_{-\pi/2}^{\pi/2} |F(\theta,\phi)|^2 \cos\theta \, d\theta \, d\phi}. \tag{2.44}$$

G is used here for directivity, although it is also often used for gain. Since conduction losses of these arrays are usually much less than radiation resistances, the gain and directivity are essentially equal, except for impedance mismatch effects. For isotropic elements, the array factor is rotationally symmetric, leading immediately to a simplification:

$$G = \frac{2|F|^2}{\int_{-\pi/2}^{\pi/2} |F(\theta)|^2 \cos\theta \, d\theta}. \tag{2.45}$$

For a uniform broadside array, the closed form pattern expression is inserted to yield

$$G = \frac{N^2}{\int_0^{\pi/2} \frac{\sin^2 \frac{1}{2} Nkdu}{\sin^2 \frac{1}{2} kdu} \cos\theta \, d\theta}. \tag{2.46}$$

A change of variable to u gives

$$G = \frac{N^2 d/\lambda}{\int_0^{d/\lambda} \frac{\sin^2 \frac{1}{2} Nkdu}{\sin^2 \frac{1}{2} kdu} du}. \tag{2.47}$$

36 BASIC ARRAY CHARACTERISTICS

This can be integrated with the help of an expansion (Whittaker and Watson, 1952):

$$\frac{N^2 d/\lambda}{G} = \int_0^{d/\lambda} \left[N + 2\sum_{n=1}^{N-1}(N-n)\cos nkdu \right] du$$

$$= N\frac{d}{\lambda} + \frac{2d}{\lambda}\sum_{n=1}^{N-1}(N-n)\operatorname{sinc} nkd. \tag{2.48}$$

The resulting directivity, where sinc x is $\sin(x)/x$, is

$$G = \frac{N^2}{N + 2\sum_{n=1}^{N-1}(N-n)\operatorname{sinc} nkd}. \tag{2.49}$$

Figure 2.27 shows array directivity versus spacing for various arrays from 2 to 24 elements. It can be noted that the directivity drops abruptly at the appear-

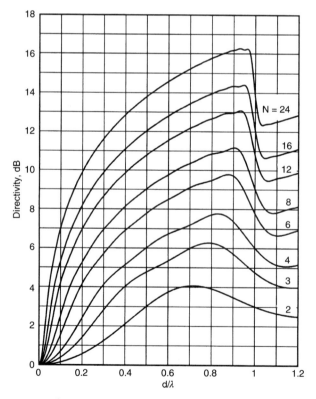

Figure 2.27 Array of isotropic elements.

ance of the first grating lobe, to about the value obtained at half-wave spacing. Although not shown on this graph, as the spacing increases further the directivity rises until a second grating lobe appears, at which point there is another drop to roughly the same half-wave spaced value (Hansen, 1972a). In fact the directivity is equal to N whenever the spacing is a multiple of 0.5λ. The minor oscillations below the directivity peak depend upon whether the boundary of real space coincides with a peak sidelobe or with a sidelobe null. This behavior can also be interpreted as the effect of mutual coupling between elements varying as the spacing changes. The summation in this equation can be recognized as a weighted sum of the mutual resistances, since the virtual mutual resistance between two isotropic radiators spaced by s is simply $120 \operatorname{sinc} ks$.

For scan angles away from broadside, the directivity becomes

$$G = \frac{N^2}{N + 2 \sum_{n=1}^{N-1} (N-n) \operatorname{sinc} nkd \cos(nkd \sin \theta_0)}. \tag{2.50}$$

This result will produce directivity changes near grating lobe incidence similar to those of Fig. 2.27. A special case is endfire, which gives directivity of

$$G = \frac{N^2}{N + 2 \sum_{n=1}^{N-1} (N-n) \operatorname{sinc} 2nkd}. \tag{2.51}$$

This replicates the results of a broadside array, but with spacing halved. Thus directivity equals N for spacing that is a multiple of quarter-wavelength.

For half-wave spacing, the sinc argument in Eqn. (2.49) is $n\pi$, which makes the directivity independent of scan angle, and equal to N. In contrast, a uniformly excited continuous line source (Hansen, 1964) has an endfire directivity of $4L/\lambda$, but a broadside directivity of $2L/\lambda$.

For large N, maximum directivity for a large array occurs at a spacing just under λ for broadside operation and just under $\lambda/2$ for endfire operation; the directivity approaches $2N$. When expressed in terms of aperture length $L = (N-1)d$, it is evident that these maximum directivities approach the same values as those of the continuous line source, viz., $2L/\lambda$ at broadside and $4L/\lambda$ at endfire.

An array of isotropic elements with an amplitude taper has a space factor of

$$F(u) = \sum_{n=1}^{N} A_n \exp j(nkdu). \tag{2.52}$$

When this is inserted into Eqn. (2.45) and integrated, the resulting directivity is

$$G = \frac{2\left|\sum_{n=1}^{N} A_n\right|^2}{\sum_{n=1}^{N}\sum_{m=1}^{N} A_n A_m^* \operatorname{sinc}(n-m)kd}. \quad (2.53)$$

Even with complex excitation coefficients, the numerator and denominator are real. For the uniform array, all $A_n = 1$, this reduces to Eqn. (2.49). Symmetric arrays allow the directivity sums to be expressed over half of the array. For example, for N even, the directivity becomes

$$G = \frac{2\left|\sum_{n=1}^{N/2} A_n\right|^2}{\sum_{n=1}^{N/2}\sum_{m=1}^{N/2} A_n A_m^* [\operatorname{sinc}(n+m-1)kd + \operatorname{sinc}(n-m)kd]}. \quad (2.54)$$

2.4.2 Directivity of Arrays of Short Dipoles

Arrays of short dipoles are of interest primarily because the directivity expressions can be integrated in closed form. Parallel short dipoles provide directivity in the ϕ dimension; the appropriate element pattern to use in the directivity equation is

$$f(\theta, \phi) = \sqrt{1 - \cos^2\theta \sin^2\phi}. \quad (2.55)$$

Collinear short dipoles provide directivity in the plane of the array; an appropriate element pattern is $\cos\theta$. The resulting directivity for a uniform broadside linear array of parallel short dipoles is

$$G = \frac{3N^2/2}{N + 3\sum_{n=1}^{N-1}(N-n)\left[\dfrac{\sin nkd}{nkd} + \dfrac{\cos nkd}{(nkd)^2} - \dfrac{\sin nkd}{(nkd)^3}\right]}. \quad (2.56)$$

For the special case of half-wave spacing, the result becomes

$$G = \frac{3N^2/2}{N + 3\sum_{n=1}^{N-1}\dfrac{(-1)^n(N-n)}{\pi^2 n^2}}. \quad (2.57)$$

Collinear short dipoles or slots yield a similar directivity:

$$G = \frac{3N^2/2}{N - 6\sum_{n=1}^{N-1}(N-n)\left[\dfrac{\cos nkd}{(nkd)^2} - \dfrac{\sin nkd}{(nkd)^3}\right]}. \qquad (2.58)$$

Again the special case of half-wave spacing yields

$$G = \frac{3N^2/2}{N - 6\sum_{n=1}^{N-1}\dfrac{(N-n)(-1)^n}{\pi^2 n^2}}. \qquad (2.59)$$

For both dipole orientations, the mutual resistance is given by the bracketed trig terms times a constant. Although these directivity expressions are simple, short elements are seldom used in arrays because of their high reactance (susceptance), low radiation resistance, and narrow bandwidth. Half-wave or resonant elements are almost always preferred.

One might infer that directivity of general dipole or slot or patch arrays could be calculated using mutual resistances, a concept pioneered by Bloch, Medhurst, and Pool (1953). This proves to be true, and will be discussed next.

2.4.3 Directivity of Arrays of Resonant Elements

Directivity will be derived for a uniform broadside linear array of half-wave dipoles in any orientation. As before, directivity is the ratio of peak power density to input power:

$$G = \frac{4\pi r^2 E_0 H_0^*}{P}. \qquad (2.60)$$

For a linear array of half-wave dipoles, the peak fields E_0 are

$$E_0 = \frac{60}{r}\sum I_n, \qquad H_0 = \frac{E_0}{120\pi}. \qquad (2.61)$$

Here the dipole currents are I_n. For uniform excitation, all $I_n = I_0$, and the numerator is

$$120\left[\sum I_n\right]^2 = 120 I_0^2 N^2. \qquad (2.62)$$

The input power is

$$P = \sum \mathrm{Re}(I_n V_n^*) = \sum I_n \sum I_m^* R_m = I_0^2 \sum\sum R_{nm}, \qquad (2.63)$$

where the mutual resistance between elements is R_{nm}. Finally, the directivity becomes

$$G = \frac{120N^2}{\sum_{n=1}^{N}\sum_{m=1}^{N} R_{nm}} = \frac{120N^2}{R_0 N + 2\sum_{n=1}^{N-1}(N-n)R_n}. \quad (2.64)$$

The self-resistance is R_0, and R_n is the mutual resistance between dipoles nd apart. Directivity for parallel and collinear linear arrays of half-wave dipoles has been calculated using compact and efficient computer algorithms for mutual impedances (Hansen, 1972b; Hansen and Brunner, 1979). Results for parallel dipole arrays are shown in Fig. 2.28 while directivity for collinear dipole arrays are shown in Fig. 2.29. It can be noticed that the parallel dipole array exhibits grating lobe directivity drops similar to those with isotropic elements. Both experience strong mutual coupling. However, for collinear arrays the mutual coupling is less and thus there is less grating lobe effect for these arrays. At the same time the parallel array gives roughly 3 dB more directivity at the peak, due to narrowing the beam in the transverse plane.

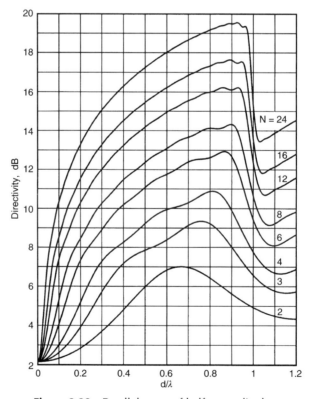

Figure 2.28 Parallel array of half-wave dipoles.

DIRECTIVITY   41

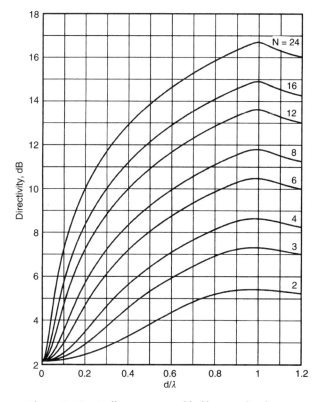

Figure 2.29 Collinear array of half-wave dipoles.

For a fixed element spacing and number of elements, maximum directivity does not occur at broadside or at endfire. This is discussed in Chapter 9 on Superdirective Arrays.

2.4.4 Planar Array Directivity

The expression for planar array directivity is

$$G = \frac{4\pi F_0^2}{\int_0^{2\pi} \int_0^{\pi} F^2(\theta) \sin\theta \, d\theta \, d\phi}. \tag{2.65}$$

For an array of isotropic elements located on a rectangular lattice, with a broadside beam, the numerator of the directivity expression becomes

$$4\pi \left[\sum_{n=1}^{N/2} \sum_{m=1}^{M/2} A_{nm} \right]^2. \tag{2.66}$$

In the denominator, F^2 is a fourfold sum:

$$F^2 = \sum_n \sum_m \sum_p \sum_q A_{nm} A_{pq} \cos(2n-1)\alpha \cos(2m-1)\beta \cos(2p-1)\alpha$$
$$\times \cos(2q-1)\beta, \qquad (2.67)$$

with $\alpha = (\pi d_x/\lambda)u$ and $\beta = (\pi d_y/\lambda)v$. Liberal and frequent use of trigonometric identities allows the four-cosine product to become a sum of 8 terms, where the integer factors A and B occur in all possible combinations. The general term is

$$\tfrac{1}{8}\cos(A\alpha \pm B\beta) \qquad (2.68)$$

and the values of A and B are

$$A = \begin{cases} n+p-1 \\ n-p \end{cases}, \qquad B = \begin{cases} m+q-1 \\ m-q \end{cases}. \qquad (2.69)$$

These can be recast into the form

$$\cos(a\cos\phi + b\sin\phi) = \cos\left(\sqrt{a^2+b^2}\sin\left(\phi + \arctan\frac{a}{b}\right)\right). \qquad (2.70)$$

This can be integrated in ϕ:

$$\int_0^{2\pi} \cos\left(\sqrt{a^2+b^2}\sin\left(\phi + \arctan\frac{a}{b}\right)\right) d\phi = 2\pi J_0(\sqrt{a^2+b^2}) \qquad (2.71)$$

and this result allows the θ integration to be performed:

$$2\pi \int_0^{\pi} J_0\left(\sqrt{a^2+b^2}\right) \sin\theta \, d\theta = 4\pi \operatorname{sinc} C, \qquad (2.72)$$

where $\sqrt{A^2+B^2} = C\sin\theta$. Finally, the directivity can be written in terms of a fourfold sum:

$$D = \frac{4\left(\sum_{n=1}^{N/2}\sum_{m=1}^{M/2} A_{nm}\right)^2}{\sum_{n=1}^{N/2}\sum_{m=1}^{M/2}\sum_{p=1}^{N/2}\sum_{q=1}^{M/2} A_{nm}A_{pq}[S_1 + S_2 + S_3 + S_4]}, \qquad (2.73)$$

where

$$S_1 = \operatorname{sinc} 2\pi\sqrt{(n+p-1)^2 d_x^2/\lambda^2 + (m+q-1)^2 d_y^2/\lambda^2},$$
$$S_2 = \operatorname{sinc} 2\pi\sqrt{(n+p-1)^2 d_x^2/\lambda^2 + (m-q)^2 d_y^2/\lambda^2}, \quad (2.74)$$
$$S_3 = \operatorname{sinc} 2\pi\sqrt{(n-p)^2 d_x^2/\lambda^2 + (m+q-1)^2 d_y^2/\lambda^2},$$
$$S_4 = \operatorname{sinc} 2\pi\sqrt{(n-p)^2 d_x^2/\lambda^2 + (m-q)^2 d_y^2/\lambda^2}.$$

Unfortunately this result does not simplify appreciably for either uniform excitation or half-wave spacing. The terms S_i are only zero for special sets of integers. This is because the cross-coupling impedance terms are not in general zero.

Although this directivity expression is for a rectangular array with rectangular lattice, it can readily be modified for other geometries. For example, a flat plane array, where the waveguide linear array sticks are of different lengths, is handled by adjusting the summation limits. Other lattices such as hexagonal can also be accommodated.

In general, it is necessary to use approximations for planar array directivity. For example, the directivity is approximately

$$G \simeq 2G_x G_y, \quad (2.75)$$

where G_x and G_y are the linear array directivities along x and along y. The factor of 2 gives a result that compares well with exact square array calculations (Hansen, 1972a); it also accounts for a pencil beam radiating on one side only.

The previous derivation was for directivity of a planar array of isotropic elements. Directivity of an array of actual elements can be computed whenever the mutual impedance between any two elements can be calculated. Following the method of Section 2.7, the general directivity formula is

$$D = \frac{120 \left[\sum_n \sum_m A_{nm} \right]^2}{\sum_n \sum_m \sum_p \sum_q A_{nm} A_{pq} R_{nmpq}}. \quad (2.76)$$

Here the sums go over all the nm elements, and R_{nmpq} is a mutual resistance between the nmth element and the pqth element.

An approximate understanding of planar array effective aperture is provided by considering the directivity of a uniformly excited planar aperture. Let the aperture be in the y,z-plane of a spherical coordinate system, but a coordinate system where $\sin\theta$ is replaced by $\cos\theta$. The aperture normal is then $\theta_0 = \pi/2$, $\phi_0 = 0$. The space factor is $F = \operatorname{sinc} \pi u \operatorname{sinc} \pi v$ where

44 BASIC ARRAY CHARACTERISTICS

$$u = \frac{L}{\lambda}(\cos\theta - \cos\theta_0), \qquad v = \frac{W}{\lambda}\sin\theta\sin\phi. \tag{2.77}$$

Scan in the $\phi_0 = 0$ plane is assumed, which gives for directivity

$$G = \frac{4\pi}{\int_0^{2\pi}\int_{-\pi/2}^{\pi/2} F^2 \sin\theta\, d\theta\, d\phi}. \tag{2.78}$$

The ϕ integral can be simplfied by changing the variable to v. The necessary cosine is inserted since it is essentially unity over the main beam for large apertures. The error is larger at angles where the sidelobe energy is low. This gives the ϕ integral approximately as

$$\int_{-\pi/2}^{\pi/2} F^2\, d\phi \simeq \frac{\lambda}{W\sin\theta}. \tag{2.79}$$

The directivity is now approximately

$$\frac{4\pi}{G} \simeq \frac{\lambda}{W}\int_0^{\pi}\frac{\sin^2\pi u}{\pi^2 u^2}\, d\theta. \tag{2.80}$$

A similar change of u variable produces a result (King and Thomas, 1960) which is valid as long as the beam is narrow in both planes:

$$G \simeq \frac{4\pi LW \sin\theta_0}{\lambda^2}. \tag{2.81}$$

When the angle θ is changed back to that measured from the array normal, the effective aperture varies as $\cos\theta_0$. However, as the beam broadens with scan, the approximations become disabling. For angles near endfire, the same ϕ approximation is used, but the θ integration is approximated differently. Let

$$(-u) = \frac{L}{2\lambda}(\theta+\theta_0)(\theta-\theta_0), \tag{2.82}$$

which gives near endfire

$$\frac{4\pi}{G} \simeq \frac{\lambda^2}{LW}\int_0^1 \frac{\sin^2\pi u}{\pi^2 u^2 \sqrt{(2\lambda u/L)^2 - \theta_0^2}}\, du. \tag{2.83}$$

At endfire the result is

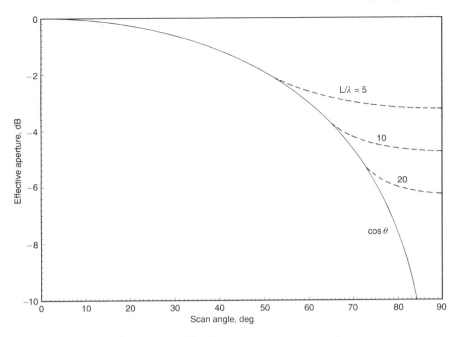

Figure 2.30 Effective aperture vs. scan angle.

$$G \simeq 3\sqrt{2}\pi \frac{W}{\lambda}\sqrt{\frac{L}{\lambda}}. \qquad (2.84)$$

Figure 2.30 shows how the effective aperture varies with principal plane scan, and from this the range of usefulness of the $\cos\theta_0$ effective aperture may be observed. The $\cos\theta_0$ value decays to the endfire value for

$$\cos\theta_0 = \frac{3}{2\sqrt{2L/\lambda}}. \qquad (2.85)$$

It should be used only for scan angles well above this critical value. Note that none of these $\cos\theta$ effective aperture results includes *scan impedance* effects; see Chapter 7.

REFERENCES

Allen J. L. et al., *Phased Array Radar Studies*, Lincoln Lab., TR-228, Aug. 1960.

Bloch, A., Medhurst, R. G., and Pool, S. D., "A New Approach to the Design of Super-Directive Aerial Arrays," *Proc. IEE*, Vol. 100, Part III, 1953, pp. 303–314.

Elliott, R. S., "The Theory of Antenna Arrays," in *Microwave Scanning Antennas*, Vol. II, R. C. Hansen, Ed., Academic Press, 1966, [Peninsula Publishing, 1985] Chapter 1.

Hansen, R. C., *Microwave Scanning Antennas*, Vol. 1, Academic Press, 1964 [Peninsula Publishing, 1985] Chapter 1.

Hansen, R. C., "Comparison of Square Array Directivity Formulas," *Trans. IEEE*, Vol. AP-20, Jan. 1972a, pp. 100–102.

Hansen, R. C., "Formulation of Echelon Dipole Mutual Impedance for Computer," *Trans. IEEE*, Vol. AP-20, Nov. 1972b, pp. 780–781.

Hansen, R. C., "Linear Arrays," in *Handbook of Antenna Design*, A. W. Rudge et al., Eds., IEE/Peregrinus, 1983, Chapter 9.

Hansen, R. C. and Brunner, G., "Dipole Mutual Impedance for Design of Slot Arrays," *Microwave J.*, Vol. 22, Dec. 1979, pp. 54–56.

King, M. J. and Thomas, R. K., "Gain of Large Scanned Arrays," *Trans. IRE*, Vol. AP-8, Nov. 1960, pp. 635–636.

Lo, Y. T. and Lee, S. W., "Affine Transformation and Its Application to Antenna Arrays," *Trans. IEEE*, Vol. AP-13, Nov. 1965, pp. 890–896.

Mailloux, R. J., *Phased Array Antenna Handbook*, Artech House, 1994.

Miller, C. J., "Minimizing the Effects of Phase Quantization Errors in an Electronically Scanned Array," *RADC Symposium Electronically Scanned Array Techniques and Applications*, April 1964, RADC-TDR-64-225, AD-448 481.

Nelson, E. A., "Quantization Sidelobes of a Phased Array with a Triangular Element Arrangement," *Trans. IEEE*, Vol. AP-17, May 1969, pp. 363–365.

Sharp, E. D., "A Triangular Arrangement of Planar-Array Elements That Reduces the Number Needed," *Trans. IRE*, Vol. AP-9, Mar. 1961, pp. 126–129.

Smith, M. S. and Guo, Y. C., "A Comparison of Methods for Randomizing Phase Quantization Errors in Phased Arrays", *Trans. IEEE*, Vol. AP-31, Nov. 1983, pp. 821–827.

Southworth, G. C., "Certain Factors Affecting the Gain of Directive Antenna Arrays," *Proc. IRE*, Vol. 18, Sept. 1930, pp. 1502–1536.

Von Aulock, W. H., "Properties of Phased Arrays," *Proc. IRE*, Vol. 48, Oct. 1960, pp. 1715–1727.

Whittaker, E. T. and Watson, G. N., *Modern Analysis*, 4th ed., Cambridge University Press, 1952.

CHAPTER THREE

Linear Array Pattern Synthesis

3.1 INTRODUCTION

Linear arrays of radiating antenna elements have been developed over many decades. This chapter brings together the important developments that provide a detailed understanding of array design and synthesis. Included are the design of narrow-beam, low-sidelobe arrays, shaped beam arrays, arrays designed by constrained optimization, and aperiodic arrays. Many older data have been recalculated using the excellent available computing power, as well as appropriate new supplementary data.

3.1.1 Pattern Formulations

The fields radiated from a linear array are a superposition (sum) of the fields radiated by each element in the presence of the other elements. Each element has an excitation parameter, which is current for a dipole, voltage for a slot, and mode voltage for a multiple-mode element. This chapter is concerned primarily with the "forced excitation" problem, where the drive of each element is individually adjusted so that the excitation is as desired. This adjustment of the drive accommodates the mutual coupling among feeds which changes the element input impedance (admittance). A more common situation is that of "free excitation": the feed network is designed to produce the desired excitations, but as the scan angle or frequency changes, the element "*scan impedance*" change will result in a change of excitation. These matters are discussed more fully in Chapter 7.

The excitation of each element will be complex, with amplitude and phase. This discrete distribution is often called an aperture distribution, where the array is the aperture. The far-field radiation pattern is the discrete Fourier transform of the array excitation. The pattern is sometimes called the space factor. A commonly used array notation is appropriate for this chapter: the angle from broadside is θ, the element spacing is d, and the array variable u is

$$u = \frac{d}{\lambda}(\sin\theta - \sin\theta_0). \tag{3.1}$$

Although the far-field pattern can be written as a simple sum of amplitude and exponential phase path length terms, it is convenient to provide a real pattern in a formulation valid for both even and odd numbers of elements. This formulation refers the phase to the array center, which results in the real pattern, and is

$$F(u) = \sum_{n=1}^{N} A_n \exp[j2\pi(n-1)u]. \tag{3.2}$$

Excitation coefficients are A_n, the array has N elements, and the beam peak is at θ_0. The array is represented in Fig. 3.1.

To understand how the element contributions combine, the unit circle approach of Schelkunoff (1943) replaces the exponential factor by a new variable:

$$w = \exp(j2\pi u). \tag{3.3}$$

This gives the pattern as a polynomial in w:

$$F(u) = \sum_{n=1}^{N} A_n w^{n-1}. \tag{3.4}$$

Physical space, commonly called real space, correponds to part of the unit circle in the complex w plane. The polar angle is $\psi = 2\pi u$; the part of the unit circle covered is for

$$-kd(1 + \sin\theta_0) \le \psi \le kd(1 - \sin\theta_0). \tag{3.5}$$

With half-wave spacing, w traverses the unit circle once. Wavelength spacing produces two traverses; and so on. The polynomial in w has $N - 1$ roots which

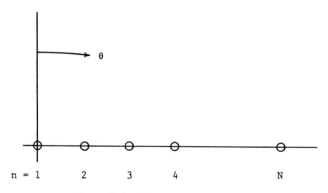

Figure 3.1 Linear array geometry.

may or may not lie on the unit circle. The pattern is given by the product of the distances from the observation point in w (on the circle) to each of the zeros (roots); see Fig. 3.2. As w moves around the circle, lobes form, then decay. When zeros are located on the unit circle, pattern nulls are produced; zeros off the unit circle may give pattern minima. A simple case is the uniform array:

$$F(u) = \frac{\sin N\pi u}{N \sin \pi u} \exp\left[j\pi(N-1)u\right] = \frac{w^N - 1}{w - 1}. \tag{3.6}$$

Roots are equally spaced on the circle by $2\pi/N$, and the exponential factor gives the phase at the center of the array. This control of pattern by movement of w-plane zeros will be utilized later, in shaped beam synthesis. Readers may recognize that the unit circle technique developed by Schelkunoff is exactly the Z transform used many years later in circuit analysis.

3.1.2 Physics Versus Mathematics

For uniform excitation, calculation of array performance is relatively easy; tapered distributions used to reduce sidelobes are less easy. In the days BC (before computers), aperture distributions were chosen for their easy integrability to closed-form solutions. For example, a cosine to the nth power, on a pedestal, allows the sidelobe level to be adjusted, and is readily calculated. There are, however, two disadvantages to these distributions. In the example there are two parameters: exponent and pedestal height. For given values of these, all important characteristics (beamwidth, aperture efficiency, sidelobe level, beam efficiency, etc.) can be calculated. But there is no easy way to choose the two parameters to optimize efficiency for a given sidelobe level, for example. More fundamental is the second disadvantage: these distributions, depending on the values of parameters selected, may be quite inefficient in

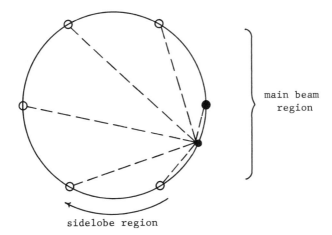

Figure 3.2 Unit circle for seven-element array.

terms of beamwidth, aperture efficiency, etc. Another example is the popular Gaussian. With a pedestal (a truncated Gaussian) it is multiple parameter distribution, and, what is worse, the Fourier transform is complex. Thus in modern antenna work easy mathematics has yielded to good physics: aperture distributions should be designed by proper placement of pattern-function zeros, preferably using a single parameter. This assures good physics, which means highly efficient, low-Q apertures. Design can be handled via the computer. Taylor was a pioneer in this approach to narrow-beamwidth low-sidelobe distributions; his design principles are discussed next.

3.1.3 Taylor Narrow-Beam Design Principles

A common pattern requirement is for high directivity and low sidelobes, useful for radar, communications, mapping, etc. *Sidelobe level* (SLL) is the amplitude of the highest sidelobe, usually that closest to the main beam, normalized to the main beam peak. It is convenient also to use *sidelobe ratio* (SLR), which is the inverse of SLL. The main beam may be fixed at broadside or at some other angle, or it may be electronically scanned to any desired angle. The terms "electronic scanning array" and "phased array" are synonymous. For maximum directivity without superdirectivity, uniform excitation is used. When lower sidelobes than those provided by uniform excitation are required, the aperture amplitude is tapered (apodized, shaded) from the center to the ends. There are some general rules, derived by Taylor and others.

Taylor (1953) and his colleagues developed some important general rules for low-sidelobe patterns. These are listed here.

1. Symmetric amplitude distributions give lower sidelobes.
2. $F(u)$ should be an entire function of u.
3. A distribution with a pedestal produces a far-out sidelobe envelope of $1/u$.
4. A distribution going linearly to zero at the ends produces a far-out sidelobe envelope of $1/u^2$.
5. A distribution that is nonzero at the ends (pedestal) is more efficient.
6. Zeros should be real (located on the w unit circle).
7. Far-out zeros should be separated by unity (in u).

In the sections that follow, these principles will be employed to produce highly efficient and robust (low-Q) excitations and the corresponding patterns.

Few methods exist for the design of efficient array distributions for prescribed patterns; most syntheses utilize continuous apertures where the aperture distribution is sampled to obtain the array excitation values. As long as the array spacing is at or near half-wave, and the number of elements is not small, this procedure works well. When the number of elements is too small, a discrete $\bar{n}$ synthesis of the type presented in Section 3.6 can be used. The cost is that a new synthesis is required for each different number of elements. A more sophisticated method matches the zeros of the desired continuous pattern to those of the discrete array pattern; see Section 3.9.2. Of course, the Dolph–Chebyshev synthesis of Section 3.2 is discrete from the start, but these designs are seldom used because of the high farther-out sidelobes.

3.2 DOLPH–CHEBYSHEV ARRAYS

3.2.1 Half-Wave Spacing

A half-wave spaced array yields maximum directivity for a given sidelobe ratio when all sidelobes are of equal height. Dolph (1946) recognized that Chebyshev polynomials were ideally suited for this purpose. In the range ±1 the polynomial oscillates with unit amplitude, while outside this range it becomes monotonically large. See, for example, Fig. 3.3 for the graph of a ninth-order Chebyshev polynomial. The polynomial can be simply expressed in terms of trig type functions as

$$T_n(x) = \begin{cases} (-1)^n \cosh(n \operatorname{arccosh} |x|), & x < -1; \\ \cos(n \arccos x), & |x| \leq 1; \\ \cosh(n \operatorname{arccosh} x), & x \geq 1. \end{cases} \quad (3.7)$$

It can also be readily calculated through recursion. A direct correspondence between the array polynomial and the Chebyshev polynomial is not feasible, because the array main beam must be symmetric and have zero slope at the

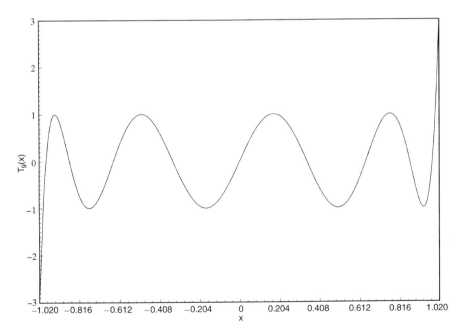

Figure 3.3 Chebyshev polynomial.

center. Since an N-element array has $N-1$ zeros, an $(N-1)$-degree Chebyshev, which has the same number of zeros, is appropriate. The oscillatory part of the Chebyshev polynomial is mapped once onto the sidelobes on one side of the main beam; the main beam up to the peak is mapped onto the $x > 1$ region of the Chebyshev polynomial. The transformation that performs this is $x = x_0 \cos(u/2)$. This transforms the Chebyshev polynomial $T_{N-1}(x)$ to the pattern function $F(u)$. The pattern variable is given by $u = kd \sin \theta$. Thus the pattern factor is simply

$$F(u) = T_{N-1}\left(x_0 \cos \frac{u}{2}\right). \tag{3.8}$$

As θ varies from $-\pi/2$ to $\pi/2$, x varies from $(x_0 \cos \pi d/\lambda)$ up to $\theta = 0$ and then back to $x_0 \cos \pi d/\lambda$. For half-wavelength spacing, the minimum value of x is zero. The maximum allowable spacing is determined by the need to prevent x falling below -1. This gives the maximum element spacing for a broadside array as

$$\frac{d}{\lambda} = \frac{1}{\pi} \arccos\left(-\frac{1}{x_0}\right), \tag{3.9}$$

which approaches one-wavelength spacing for large arrays.

The voltage sidelobe ratio is given by

$$\text{SLR} = T_{N-1}(x_0), \qquad \text{SLR} > 1. \tag{3.10}$$

Given the SLR, the mapping constant x_0 is given by

$$\begin{aligned} x_0 &= \cosh \frac{\operatorname{arccosh} \text{SLR}}{N-1} \\ &= \tfrac{1}{2}\left[\text{SLR} + \sqrt{\text{SLR}^2 - 1}\right]^{1/N} + \tfrac{1}{2}\left[\text{SLR} - \sqrt{\text{SLR}^2 - 1}\right]^{1/N}. \end{aligned} \tag{3.11}$$

The second form is given because some computers do not have inverse hyperbolic function subroutines. The Dolph formulation is valid only for array spacings of 0.5λ and larger. A typical array pattern is given in Fig. 3.4. Using the array center as phase reference, the space factor may be written as a series:

$$F(u) = \sum_{n=1}^{N} A_n \exp\left\{j\left[(2n - N - 1)\frac{u}{2}\right]\right\}. \tag{3.12}$$

This finite series is inverted to produce the coefficients:

$$A_n = \frac{1}{N} \sum F\left(\frac{2\pi m}{N}\right) \exp\left\{-j\left[(2n - N - 1)\frac{\pi m}{N}\right]\right\}. \tag{3.13}$$

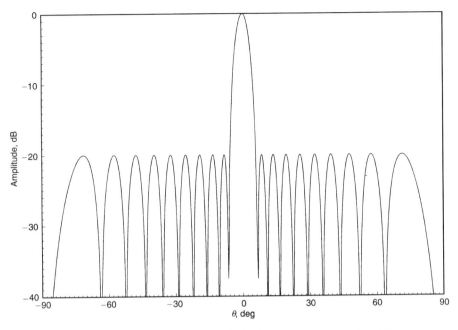

Figure 3.4 20-element Dolph–Chebyshev array pattern, $d = 0.5\lambda$.

When the Chebyshev polynomial is substituted in, the coefficients are given as a sum of Chebyshev polynomials:

$$A_n = \frac{1}{N} \sum_m T_{N-1}\left(x_0 \cos \frac{\pi m}{N}\right) \exp\left\{-j\left[(2n - N - 1)\frac{\pi m}{N}\right]\right\}. \tag{3.14}$$

Because the pattern is symmetric, additional simplifications can be made. The formulas developed by Stegen (1953) are convenient for calculating the excitation coefficients. In these formulas the elements are numbered from the array center, and are different for odd and even N. For even $N = 2M$:

$$A_n = \frac{2}{N}\left[\text{SLR} + 2\sum_{m=1}^{M-1} T_{N-1}\left(x_0 \cos \frac{\pi m}{N}\right) \cos\left((2n-1)\frac{\pi m}{N}\right)\right], \quad 1 \le n \le M \tag{3.15}$$

And for odd $N = 2M + 1$:

$$A_n = \frac{2}{N}\left[\text{SLR} + 2\sum_{m=1}^{M} T_{N-1}\left(x_0 \cos \frac{m\pi}{N}\right) \cos \frac{2nm\pi}{N}\right], \tag{3.16}$$

These results are valid for spacings $\ge \lambda/2$.

These formulas are awkward for large arrays; simplified approximate versions have been developed by Van der Maas (1954) and Barbiere (1952).

54 LINEAR ARRAY PATTERN SYNTHESIS

However, with current computer capability, such approximations are unnecessary. Stegen's formulas can be implemented directly. An extensive table of coefficients, directivity, and beamwidth has been prepared by Brown and Scharp (1958), giving $N = 3(1)40$ and $SLR = 10(1)40\,dB$ for $d = \lambda/2$. However, there are significant errors near the upper values of N and SLR, presumably due to roundoff error. Users should compute their own values using the formulas above, or those that follow.

The array coefficients for a typical 30-element array are shown in Fig. 3.5. It may be observed that none of the three distributions is monotonic, and that the 20 dB SLR values are larger at the ends than at the center. In general the high edge values increase with larger arrays, and with lower SLR. This curve points up a major difficulty in the utilization of Dolph–Chebyshev arrays: the nonmonotonic distributions and end spikes are difficult to achieve. A further disadvantage is that the equal level far-out sidelobes tend to pick up undesired interference and clutter. For these reasons, Dolph–Chebyshev arrays are seldom used. Figure 3.6 gives values of the maximum N for a given sidelobe ratio that produces a monotonic distribution, and also the maximum N versus sidelobe ratio that produces an end spike no larger than the center excitation value.

Beamwidth is simply calculated from the pattern Eqn. (3.12); however, a root finder needs to be used to find the beamwidth values from the equation

$$\sqrt{2}T_{N-1}\left(x_0 \cos\frac{u_3}{2}\right) = T_{N-1}(x_0). \tag{3.17}$$

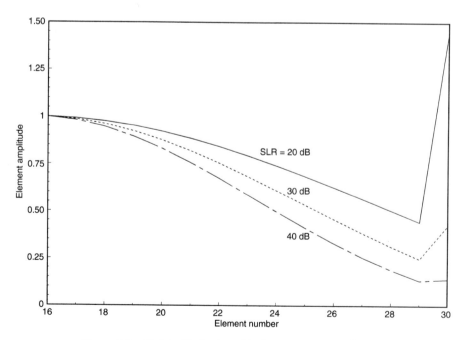

Figure 3.5 Dolph–Chebyshev 30-element array coefficients.

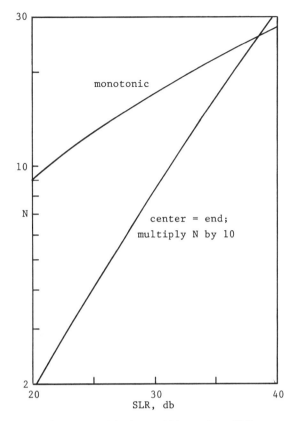

Figure 3.6 Maximum N for a given SLR.

Figure 3.7 gives normalized beamwidth in degrees for a half-wave spaced array with number of elements from 10 to 100, versus sidelobe ratio. It can be seen that the increase in beamwidth with SLR is roughly linear as expected.

Directivity of a Dolph–Chebyshev array is calculated from the directivity formulas of Chapter 2 using the excitation coefficients given above. Rather than show directivity, it is more useful to show array efficiency, which is the ratio of array directivity to that of a uniformly excited array of the same number of elements and spacing. Figure 3.8 shows array efficiency for half-wave spaced arrays. It is interesting to note that the efficiency peaks for a particular sidelobe ratio for each number of array elements. This is because the sidelobe energy is a function primarily of the sidelobe ratio, while the mainbeam energy primarily depends upon the number of elements. Thus as the sidelobe energy becomes a larger fraction of the total, the efficiency decreases. The values of number of elements and sidelobe ratio that provide maximum efficiency are shown in Fig. 3.9. This curve is fitted approximately by $\log N \simeq 0.08547 \, \text{SLR}_{db} - 0.4188$.

56 LINEAR ARRAY PATTERN SYNTHESIS

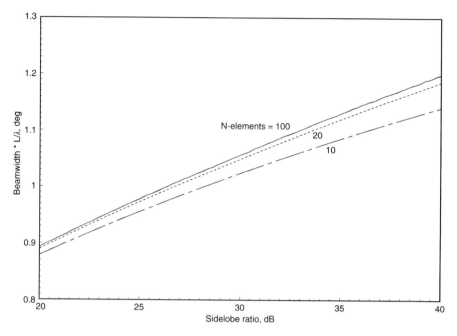

Figure 3.7 Dolph–Chebyshev distribution beamwidth.

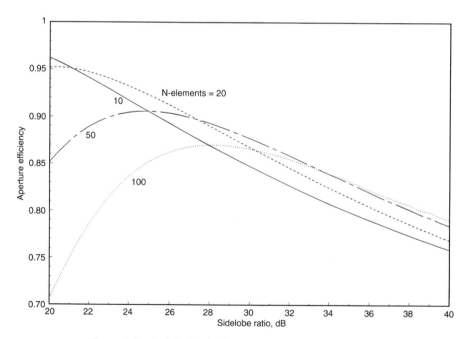

Figure 3.8 Dolph–Chebyshev array efficiency, $d = 0.5\lambda$.

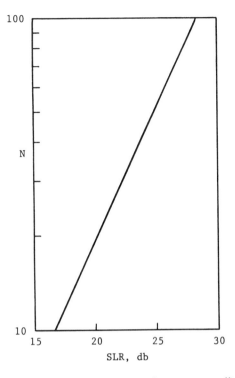

Figure 3.9 Dolph–Chebyshev array with maximum efficiency.

3.2.2 Spacing Less Than Half-Wave

The derivations in the previous section are limited to half-wave spacing or greater, because smaller spacings only utilize a smaller part of the oscillatory region of the polynomial. For N odd, Riblet (1947) showed that this restriction could be removed; the pattern factor is formed by starting near the end of the Chebyshev $+1$ region, tracing the oscillatory region to the other end, and then retracing back to the start end and up the monotonic portion to form the main beam half. The exact starting point depends on Nkd. Since the M-order Chebyshev has $M-1$ oscillations, which are traced twice, and since the trace from zero to 1 and back forms the sidelobes, the pattern factor always has an odd number of sidelobes each side, and thus an even number of zeros. As a result, only an odd number of elements can be formed into a Chebyshev array for spacing less than half-wave using the Riblet synthesis. The pattern function is given by

$$F(u) = T_M(a\cos(u) + b),$$
$$a = \frac{z_0 + 1}{1 - \cos(kd)}, \quad b = \frac{z_0 \cos(kd) + 1}{\cos(kd) - 1}. \quad (3.18)$$

As before, $u = kd \sin \theta$ and $M = (N-1)/2$. The value of z_0 is

$$z_0 = \cosh \frac{\text{arccosh}(\text{SLR})}{M}. \tag{3.19}$$

Formulas have been developed by many, including DuHamel (1953), Brown (1957, 1962), Drane (1963, 1964), and Salzer (1975). The formulas of Drane are suitable for computer calculation of superdirective arrays (see Chapter 9) and thus are used here. Again the elements are numbered starting with zero at the center; the array excitation coefficients are

$$A_n = \frac{\epsilon_n}{4M} \sum \epsilon_m \epsilon_{M_2-m} T_n(x_n) T_M(ax_n + b) + (-1)^n T_M(b - ax_n), \tag{3.20}$$

where $\epsilon_i = 1$ for $i = 0$ and $\epsilon_i = 2$ for $i > 0$. The variable $x_n = \cos n\pi/M$. The integers M_1 and M_2 are the integer parts of $M/2$ and $(M+1)/2$, respectively. Small spacings may require double-precision arithmetic due to the subtraction of terms.

For half-wave spacing, the a and b above reduce to

$$a = \tfrac{1}{2}(z_0 + 1), \qquad b = \tfrac{1}{2}(z_0 - 1). \tag{3.21}$$

For half-wave spacing the two approaches give identical results. For this case the pattern factor of Eqn. (3.8) is equal to the pattern factor of Eqn. (3.18):

$$T_{N-1}\left(x_o^{N-1} \cos \frac{u}{2}\right) \equiv T_M\left(\frac{z_o^M(\cos(u) + 1) + \cos(u) - 1}{2}\right). \tag{3.22}$$

DuHamel (1953) extended the Dolph–Chebyshev design to endfire arrays, but only for spacing less than half-wave. This is not a serious restriction, as spacing is customarily made quarter-wave at endfire to avoid a large back lobe. For any scan angle, u is modified as usual for scanned arrays:

$$u = ka(\sin\theta - \sin\theta_0), \tag{3.23}$$

where θ_0 is the main beam scan angle. The interelement phase shift is $kd \sin\theta_0$, and the coefficients a and b become

$$a = \frac{3 + z_0 + 2\sqrt{2(z_o + 1)}\cos kd}{2 \sin^2 kd}, \tag{3.24}$$

$$b = \frac{(\sqrt{z_o + 1} + \sqrt{2}\cos kd)^2}{2 \sin^2 kd}. \tag{3.25}$$

Drane (1968) has shown that a large Dolph–Chebyshev array has a directivity independent of scan angle for any spacings between half-wave and full-wave, as long as the spacing admits no grating lobes.

3.3 TAYLOR ONE-PARAMETER DISTRIBUTION

3.3.1 One-Parameter Design

Taylor (1955) recognized that to produce a linear aperture distribution with a sidelobe envelope approximating a $1/u$ falloff, the uniform amplitude $\sin(x)/x$ pattern could be used as a starting point. He understood that the height of each sidelobe is controlled by the spacing between the aperture pattern factor zeros on each side of the sidelobe. Since the sinc pattern has a $1/u$ sidelobe envelope, it was necessary only to modify the close-in zeros to reduce the close-in sidelobes. The far-out zeros were left at the integers. The shifting was accomplished by setting zeros equal to

$$u = \sqrt{n^2 + B^2}, \tag{3.26}$$

where B is a positive real parameter. The pattern function is obtained by writing the sinc function as the ratio of two infinite products. Then the zeros are shifted as mentioned. In this form the pattern function, where C is a constant, is

$$F(u) = \frac{C \prod_{n=1}^{\infty}\left[1 - \frac{u^2 - B^2}{n^2}\right]}{\prod_{n=1}^{\infty}\left[1 + \frac{B^2}{n^2}\right]}. \tag{3.27}$$

This is more simply written as

$$\begin{aligned} F(u) &= \frac{\sinh \pi\sqrt{B^2 - u^2}}{\pi\sqrt{B^2 - u^2}}, & u \leq B; \\ F(u) &= \frac{\sin \pi\sqrt{u^2 - B^2}}{\pi\sqrt{u^2 - B^2}}, & u \geq B. \end{aligned} \tag{3.28}$$

Now the constant $C = \sinh(\pi B)/\pi B$. This pattern function then is a modified sinc pattern, with the single parameter B controlling all characteristics: sidelobe level, beamwidth, directivity efficiency, beam efficiency, etc. A transition from the sinc form to the hyperbolic form occurs on the side of the main beam at $u = B$. For smaller u the hyperbolic form provides the central part of the main beam, with a peak value of $\sinh(\pi B)/\pi B$. The trig form provides the lower part of the main beam and the sidelobes. The sidelobe ratio is immediately that of the sinc function multiplied by the beam peak value. In dB this is

$$\text{SLR} = 20\log\frac{\sinh \pi B}{\pi B} + 13.2614 \quad \text{dB}. \tag{3.29}$$

A typical Taylor one-parameter pattern, plotted versus u, is shown in Fig. 3.10.

60 LINEAR ARRAY PATTERN SYNTHESIS

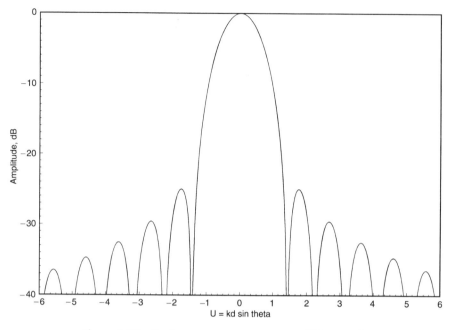

Figure 3.10 Taylor one-parameter pattern, SLR = 25 dB.

The beamwidth is found by applying a root finder to the equation

$$\sqrt{2}\, F(u_3) = \frac{\sinh \pi B}{\pi B}. \qquad (3.30)$$

These values are given in Table 3.1 for sidelobe ratios up to 50 dB. The u_3 beamwidth is converted to physical angle θ_3 using

$$\frac{\theta_3}{\lambda/L} = \frac{2L}{\lambda} \arcsin \frac{u_3}{L/\lambda}. \qquad (3.31)$$

TABLE 3.1 Taylor One-Parameter Characteristics

SLR (dB)	B	u_3 (rad)	η_t	$\eta_t u_3$	Edge taper (dB)
13.26	0	0.4429	1	0.443	0
15	0.3558	0.4615	0.993	0.457	2.5
20	0.7386	0.5119	0.933	0.478	9.2
25	1.0229	0.5580	0.863	0.481	15.3
30	1.2762	0.6002	0.801	0.481	21.1
35	1.5136	0.6391	0.751	0.480	26.8
40	1.7415	0.6752	0.709	0.479	32.4
45	1.9628	0.7091	0.674	0.478	37.9
50	2.1793	0.7411	0.645	0.478	43.3

For long linear apertures the normalized beamwidth is independent of L/λ and is

$$\frac{\theta_3}{\lambda/L} \simeq 2u_3. \qquad (3.32)$$

Figure 3.11 shows normalized beamwidth for apertures of length 5, 10, and 50 wavelengths. It can be seen that there is only a small difference between 5 and 10 wavelength apertures for the larger sidelobe apertures. For most cases then the approximate formula is satisfactory.

For some mapping and radiometer applications, the fraction of energy radiated in the main beam null-to-null is important. This is called beam efficiency η_b. The beam efficiency is calculated by integrating the square of the pattern function over the main beam, and dividing by the integral of the square of the pattern function over the entire pattern. This would normally require numerical integration over all of the pattern sidelobes, an expensive and relatively inaccurate task. However, it will be seen later that the taper efficiency η_t also involves this integration over all the pattern sidelobes. This allows the beam efficiency to be written in terms of the taper efficiency and an integral only over the main beam, a relatively simple and fast integration. This result is

$$\eta_b = \frac{2L\eta_t}{F_0^2 \lambda} \int_0^{\theta_1} |F|^2 \cos\theta \, d\theta \qquad (3.33)$$

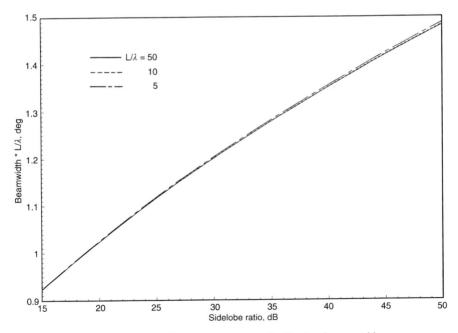

Figure 3.11 Taylor one-parameter distribution beamwidth.

LINEAR ARRAY PATTERN SYNTHESIS

Figure 3.12 shows beam efficiency versus sidelobe ratio. The curve is independent of L/λ for L/λ as small as 5.

The aperture distribution is the inverse transform of Eqn. (3.28):

$$g(p) = I_0(\pi B\sqrt{1 - p^2}), \qquad (3.34)$$

where p is the distance from aperture center to aperture end, with $p = 1$ at the end. I_0 is the zero-order modified Bessel function. The pedestal height is $1/I_0(\pi B)$. Aperture amplitude distributions from center to end are given for several SLRs in Fig. 3.13. Aperture (taper) efficiency is given by

$$\eta_t = \frac{|F(0)|^2}{\displaystyle\int_{-L/2\lambda}^{L/2\lambda} |F(u)|^2 \, du}. \qquad (3.35)$$

Because of the rapid sidelobe decay with u, the integral limits may be approximated by infinity even for modest apertures. The integration can then be reduced to a tabulated integral (Rothman, 1949):

$$\eta_t = \frac{2\sinh^2 \pi B}{\pi B \, \bar{I}_0(2\pi B)}, \qquad (3.36)$$

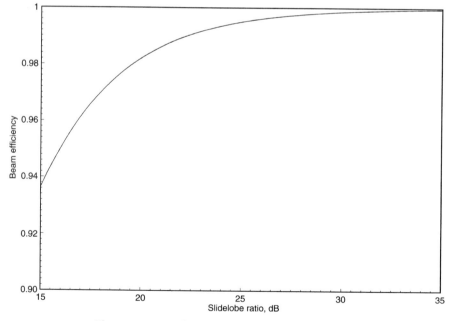

Figure 3.12 Taylor one-parameter beam efficiency.

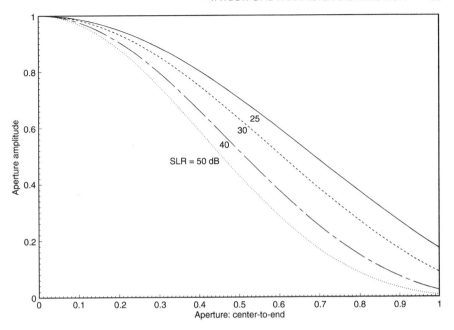

Figure 3.13 Taylor one-parameter aperture distribution.

where $\bar{I}_0$ is the integral of I_0 from zero to the argument. Values of aperture efficiency are shown in Table 3.2.

3.3.2 Bickmore–Spellmire Two-Parameter Distribution

Occasionally it is desirable to provide an envelope taper more rapid than $1/u$. The two-parameter family of distributions developed by Bickmore and Spellmire (1956) allows this. The two-parameter pattern factor is

TABLE 3.2 Taylor One-Parameter Efficiencies

SLR (dB)	B	η_t	η_b
13.26	0	1	0.9028
15	0.3558	0.9931	0.9364
20	0.7386	0.9329	0.9820
25	1.0229	0.8626	0.9950
30	1.2762	0.8014	0.9986
35	1.5136	0.7509	0.9996
40	1.7415	0.7090	0.9999
45	1.9628	0.6740	1.0000
50	2.1793	0.6451	1.0020

$$F(u) = \frac{J_\nu \pi \sqrt{u^2 - C^2}}{(\pi \sqrt{u^2 - C^2})^\nu}. \tag{3.37}$$

A value of $\nu = 1/2$ yields the one-parameter pattern while $\nu > 1/2$ gives envelope tapers more rapid than $1/u$. When $\nu = -1/2$, the "ideal" pattern of Eqn. (3.28) results. The parameter C controls initial sidelobe level. Note that these two parameters are independent, unlike the obsolete multiple-parameter distributions previously discussed. The aperture distribution is

$$g(p) = \frac{J_{\nu-1/2}(j\pi C\sqrt{1-p^2})}{(j\pi C\sqrt{1-p^2})^{\nu-1/2}} \tag{3.38}$$

When $\nu < 1/2$, singularities occur at the aperture ends. A value of $\nu = 3/2$, for example, gives a $1/u^2$ sidelobe envelope taper. These more rapid tapers are seldom employed because of their low aperture efficiency.

3.4 TAYLOR N-BAR APERTURE DISTRIBUTION

The Taylor $\bar{n}$ *distribution* was developed as a compromise between the Dolph–Chebyshev or "ideal" aperture with its constant-level sidelobes, and the $1/u$ sidelobe envelope falloff of the $\sin(x)/x$ pattern. The goal of this $\bar{n}$ distribution is to obtain higher efficiency while retaining most of the advantages of a tapered distribution. To accomplish this, Taylor started from a different point from that used for the one-parameter distribution. Here his starting point is the "ideal" linear aperture, which has a pattern of equal-level sidelobes like the Dolph–Chebyshev array distribution. Thus for any linear phase uniform aperture it provides the highest directivity for a given sidelobe ratio. The equal-level sidelobes are provided by a cosine function, with the main beam provided by a hyperbolic cosine:

$$\begin{aligned} F(u) &= \cosh \pi \sqrt{A^2 - u^2}, & u \leq A; \\ F(u) &= \cos \pi \sqrt{u^2 - A^2}, & u \geq A. \end{aligned} \tag{3.39}$$

The single parameter A controls the distribution and pattern characteristics. Pattern zeros $z_n = \pm\sqrt{A^2 + (n - 1/2)^2}$. The sidelobe ratio is the value at $u = 0$:

$$\text{SLR} = \cosh A \tag{3.40}$$

and the parameter in terms of SLR is

$$A = \frac{1}{\pi} \ln \left(\text{SLR} + \sqrt{\text{SLR}^2 - 1} \right). \tag{3.41}$$

The 3 dB beamwidth is given by

$$u_3 = \frac{1}{\pi}\left[\ln^2\left(\text{SLR} + \sqrt{\text{SLR}^2 - 1}\right) - \ln^2\left(\text{SLR}/\sqrt{2} + \sqrt{\text{SLR}^2/2 - 1}\right)\right]^{1/2}. \quad (3.42)$$

There is a transition on the side of the main beam at $u = A$ between the trig and hyperbolic forms. This canonical distribution is never used, as it not only has high far-out sidelobes but the distribution has an infinite spike at the ends. Of course this is the limiting value of the Dolph–Chebyshev distribution as the number of elements increases. However the "ideal" distribution led to a widely used distribution, the Taylor $\bar{n}$.

Taylor utilized the first few equal sidelobes of the "ideal" distribution to provide increased efficiency, and shifted zeros of the far-out sidelobes to produce a $1/u$ envelope to reduce interference and clutter, and to provide a low-Q distribution. To do this, the u scale is stretched slightly by a factor σ, with σ slightly greater than unity. Thus the close-in zero locations are not changed significantly. At some point a zero will fall at an integer; from this transition point on, the zeros occur at $\pm n$. The pattern then has $\bar{n}$ roughly equal sidelobes; beyond $u = \bar{n}$ the sidelobe envelope decays as $1/u$. The first $\bar{n}$ lobes are not of precisely equal heights, as the transition region allows some decay of the sidelobes near the transition point. The envelope beyond the transition point is slightly different from $1/u$. The zeros are

$$\begin{aligned} z_n &= \pm\sigma\sqrt{A^2 + (n - \tfrac{1}{2})^2}, & 1 \le n \le \bar{n}; \\ z_n &= \pm n, & \bar{n} \le n. \end{aligned} \quad (3.43)$$

The dilation parameter is

$$\sigma = \frac{\bar{n}}{\sqrt{A^2 + (\bar{n} - \tfrac{1}{2})^2}}. \quad (3.44)$$

The pattern function incorporating these zeros was first written as an infinite product on the zeros, starting with the expansion of the "ideal" aperture. For both conceptual and computational reasons it is advantageous to incorporate the regularly spaced zeros into a trig function, with the $\bar{n} - 1$ modified zeros in an explicit finite product:

$$F(u) = \frac{\sin \pi u}{\pi u} \prod_{n=1}^{\bar{n}-1} \frac{1 - u^2/z_n^2}{1 - u^2/n^2}. \quad (3.45)$$

This pattern function can also be written as a superposition of sinc beams each side of the central beam spaced at unit intervals in u:

$$F(u) = \sum_{n=-(\bar{n}-1)}^{\bar{n}-1} F(n, A, \bar{n}) \,\text{sinc}\,[\pi(u+n)]. \tag{3.46}$$

Figure 3.14 shows a typical Taylor $\bar{n}$ pattern in u. It may be observed that the first four sidelobes have an envelope closer to constant level than to $1/u$; the $\bar{n}$ transition forces a gradual change in sidelobe height. For large $\bar{n}$, the nearly constant sidelobes are more pronounced.

The beamwidth is accurately given by σu_3^i, where u_3^i is the "ideal" beamwidth as given in Eqn. (3.42). Table 3.3 gives the parameter A versus sidelobe

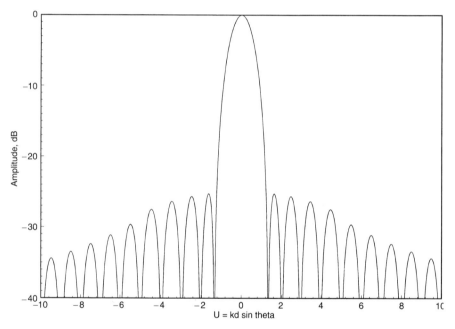

Figure 3.14 Taylor $\bar{n}$ pattern, SLR = 25 dB, $\bar{n}$ = 5.

TABLE 3.3 Taylor $\bar{n}$ Characteristics

SLR (dB)	A	u_3	σ				
			$\bar{n}=2$	$\bar{n}=4$	$\bar{n}=6$	$\bar{n}=8$	$\bar{n}=10$
20	0.9528	0.4465	1.1255	1.1027	1.0749	1.0582	1.0474
25	1.1366	0.4890		1.0870	1.0683	1.0546	1.0452
30	1.3200	0.5284		1.0693	1.0608	1.0505	1.0426
35	1.5032	0.5653			1.0523	1.0459	1.0397
40	1.6865	0.6000			1.0430	1.0407	1.0364
45	1.8697	0.6328				1.0350	1.0328
50	2.0530	0.6639					1.0289

ratio, along with the beamwidth and dilation parameter. Normalized beamwidth, BW × L/λ, is shown in Fig. 3.15. A continuous curve is not used as the $\bar{n}$ value for each SLR is different. Also shown is the beamwidth curve for the Taylor *one-parameter distribution*. As expected, the $\bar{n}$ distribution is significantly better, especially for high SLR.

The aperture distribution is best expressed as a finite Fourier series:

$$g(p) = 1 + 2 \sum_{n=1}^{\bar{n}-1} F(n, A, \bar{n}) \cos n\pi p. \tag{3.47}$$

Again the aperture variable p is zero at the center and unity at the ends. The coefficient used in this formula, and in the previous pattern formula is

$$F(n, A, \bar{n}) = \frac{[(\bar{n} - 1)!]^2}{(\bar{n} - 1 + n)!(\bar{n} - 1 - n)!} \prod_{m=1}^{\bar{n}-1}\left[1 - \frac{n^2}{z_m^2}\right], \tag{3.48}$$

where $F(0, A, \bar{n}) = 1$. The amplitude distribution for half of the $\bar{n}$ aperture is given in Fig. 3.16. In comparison with Fig. 3.13 for the *one-parameter distribution*, the $\bar{n}$ "tails" are much higher. That is, the distribution pedestal is higher, leading to better efficiency in accordance with Taylor's rules. Tables of aperture distribution and coefficients are given for SLR = 20(5)40 dB and $\bar{n} = 3(1)10$ by Hansen (1964). More extensive tables are available in Brown and Scharp (1958).

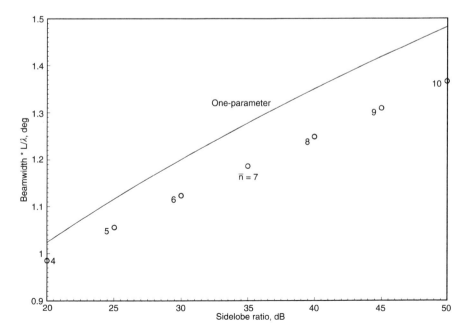

Figure 3.15 Taylor distribution beamwidth.

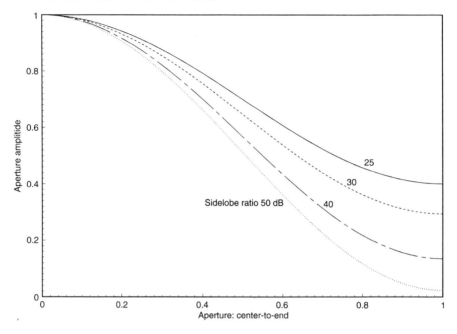

Figure 3.16 Taylor n̄ aperture distribution.

Directivity and aperture efficiency are easily calculated. It is easier to integrate the aperture distribution, since it is expressed as a Fourier series of orthogonal terms, than to integrate the pattern function:

$$\eta = \frac{\left| \int_0^1 g(p)\, dp \right|^2}{\int_0^1 g^2(p)\, dp}. \tag{3.49}$$

Using Eqn. (3.47) for the distribution function, the numerator of the efficiency expression is unity since the cosine terms integrate to zero. In the denominator the g^2 product gives unity plus two single series plus a double series. Each term of the single series integrates to zero. The double series integral is zero for n not equal to m, and gives one-half for $n = m$. Thus the aperture efficiency is

$$\frac{1}{\eta} = 1 + 2 \sum_{n=1}^{\bar{n}-1} F^2(n, A, \bar{n}). \tag{3.50}$$

Calculation is easy even for large values on n̄. For each sidelobe ratio there is a value of n̄ that gives maximum efficiency because, as n̄ increases from a small value, the main beam energy decreases faster than the sidelobe energy increases. For large n̄, however, the main beam energy changes more slowly. Typical n̄ aperture efficiency values are shown in Fig. 3.17. Since each SLR value has a different n̄ value, a continuous curve is not used. For comparison, the efficiency of the *one-parameter distribution* is also shown. Table 3.4 gives the values of n̄ that yield maximum efficiency. Also shown are the largest values of n̄ that allow a monotonic distribution, along with the corresponding aperture efficiencies. The efficiency penalty for choosing a monotonic n̄ over the maximum efficiency n̄ is modest, as may be observed.

Beam efficiency is calculated using Eqn. (3.33), with the aperture efficiency obtained above. Table 3.5 gives values for SLR from 20 to 35 dB with typical n̄

TABLE 3.4 Maximum Efficiency and Monotonic n̄ Values

SLR	Max η_t values		Monotonic	
	n̄	η_t	n̄	η_t
20	6	0.9667	3	0.9535
25	12	0.9252	5	0.9105
30	23	0.8787	7	0.8619
35	44	0.8326	9	0.8151
40	81	0.7899	11	0.7729
45				
50				

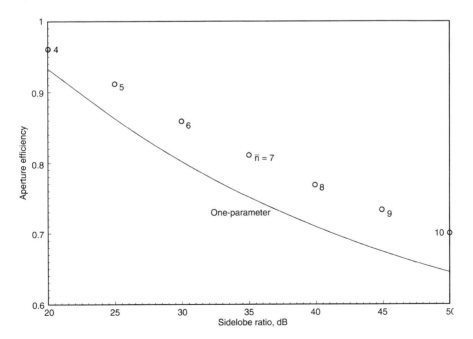

Figure 3.17 Taylor distribution aperture efficiency.

70 LINEAR ARRAY PATTERN SYNTHESIS

TABLE 3.5 Taylor n̄ Beam Efficiency

SLR (dB)	n̄	η_b
20	4	0.9658
25	5	0.9861
30	6	0.9947
35	7	0.9980

values for each. These results may be compared with those of Fig. 3.12 for the *one-parameter distribution*.

It is important to discuss the proper range of values of n̄. Too large a value of n̄ for a particular sidelobe ratio will, as shown, give a nonmonotonic aperture distribution; a large peak may even be produced at the ends of the aperture. Too large a value of n̄ will also not allow the transition zone zeros to behave properly. Figure 3.18 shows the behavior of zero spacings on n̄ for an SLR of 30 dB. The largest overshoot occurs for n̄ = 9, just above the largest monotonic value of n̄ = 8. Overshoots occur out to (and beyond) the maximum efficiency n̄ = 23. However an n̄ = 4 shows no overshoot; n̄ = 5 and 6 give small overshoots. For other values of SLR these characteristics occur also, but for different n̄ value.

In summary, the Taylor n̄ *distribution* is widely used because it gives slightly better efficiency and beamwidth than the Taylor *one-parameter distribution*, for the same sidelobe level.

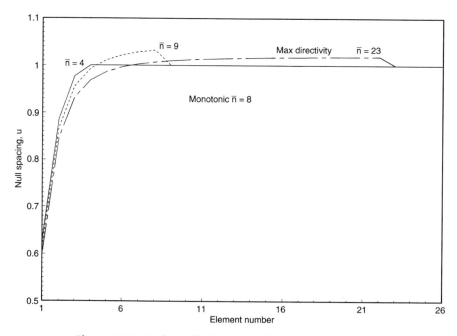

Figure 3.18 Taylor n̄ distribution null spacing, SLR = 30 dB.

3.5 LOW-SIDELOBE DISTRIBUTIONS

3.5.1 Comparison of Distributions

Low sidelobes, which might be roughly defined as -30 to -60 dB, are of interest for several reasons: reduction of radar and communications intercept probability, reduction of radar clutter and jammer vulnerability, and increasing spectrum congestion in satellite transmissions. From the previous sections it is clear that a low sidelobe distribution should be heavily tapered in amplitude, and from Taylor's rules have a reasonable pedestal height. The $\bar{n}$ space factor appears to be ideal for these applications.

The design of a high-performance low-sidelobe pattern should again emphasize the pattern zeros. The close-in zeros are adjusted to obtain a few nearly equal sidelobes at the design sidelobe level, while farther-out zeros are placed to give a $1/u$ envelope. This scientifically designed pattern and distribution is in stark contrast to the classical World War II era distributions, which were all chosen for easy integrability. For example, the Hamming distribution is a cosine of a doubled argument, on a pedestal (Blackman and Tukey, 1958). This distribution is given by

$$g(p) = 0.54 + 0.46 \cos \pi p = a + b \cos \pi p, \qquad (3.51)$$

giving a value of unity in the center and an end value of 0.08. The excitation efficiency is

$$\eta_t = \frac{a^2}{a^2 + 2b^2} = 0.7338, \qquad (3.52)$$

and the half-power beamwidth is $u_3 = 0.6515$. The pattern function has zeros at $u = 2, 3, 4, \ldots$, with another zero at $u = a/(a-b) = 2.5981$. The pattern function is

$$F(u) = \frac{\left[(a-b)u^2 - a\right] \sin \pi u}{\pi a u (u^2 - 1)}. \qquad (3.53)$$

The close spacing of the first two zeros produces an irregular sidelobe envelope; the highest sidelobe is the fourth at -42.7 dB. The first and second sidelobes are -44.0 dB and -56.0 dB.

A comparison of the Taylor one-parameter, Taylor $\bar{n}$, and Hamming distributions is illuminating. Since the Hamming SLR = 42.7 dB, this is used for the others as well. Table 3.6 gives the first 10 pattern zeros for these distributions, with values of $\bar{n} = 6$ and $\bar{n} = 10$ used. All of the Taylor distributions exhibit a relatively smooth progression of zeros, but the Hamming has a singular discontinuity at the second zero. Table 3.7 gives the spacing between zeros corresponding to Table 3.6. Here the smooth progression of all the Taylor distributions is again evident, although there is a slight oscillation at the transition $\bar{n}$ for the $\bar{n} = 10$ case. Note that the Hamming null spacings

72 LINEAR ARRAY PATTERN SYNTHESIS

TABLE 3.6 Zeros for SLR = 42.7 dB

n	Taylor One-Parameter	Taylor		Hamming
		n̄ = 6	n̄ = 10	
1	2.112	1.894	1.897	2
2	2.732	2.398	2.396	2.598
3	3.550	3.173	3.166	3
4	4.412	4.069	4.056	4
5	5.335	5.020	5.002	5
6	6.282	6	5.978	6
7	7.243	7	6.970	7
8	8.214	8	7.974	8
9	9.190	9	8.984	9
10	10.172	10	10	10

TABLE 3.7 Null Spacings for SLR = 42.7 dB

n	Taylor One-Parameter	Taylor		Hamming
		n̄ = 6	n̄ = 10	
1	0.619	0.503	0.499	0.598
2	0.799	0.776	0.770	0.402
3	0.881	0.895	0.890	1
4	0.923	0.951	0.946	1
5	0.947	0.980	0.976	1
6	0.961	1.0	0.993	1
7	0.970	1.0	1.003	1
8	0.977	1.0	1.011	1
9	0.981	1.0	1.016	1
10	0.985	1.0	1.0	1

change from roughly 0.6 to 0.4 to 1, clearly an inefficient way to design an aperture distribution. Finally, in Table 3.8 are given the beamwidth in u and the aperture efficiency. The Hamming is roughly 3% poorer in both indices. Although no data are given, the Taylor distributions have an intrinsically lower Q. That is, errors in aperture excitation will affect the Hamming distribution more, owing to its higher Q.

Because phase errors affect low-sidelobe patterns strongly, it may be expected that measurement distance will change such patterns. The conventional far-field distance of $2L^2/\lambda$, where L is array length, for measurement and operation, may not be sufficient for low-sidelobe designs. As the observation distance moves in from infinity, the first sidelobe rises and the null starts filling. Then the sidelobe becomes a shoulder on the now wider main beam, and the second null rises. This process continues as the distance decreases. To first order the results are dependent only on design sidelobe level. Details are given in Chapter 12.

TABLE 3.8 Comparison of Distributions for SLR = 42.7 dB

	Taylor One-Parameter	Taylor		Hamming
		n̄ = 6	n̄ = 10	
u_3	0.694	0.637	0.635	0.651
η_t	0.690	0.754	0.755	0.734
$u_3 \eta_t$	0.478	0.480	0.480	0.478

3.5.2 Average Sidelobe Level

It is sometimes desirable to be able to estimate the average sidelobe level for an antenna. Because the average level and the pattern envelope are approximately related to the antenna directivity, this can be done; the Taylor one-parameter distribution will be used as an example. Since most antenna patterns have a sidelobe envelope with $1/u$ asymptotic decay, this example is widely applicable. The average power sidelobe level is defined here as the integral from the first null to $\pi/2$:

$$\begin{aligned} \text{SLL}_{\text{ave}} &= \frac{\lambda}{\pi^2 F_0^2 L} \int_{u_0}^{L/\lambda} \frac{du}{u^2 - B^2} \\ &= \frac{\lambda}{2\pi^2 F_0^2 LB} \ln \frac{(L/\lambda - B)(u_0 + B)}{(L/\lambda + B)(u_0 - B)}. \end{aligned} \tag{3.54}$$

The first null is at $u_0 = \sqrt{1 + B^2}$. For long apertures, $L/\lambda \gg B$, which gives

$$\text{SLL}_{\text{ave}} \simeq \frac{\lambda}{\pi^2 F_0^2 LB} \ln \frac{u_0 + B}{u_0 - B}. \tag{3.55}$$

The actual sidelobe power is less because of the sidelobe shape. A typical shape, independent of the envelope, is $\sin^2 \pi u$, giving an integral of $1/2$. Thus the sidelobe power is 3 dB below the envelope. The directivity is $G = 2\eta_t L/\lambda$, with η_t the aperture efficiency, and average sidelobe level may be written in terms of directivity:

$$\text{SLL}_{\text{ave}} \simeq \frac{\eta_t}{2\pi^2 F_0^2 GB} \ln \frac{u_0 + B}{u_0 - B}. \tag{3.56}$$

Given an SLR, B is fixed, and the right-hand factor can be calculated; it is shown in Fig. 3.19 and represents a correction of less than 3 dB for high SLR. In dB terms, the average sidelobe level becomes

$$\text{SLR}_{\text{ave}} \simeq G_{\text{dB}} + (SLR - 13.26 \text{ dB}) - 10 \log \left[\frac{\eta_t}{2B} \ln \frac{u_0 + B}{u_0 - B} \right] + 9.95 \text{ dB}. \tag{3.57}$$

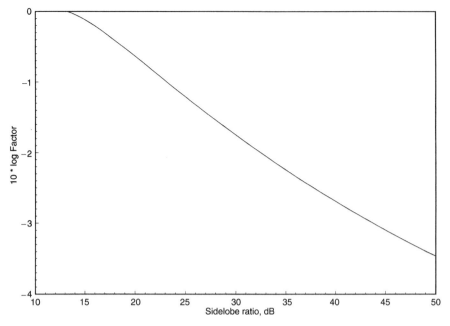

Figure 3.19 Average sidelobe level factor.

As expected, higher directivity yields lower average sidelobe level, as does SLR.

3.6 VILLENEUVE N-BAR ARRAY DISTRIBUTION

For most linear array designs the number of elements is sufficiently large that the continuous Taylor distributions previously discussed can be sampled. Villeneuve (1984) developed an elegant method for the design of Taylor $\bar{n}$ arrays for a small number of elements. This was accomplished by adjusting $\bar{n} - 1$ close-in zeros of the Dolph–Chebyshev array polynomial (instead of the close-in zeros of the continuous aperture function). The Chebyshev polynomial is written in product form so that the close-in zeros can be shifted. The Taylor dilation factor σ is also used. The result is the pattern function

$$F_m(w) \equiv E(\psi) = \exp jN\psi/2 \frac{\sin\left(\frac{2N+1}{2}\psi\right)}{\sin(\frac{1}{2}\psi)}$$

$$\times \frac{\prod_{m=1}^{\bar{n}-1} \sin\left(\frac{\psi - \psi'_m}{2}\right) \sin\left(\frac{\psi + \psi'_m}{2}\right)}{\prod_{m=1}^{\bar{n}-1} \sin\left(\frac{\psi - \frac{m2\pi}{2N+1}}{2}\right) \sin\left(\frac{\psi + \frac{m2\pi}{2N+1}}{2}\right)}, \quad (3.58)$$

where $\psi = kd \sin \theta$, $u = u_0 \cos \psi/2$, and $\psi' = \sigma\psi$.

This pattern is the discrete array equivalent of the Taylor $\bar{n}$ for a continuous aperture. The dilation factor is

$$\sigma = \frac{2\pi\bar{n}}{(2N+1)\psi_{\bar{n}}}. \tag{3.59}$$

The amplitude excitation coefficients are found by applying a finite transform to the pattern function. The coefficients are slightly different for even and odd numbers of elements. For an even number of elements $N = 2M$, the following relations hold:

$$u_0 = \cosh\left[\frac{1}{2N-1} \ln\left(\eta + \sqrt{\eta^2 - 1}\right)\right],$$

$$\psi_p = 2\arccos\left[\frac{1}{u_0} \cos(2p-1)\frac{\pi}{2(2N-1)}\right], \quad p = 1, \ldots, 2N-1, \tag{3.60}$$

$$\sigma = \frac{\bar{n}2\pi}{2N\psi_{\bar{n}}} = \frac{\bar{n}\pi}{2N \arccos\left[\frac{1}{u_0} \cos\left((2\bar{n}-1)\frac{\pi}{2(2N-1)}\right)\right]}.$$

The coefficients are now given by

$$a_p = \frac{1}{2N} \sum_{m=-(\bar{n}-1)}^{\bar{n}-1} E\left(\frac{m2\pi}{2N}\right) \exp\{-j[(p-1/2)m2\pi/(2N)]\}, \quad -(N-1) < p < N$$

$$= \frac{1}{2N} \left\{ E(0) + 2 \sum_{m=1}^{\bar{n}-1} E\left(\frac{m2\pi}{2N}\right) \cos\left[(p-\tfrac{1}{2})\frac{m\pi}{N}\right] \right\}, \tag{3.61}$$

where

$$E\left(\frac{m2\pi}{2N}\right) = \frac{2N(-1)^m \prod_{q=1}^{\bar{n}-1} \sin\left(\frac{m\pi}{2N} - \frac{\psi'_q}{2}\right) \sin\left(\frac{m\pi}{2N} + \frac{\psi'_q}{2}\right)}{\sin\left(\frac{m\pi}{2N}\right) \sin\left(\frac{2m\pi}{2N}\right) \prod_{\substack{q=1 \\ q\neq m}}^{\bar{n}-1} \sin\left(\frac{(m-q)\pi}{2N}\right) \sin\left(\frac{(m+q)\pi}{2N}\right)},$$

$$E(0) = \frac{2N \prod_{q=1}^{\bar{n}-1} \sin^2\left(\frac{\psi'_q}{2}\right)}{\prod_{q=1}^{\bar{n}-1} \sin^2\left(\frac{q\pi}{2N}\right)}.$$

$$\tag{3.62}$$

For an odd number of elements, where $N = 2M + 1$, the relations are

$$\sigma = \frac{\bar{n}\pi}{(2N+1)\arccos\left[\frac{1}{u_0}\cos\left((2\bar{n}-1)\frac{\pi}{4N}\right)\right]},$$

$$u_0 = \cosh\left[\frac{1}{2N}\ln\left(\eta + \sqrt{\eta^2 - 1}\right)\right], \quad (3.63)$$

$$\psi_p = 2\arccos\left[\frac{1}{u_0}\cos(2p-1)\frac{\pi}{4N}\right], \quad p = 1, \ldots, 2N.$$

The coefficients become

$$a_p = \frac{1}{2N+1}\sum_{m=-(\bar{n}-1)}^{\bar{n}-1} E\left(\frac{m2\pi}{2N+1}\right)\exp\{-j[pm2\pi/(2N+1)]\}, \quad -N \leq p \leq N$$

$$= \frac{1}{2N+1}\left[E(0) + 2\sum_{m=1}^{\bar{n}-1} E\left(\frac{m2\pi}{2N+1}\right)\cos\left(\frac{pm2\pi}{(2N+1)}\right)\right],$$

$$-N \leq p \leq N, \quad (3.64)$$

where

$$E\left(\frac{m2\pi}{2N+1}\right) = \frac{(2N+1)(-1)^m \prod_{q=1}^{\bar{n}-1} \sin\left(\frac{m\pi}{2N+1} - \frac{\psi'_q}{2}\right) \sin\left(\frac{m\pi}{2N+1} + \frac{\psi'_q}{2}\right)}{\sin\left(\frac{m\pi}{2N+1}\right)\sin\left(\frac{2m\pi}{2N+1}\right) \prod_{\substack{q=1 \\ q \neq m}}^{\bar{n}-1} \sin\left(\frac{[m-q]\pi}{2N+1}\right)\sin\left(\frac{[m+q]\pi}{2N+1}\right)},$$

$$E(0) = \frac{(2N+1)\prod_{q=1}^{\bar{n}-1}\sin^2\left(\frac{\psi'_q}{2}\right)}{\prod_{q=1}^{\bar{n}-1}\sin^2\left(\frac{q\pi}{2N+1}\right)}. \quad (3.65)$$

These expressions appear complex but are readily programmed for computer.

The aperture efficiency of Villeneuve arrays has been computed by summing virtual mutual resistances (Hansen, 1985), using the formula

$$G = \frac{\left(\sum_{n=1}^{N} A_n\right)^2}{\sum_{n=1}^{N}\sum_{m=1}^{N} A_n A_m \operatorname{sinc}[(n-m)kd]}. \quad (3.66)$$

TABLE 3.9 Villeneuve ñ Array Efficiency

N	\multicolumn{8}{c}{SLR (dB)}

	25		30		35		40	
N	ñ	η_t	ñ	η_t	ñ	η_t	ñ	η_t
41	5	0.9026	7	0.8551	9	0.8098	12	0.7701
31	5	0.9000	7	0.8528	9	0.8078	12	0.7693
21	5	0.8947	7	0.8479	11	0.8060	11	0.7630
15	6	0.8917	7	0.8409	8	0.7949	8	0.7540
11	6	0.8807	6	0.8276	6	0.7825	6	0.7447
9	5	0.8674	5	0.8158	5	0.7732	5	0.7384
7	4	0.8468	4	0.7987	4	0.7608	4	0.7310
5	3	0.8182	3	0.7725	3	0.7448		

TABLE 3.10 Taylor ñ Array Efficiency

SLR (dB)	ñ	η_t
25	5	0.9105
30	7	0.8619
35	9	0.8151
40	11	0.7729

Table 3.9 gives aperture efficiencies for Villeneuve odd arrays from 5 to 41 elements, with sidelobe ratios of 25 to 40 dB. For each case ñ is chosen to be the largest that allows a monotonic distribution. These values can be compared with the corresponding Taylor ñ efficiencies in Table 3.10. These results show that there is only a modest penalty in arrays with as few as 10 elements. The tables allow the designer to determine when the size of the array requires use of a Villeneuve distribution.

3.7 DIFFERENCE PATTERNS

3.7.1 Canonical Patterns

A difference pattern typically has a null at broadside; these patterns are used for accurate determination of angle of arrival. The widely used monopulse radar antenna employs a difference pattern. The aperture distributions considered here are those with a biphasal distribution, where one-half of the aperture is inphase, while the other half is 180 deg out of phase. The uniformly excited linear array, with the two halves of the array fed through a 180 deg hybrid, is a simple linear difference antenna. The hybrid provides a sum pattern which is sinc πu, and a difference pattern which is

$$F(u) = \frac{\sqrt{2}\,(1 - \cos \pi u)}{\pi u}. \tag{3.67}$$

78 LINEAR ARRAY PATTERN SYNTHESIS

The difference distribution is normalized to make the integral of the aperture distribution squared equal to unity. For this simple difference pattern, the peak value is 1.0248, and the slope in the center is 0.7071. The first sidelobe is $-10.57\,\mathrm{dB}$ below the peak. These high sidelobes make this uniform difference pattern unattractive.

The excitation efficiency is obtained from the formula for excitation efficiency for equiphase apertures (Friis and Lewis, 1947), modified to include the pattern peak location u_0:

$$\eta_t = \frac{\left(\int g(p)\exp(j\pi p u_0)\,dp\right)^2}{2\int g^2(p)\,dp}. \tag{3.68}$$

The result for the uniform difference aperture is $\eta_t = 0.7420$.

The sidelobes of the maximum slope difference pattern (Powers, 1967; Kirkpatrick, 1953) are even higher. This distribution is triangular, zero at the center and maximum at each edge. Using the same normalization, the peak (edge) value is 1.2247. The slope at the center is 0.8165 and the first sidelobe is $-8.28\,\mathrm{dB}$ below the main beam. The pattern function is given by

$$F(u) = \frac{\sqrt{6}\,(\operatorname{sinc}\pi u - \cos\pi u)}{\pi u}. \tag{3.69}$$

The excitation efficiency is $\eta_t = 0.5708$.

A third distribution is that providing maximum directivity. This maximum directivity pattern function is

$$F(u) = \operatorname{sinc}[\pi(u - u_m)] - \operatorname{sinc}[\pi(u + u_m)],$$

where the constant $u_m = 0.715148$. The aperture distribution that corresponds is given by $g = \sin u_m \pi y$. The first sidelobe is $-12.59\,\mathrm{dB}$ below the peak. The aperture efficiency is 0.6086. Figure 3.20 gives uniform and maximum directivity difference patterns. As might be expected, the latter has slightly narrower main difference beams and lower sidelobes. Of course the price is a complex aperture distribution, as shown in Fig. 3.21. Each half of the distribution is a portion of a sine function. Here the peak of the aperture distribution has been normalized to unity for convenience in plotting.

3.7.2 Bayliss Patterns

The high sidelobes of the preceding distributions can be reduced by applying the principles of Taylor to the difference pattern. This was done by Bayliss (1968), who produced a difference pattern analogous to the Taylor ñ pattern. Bayliss started with the Taylor "ideal" pattern function (see Section 3.4), then differentiated this to obtain a difference pattern. It was perhaps surprising that the resulting sidelobes were of irregular levels, with a nonmonotonic sidelobe

DIFFERENCE PATTERNS 79

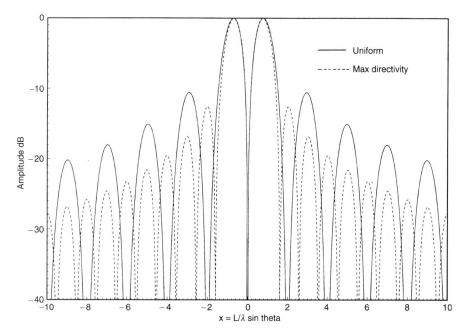

Figure 3.20 Uniform and maximum directivity difference patterns.

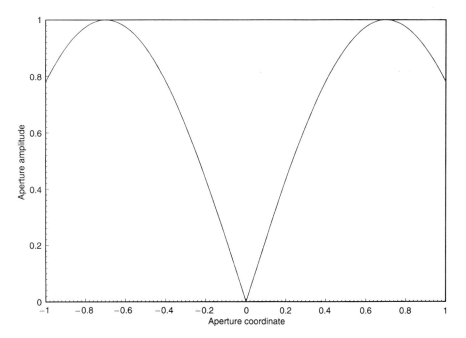

Figure 3.21 Maximum directivity difference pattern distribution.

80 LINEAR ARRAY PATTERN SYNTHESIS

envelope. An iterative procedure was used to adjust these zeros to yield equal-level sidelobes. Only four needed adjustment, but these four zeros depend upon the SLR. Bayliss gives the four zeros, the Taylor parameter A, and the difference peak location u_0 in terms of fourth-order polynomials in SLR (in dB). Table 3.11 gives the corrected Bayliss parameters, using the polynomial coefficients in Table 3.12. The other zeros are the same as for the linear Taylor $\bar{n}$ case:

$$z_n = \pm\sqrt{A^2 + n^2}, \qquad n > 4. \tag{3.71}$$

The dilation factor is given by

$$\sigma = \frac{\bar{n} + \frac{1}{2}}{z_{\bar{n}}}. \tag{3.72}$$

The pattern is provided by replacing the first $\bar{n}$ zeros of $\cos \pi u$ by the modified set. This results in a ratio of finite products for the pattern function:

$$F(u) = u \cos \pi u \, \frac{\prod_{n=1}^{\bar{n}-1}\left(1 - \dfrac{u^2}{\sigma^2 z_n^2}\right)}{\prod_{n=0}^{\bar{n}-1}\left(1 - \dfrac{u^2}{(n+1/2)^2}\right)}. \tag{3.73}$$

TABLE 3.11 Bayliss Roots and Parameters

	SLR (dB)					
	15	20	25	30	35	40
A	1.00790	1.22472	1.43546	1.64126	1.84308	2.04154
z_1	1.51240	1.69626	1.88266	2.07086	2.26025	2.45039
z_2	2.25610	2.36980	2.49432	2.62754	2.76748	2.91234
z_3	3.16932	3.24729	3.33506	3.43144	3.53521	3.64518
z_4	4.12639	4.18544	4.25273	4.32738	4.40934	4.49734
u_0	0.66291	0.71194	0.75693	0.79884	0.83847	0.87649

TABLE 3.12 Bayliss Polynomial Coefficients

	C_1	$10 C_2$	$10^3 C_3$	$10^5 C_4$	$10^7 C_5$
A	0.30387530	−0.5042922	−0.27989	−0.343	−0.2
u_1	0.98583020	−0.3338850	0.14064	0.190	0.1
u_2	2.00337487	−0.1141548	0.41590	0.373	0.1
u_3	3.00636321	−0.0683394	0.29281	0.161	0.0
u_4	4.00518423	−0.0501795	0.21735	0.088	0.0
u_0	0.47972120	−0.1456692	−0.18739	−0.218	−0.1

DIFFERENCE PATTERNS   81

The pattern function can also be written as a finite sum over $\bar{n}$ zeros:

$$F(u) = u \cos \pi u \sum_{n=0}^{\bar{n}-1} \frac{(-1)^n B_n}{(n+\frac{1}{2})^2 - u^2}. \tag{3.74}$$

Figure 3.22 shows a typical Bayliss pattern for SLR $= 25$ dB and $\bar{n} = 5$. As in the case of the Taylor $\bar{n}$ pattern, there is a smooth transition between the nearly equal-level sidelobes and the $1/u$ envelope sidelobes. The aperture distribution is also given by a finite $\bar{n}$ sum:

$$g(p) = \sum_{n=0}^{\bar{n}-1} B_n \sin \frac{\pi p}{\eta + \frac{1}{2}}, \tag{3.75}$$

where the coefficients are

$$B_m = -(-1)^m (m-\tfrac{1}{2})^2 \frac{\prod_{n=1}^{\bar{n}-1}\left(1 - \frac{(m+\frac{1}{2})^2}{\sigma^2 z_n^2}\right)}{\prod_{\substack{n=0 \\ n \neq m}}^{\bar{n}-1}\left(1 - \frac{(m+\frac{1}{2})^2}{(n+\frac{1}{2})^2}\right)}. \tag{3.76}$$

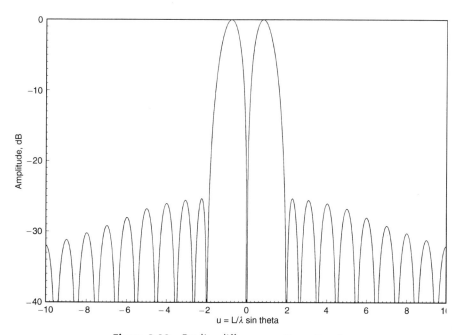

Figure 3.22 Bayliss difference pattern, $\bar{n} = 5$.

82 LINEAR ARRAY PATTERN SYNTHESIS

Figure 3.23 shows the aperture distribution corresponding to the pattern of Figure 3.22. As expected from the Taylor principles, there is an appreciable pedestal at the aperture edges. The excitation efficiency for the Bayliss aperture is

$$\eta_t = 2u_0^2 \cos^2 \pi u_0 \frac{\left[\sum_{n=0}^{\bar{n}-1} \frac{(-1)^n B_n}{\left[u_0^2 - \left(n+\frac{1}{2}\right)^2\right]^2}\right]}{\sum_{n=0}^{\bar{n}-1} B_n^2}. \qquad (3.77)$$

The slope at the center is also given in terms of the coefficients:

$$S = \frac{2}{\pi} \sum_{n=0}^{\bar{n}-1} \frac{(-1)^n B_n}{\left(n+\frac{1}{2}\right)^2}. \qquad (3.78)$$

Table 3.13 gives Bayliss aperture efficiency and slope.

Bayliss difference patterns possess the same type of extended far-field distances as the low-sidelobe designs of Section 3.5. Again the first sidelobe height and null depth deteriorate as the distance decreases from $2L^2/\lambda$. The progressive raising and shouldering of sidelobes continues as R decreases. Details are given in Chapter 12.

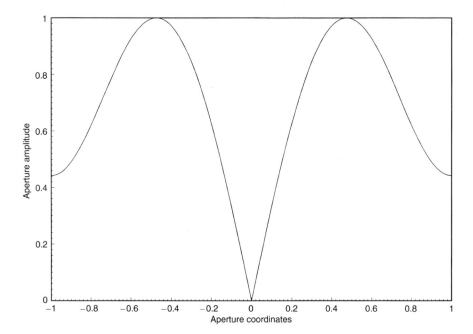

Figure 3.23 Bayliss difference pattern distribution.

TABLE 3.13 Bayliss Efficiency and Pattern Slope

SLR (dB)	n̄	Aperture Efficiency	Normalized Slope
15	4	0.5959	0.9567
20	4	0.5846	0.8974
25	5	0.5633	0.8427
30	6	0.5393	0.7912
35	7	0.5162	0.7448
40	8	0.4951	0.7037

3.7.3 Sum and Difference Optimization

From the preceding discussions it is clear that the array and aperture distributions that produce good sum pattern sidelobes are different from those that produce good difference pattern sidelobes. For example, the Taylor n̄ and Bayliss n̄ distributions are quite different. Ways have been developed to allow the good features of both sum and difference patterns to be realized. The two most important of these methods are the tandem feed, and the use of subarrays.

The tandem feed was invented by Kinsey (1970) and typically consists of a ladder network with a directional coupler at the end of each rung. The inboard arms of the couplers are connected along the rungs, while the outboard arms on one side are connected to the array elements, with the outboard arms on the other side terminated in loads. At the centre of each of the ladder rails is a hybrid junction, whose outputs give the sum and difference patterns. The rail of couplers adjacent to the array provides the sum excitation, while both rails of couplers provide the difference excitation. See Fig. 3.24 for a sketch of the configuration. This arrangement gives excellent control of both sum and difference patterns (Lopez, 1986; Jones and DuFort, 1971), although it is complex.

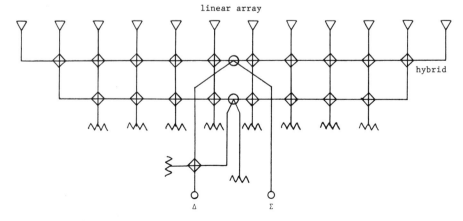

Figure 3.24 Ladder sum and difference pattern optimization.

84 LINEAR ARRAY PATTERN SYNTHESIS

The subarray configuration utilizes a linear array divided into subarrays, which need not be of equal length, although the array is symmetric about its center. Symmetric pairs of subarrays are connected to a hybrid junction; see Fig. 3.25. The sum ports of the hybrids are connected to a power combiner (divider) to provide the sum pattern, and similarly the difference ports of the hybrids are connected to a combiner to provide the difference pattern. Thus a stair-step approximation is provided to the desired sum pattern distribution, and a different stair-step approximation is provided to the desired difference pattern distribution (Kinsey, 1974). As more subarrays of smaller size are used, the stair-step approximations become better, and the resulting patterns are closer to the desired patterns. The same procedure has been applied to planar disk arrays, and is discussed in Chapter 4. Figure 3.26 shows a Lopez feed; the two waveguide rails are visible along the bottom (Rao et al., 1996).

3.7.4 Discrete Zolotarev Distributions

Zolotarev, a student of Chebyshev, developed a class of odd polynomials that gave an equi-ripple approximation in a stated interval, just as the even Chebyshev polynomials do. These were applied to microwave filter design by Levy (1970, 1971), who also gives historical and derivational details. It was recognized by McNamara (1985, 1993) that these polynomials could also be used to synthesize a difference pattern with equal-level sidelobes. The polynomial is defined as a hyperbolic cosine of an argument that is N times a function, giving a polynomial of order $2N+1$. Unfortunately, the argument function contains elliptic integrals and Jacobi η and ζ functions. The synthesis procedure will be summarized briefly; for details consult McNamara (1993). Given an

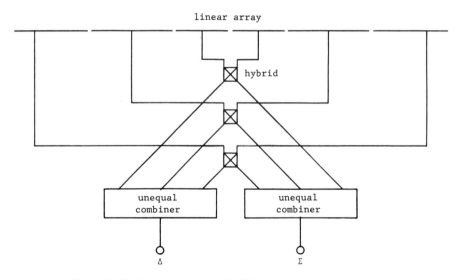

Figure 3.25 Subarray sum and difference pattern optimization.

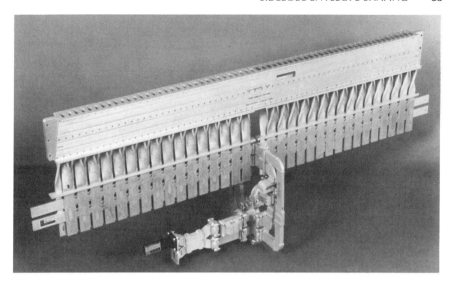

Figure 3.26 Lopez monopulse beamformer. (Courtesy Naval Research Lab.)

SLR, and an array of $2N$ elements, the order $2N - 1$ polynomial at the peak is set equal to SLR. The solution of this gives the Jacobi modulus, and the polynomial is completely determined. For half-wave spacing the polynomial is expanded into a conventional polynomial; from these coefficients the array excitation coefficients may be found. Specific computational details for the special functions are given in the reference above. Figure 3.27 shows a ninth-order Zolotarev polynomial, where x_1 and x_3 are the starts of the equal sidelobe regimes. A typical difference pattern is shown in Fig. 3.28, for a 20-element array with 30 dB SLR; the amplitudes over half of the array are shown in Fig. 3.29.

A discrete $\bar{n}$ difference pattern, analogous to the discrete Villeneuve $\bar{n}$ sum pattern, has also been developed (McNamara, 1994). The starting point, as expected from the analog, is the equal-ripple Zolotarev polynomial. The first $\bar{n}$ zeros are kept, while farther-out zeros are replaced by those of a generic pattern, chosen to allow a proper correspondence between the two patterns. A maximum-slope difference pattern is used as the generic pattern. A transition region between the $\bar{n}$ Zolotarev zeros and the generic zeros is used as before, with a dilation factor. The procedure is somewhat tedious, but the results are excellent. Figure 3.30 shows the 20-element array pattern, with nominal 25 dB SLR, and $\bar{n} = 4$. The dilation factor is $\sigma = 1.01051$.

3.8 SIDELOBE ENVELOPE SHAPING

The sidelobe envelope of a pencil beam array may be controlled through adjustment of individual sidelobe heights, using techniques developed by

86 LINEAR ARRAY PATTERN SYNTHESIS

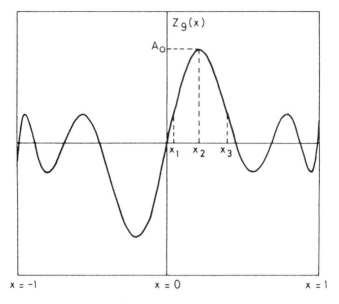

Figure 3.27 Z_9 Zolotarev polynomial. (Courtesy McNamara, D. A., "Direct Synthesis of Optimum Difference Patterns for Discrete Linear Arrays Using Zolotarev Distributions," *Proc. IEE*, Vol. MAP-140, Pt. H, Dec. 1993, pp. 495–500.)

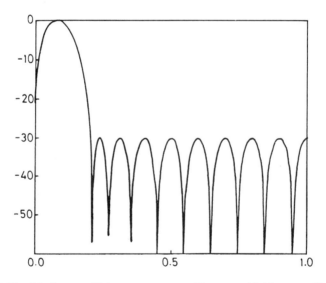

Figure 3.28 20-element Zolotarev pattern. (Courtesy McNamara, D. A., "Direct Synthesis of Optimum Difference Patterns for Discrete Linear Arrays Using Zolotarev Distributions," *Proc. IEE*, Vol. MAP-140, Pt. H, Dec. 1993, pp. 495–500.)

SIDELOBE ENVELOPE SHAPING   87

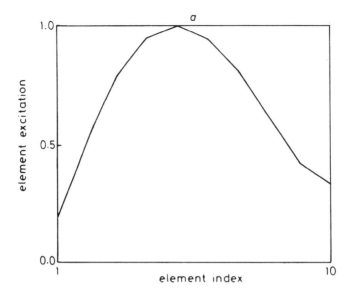

Figure 3.29 Array excitations. (Courtesy McNamara, D. A., "Direct Synthesis of Optimum Difference Patterns for Discrete Linear Arrays Using Zolotarev Distributions," *Proc. IEE*, Vol. MAP-140, Pt. H, Dec. 1993, pp. 495–500.)

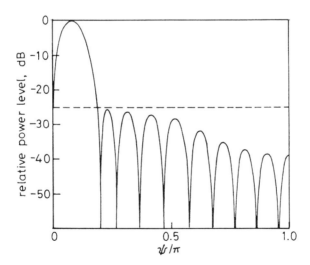

Figure 3.30 ñ Zolotarev pattern. (Courtesy McNamara, D. A., "Performance of Zolotarev and Modified-Zolotarev Difference Pattern Array Distributions," *Proc. IEE*, Vol. MAP-141, Feb. 1994, pp. 37–44.)

LINEAR ARRAY PATTERN SYNTHESIS

Elliott (1976). The principle is based on the fact that the spacing between pattern function zeros controls the heights of the sidelobes. Moving two adjacent zeros closer together reduces that sidelobe. However, as each array has only a fixed number of zeros, moving zeros closer together means that some other zeros will be farther apart. In this technique, a canonical pattern whose beamwidth and average sidelobe level are roughly the same as those of the desired pattern is selected. This canonical pattern must have the same number of pattern zeros as the desired pattern; the pattern is then written as a product of zeros. For example, a Taylor one-parameter pattern or a Taylor $\bar{n}$ pattern might be used to start.

The shaping procedure begins by writing each zero of the desired pattern as that of the corresponding zero of the canonical pattern plus a small shift. With higher-order effects discarded, the desired pattern can be expressed as the canonical pattern times a factor, which is 1 plus a sum over the pattern zeros. This sum involves the canonical pattern zeros and the unknown shifts in zeros. For narrow sidelobes, the sidelobe peak position is approximately halfway between the adjacent zeros. After inserting the sidelobe positions into the equation, a result among the desired and canonical levels of each sidelobe, and the various zeros, is obtained. This is a set of simultaneous equations, with the differences in sidelobe heights specified, and with unknown zero shifts. These equations are solved iteratively, with each resulting pattern used as a

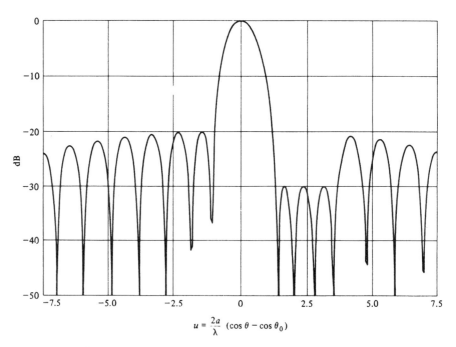

$$u = \frac{2a}{\lambda}(\cos\theta - \cos\theta_0)$$

Figure 3.31 Shaped sidelobe pattern. (Courtesy Elliott, R. S., "Design of Line-Source Antennas for Sum Patterns with Sidelobes of Individually Arbitrary Heights," *Trans. IEEE*, Vol. AP-24, 1976, pp. 76–83.)

canonical pattern for the next step. Usually fewer than 10 iterations are needed. A gradient scheme such as Newton–Raphson (Stark, 1970) provides a rapid way of finding the null shifts.

The resulting pattern is a product on the zeros which allows easy computation. The array excitation coefficients are less easily obtained. The product pattern function is multiplied out to obtain a polynomial. The coefficients are now the array excitation values. An example is a Taylor $\bar{n} = 8$ pattern with SLR = 20 dB, except for the three closest sidelobes on one side which should be at -30 dB. Figure 3.31 shows the resulting pattern (Elliott, 1977) that was obtained in three iterations; Fig. 3.32 shows the amplitude and phase of the resulting aperture distribution. The latter is of course complex, as the pattern is no longer symmetric. This pattern shaping by zero adjustment has proved to be a powerful and useful technique.

Another way of looking at this procedure is to write the array pattern in Z-transform form, where the pattern zeros are located on the unit circle; see Chapter 2. Moving two zeros closer together lowers the sidelobe that they control, but raises adjacent sidelobes. Maintaining the same beamwidth with

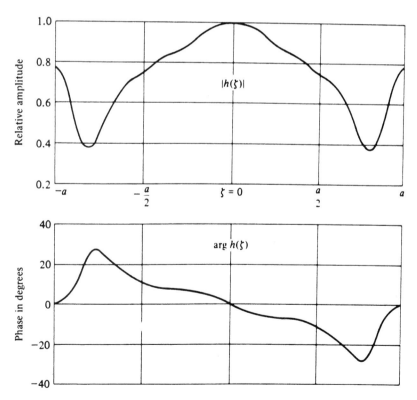

Figure 3.32 Shaped sidelobe pattern distribution. (Courtesy Elliott, R. S., "Design of Line-Source Antennas for Sum Patterns with Sidelobes of Individually Arbitrary Heights," *Trans. IEEE*, Vol. AP-24, 1976, pp. 76–83.)

90 LINEAR ARRAY PATTERN SYNTHESIS

lower overall sidelobes requires more zeros, hence a long array. Sidelobe nulls may be partially or completely filled by moving zeros off the unit circle, either inside or outside. In general, this allows multiple zero arrangements (and array excitations) that produce the same sidelobes. However, some excitations may be more robust than others, so all cases should be computed and compared.

When the array distribution is required to be real, as for resonant arrays or for some microstrip arrays, the same general synthesis technique can be used, except that additonal complex roots are introduced (Ares et al., 1994).

3.9 SHAPED BEAM SYNTHESIS

3.9.1 Woodward–Lawson Synthesis

Earlier parts of this chapter concerned pencil beam (high-gain) pattern synthesis. Often, broad, shaped beams are desired. For example, the cosecant radar pattern, or a flat-top earth overage pattern. The prototype shaped beam synthesis method was developed by Woodward (1947) and Woodward and Lawson (1948); it uses an assemblage of virtual arrays that produce orthogonal sinc beams. Thus each sinc beam has its amplitude adjusted such that the beam peak fits a point on the pattern, with the fitting points equally spaced. Each constituent beam is formed from a virtual linear array with constant amplitude and linear phase. The nth beam is given by $\sin(N\pi u)/(N \sin \pi u)$, where the array has N elements with spacing d; $u = kd(\sin\theta - \sin\theta_n)$. The beams are spaced (in $\sin\theta$) apart by $n\pi/Nd$, so all beams except the nth have zeros at the peak of that beam. The desired pattern is then sampled at approximately $N\lambda/2d$ points, with the array excitations equal to the sample amplitude. The total array distribution is simply the sum of the constituent array excitations. Given a desired pattern $F(u)$ the synthesized pattern is

$$WL(u) = \sum_{n=1}^{N} F(u_n) \frac{\sin N\pi u}{N \sin \pi u}. \tag{3.79}$$

The amount of pattern ripple is controlled by the element spacing. Figure 3.33 shows an example of a cosecant pattern synthesized from 21 beams, using a 21-element array with half-wave spacing. One zero is placed at $\theta = 0$. By shifting the zeros to allow the pattern dropoff to cross the axis, smaller ripples are obtained, as seen in Fig. 3.34. Larger spacing than half-wave will allow grating lobes, while smaller spacing tends toward superdirectivity. Thus, half-wave spacing is desirable.

The Woodward–Lawson synthesis uses constituent beams derived from a uniform distribution, giving a -13.3 dB sidelobe level for long arrays. It is possible to start with the zeros of a low-sidelobe distribution, then to move the zeros off the unit circle (in or out) to shape the beam. However, the orthogonal beam spacing here is larger; the cost of achieving lower sidelobes (and higher directivity) on the opposite side of the cosecant pattern, for example, is larger pattern ripple or a larger aperture.

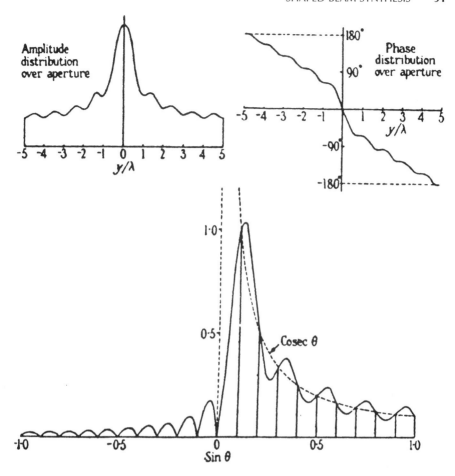

Figure 3.33 Shaped beam pattern. (Courtesy Woodward, P. M. and Lawson, J. D., "The Theoretical Precision with Which an Arbitrary Radiation Pattern May Be Obtained from a Source of Finite Size," *J. IEE*, Vol. 95, Part III, 1948, pp. 363–370.)

The Woodward–Lawson synthesis is important as it is applicable to multiple-beam antennas; see Chapter 10. It is also a good starting point for an iterative adjustment procedure of the type discussed in the previous section. The array excitations can be adjusted to minimize ripple, and to control the sidelobe envelope outside the shaped region. Intuitively one might expect the optimum cosecant pattern to have nulls in the low part of the pattern just adequate to give the prescribed sidelobe level. When the entire pattern can be specified, for example, in broadcasting antennas, a least-squares error function is then minimized by an iterative scheme: Newton–Raphson (see Section 3.9) or gradient (Perini and Idselis, 1972). Of gradient algorithms, the Fletcher–Powell (Fletcher and Powell, 1963) and Fletcher–Reeves (Fletcher and Reeves, 1964) are widely available; significant improvements have been made by

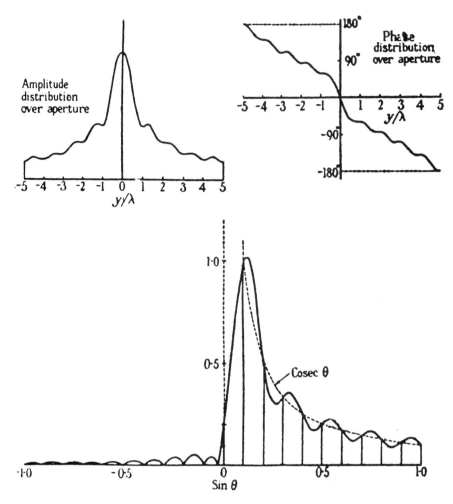

Figure 3.34 Improved shaped beam. (Courtesy Woodward, P. M. and Lawson, J. D., "The Theoretical Precision with Which an Arbitrary Radiation Pattern May Be Obtained from a Source of Finite Size," *J. IEE*, Vol. 95, Part III, 1948, pp. 363–370.)

Fletcher (1970) and Gill and Murray (1972). Least-squares synthesis tends to produce unequal ripples; an approach to a true minimax synthesis results from a least-Pth synthesis (Temes and Zai, 1969; Bandler and Charalambous, 1972). These techniques are not useful when sidelobe regions are involved because the number and position of sidelobes is usually not important; only the sidelobe envelope is important. Thus, the individual sidelobes cannot be effectively specified.

A major disadvantage of the Woodward–Lawson synthesis techniques is that in unit circle form, with the pattern written as a product of zeros, the roots in the shaped beam region are moved off the circle in pairs. Thus the available degrees of freedom are halved; these affect the number of ripples in the shaped region, hence the fidelity of the pattern match to the desired pattern. Next, a technique that utilizes all available degrees of freedom will be discussed.

3.9.2. Elliott Synthesis

Efficient synthesis techniques for shaped beams have been developed by Elliott (1982), where all zeros of the array polynomial (in unit circle terminology) are adjusted to optimize the pattern. Since factorization of complex polynomials is difficult, it is better to start the iterative synthesis process with a suitable canonical pattern that can be expressed in product form. A uniform linear array is a good choice; the roots are equally spaced on the unit circle (for half-wave spacing). The first step in the synthesis procedure is to iteratively perturb the roots on the unit circle, such that the sidelobes match the desired envelope in the sidelobe region, and such that the sidelobes are raised to the desired shaped envelope in the shaped beam region. This iterative process assumes that the peak of each lobe occurs midway between the adjacent zeros. Each zero is perturbed by a delta increment, and the resulting pattern is calculated at each lobe peak. The differences between these lobe peaks and the desired values give differential derivatives; a set of simultaneous equations relates the deltas and the derivatives. This set is best solved by the iterative Newton–Raphson method (Stark, 1970). This process typically needs on the order of 10 iterations or fewer. This first-stage result is shown in an example: a 16-element array with half-wave spacing, designed for a shaped beam of $cosec^2 \theta \cos\theta$ over a 40 deg sector, and for a sidelobe region consisting of four close sidelobes at -30 dB, with the rest at -20 dB. Figure 3.35 shows the resulting pattern (Elliott and Stern, 1984).

In the second step, the unit circle roots in the shaped region are moved radially off the circle; this fills in the nulls, producing a shaped beam with ripples. Again an iterative procedure is used, with the ripple peaks and dips assumed to occur at the unit circle points. However, when the roots are displaced from the unit circle the results show that the ripples are not symmetric, and that the sidelobes are affected; see Fig. 3.36. The iterative process must be re-run, to re-adjust the sidelobes, and balance the ripples. A final result is shown in Fig. 3.37, which shows four symmetric ripple oscillations. Table 3.14 gives the root positions; note that two roots are outside and two roots are inside the unit circle. Array excitations are found by multiplying the factors to get the array polynomial; the coefficients are the excitations. These are given for the example in Table 3.15. An interesting feature of this synthesis is that each shaped region root can be moved in or out. With M such roots, there are 2^M designs that produce the same pattern. The amplitude and phase of the element excitations will be different, and some values may be difficult to realize. For example, waveguide slots have practical limits on admittance values. The ratio of largest to smallest coupling should not be too large in order to have a robust design. Mutual coupling will also be different owing to the

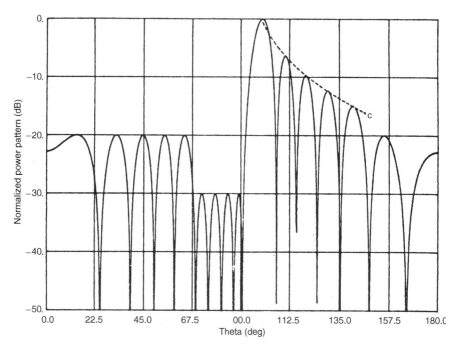

Figure 3.35 Shaped beam with prescribed sidelobes. (Courtesy Elliott, R. A. and Stern, G. J., "A New Technique for Shaped Beam Synthesis of Equispaced Arrays," *Trans. IEEE*, Vol. AP-32, Oct. 1984, pp. 1129–1133.)

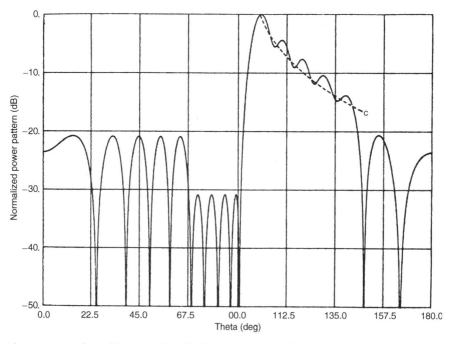

Figure 3.36 Shaped beam with null filling. (Courtesy Elliott, R. S. and Stern, G. J., "A New Technique for Shaped Beam Synthesis of Equispaced Arrays," *Trans. IEEE*, Vol. AP-32, Oct. 1984, pp. 1129–1133.)

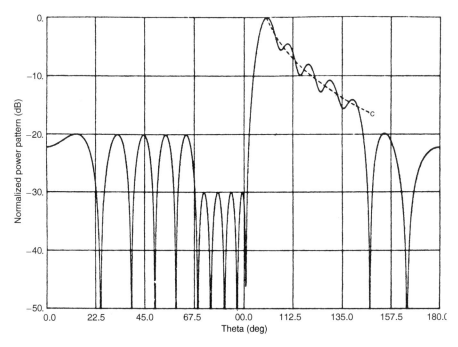

Figure 3.37 Shaped beam with readjusted sidelobes. (Courtesy Elliott, R. S. and Stern, G. J., "A New Technique for Shaped Beam Synthesis of Equispaced Arrays," *Trans. IEEE*, Vol. AP-32, Oct. 1984, pp. 1129–1133.)

TABLE 3.14 Unit Circle Root Locations

Radial Distance	Angular Distance (deg)
1.0000	−8.76
1.3219	−64.12
0.7565	−108.15
1.3588	−108.15
0.7360	−143.76
1.0000	−166.26
1.0000	171.24
1.0000	148.74
1.0000	126.24
1.0000	103.74
1.0000	81.24
1.0000	58.74
1.0000	36.24
1.0000	13.74

TABLE 3.15 Array Element Excitation

Element	Amplitude	Phase (deg)
1	0.300	231.71
2	0.242	211.46
3	0.307	198.35
4	0.378	161.79
5	0.337	132.12
6	0.431	123.71
7	0.753	85.44
8	1.000	29.68
9	1.000	−29.68
10	0.753	−85.44
11	0.431	−123.71
12	0.337	−132.12
13	0.378	−161.79
14	0.307	−198.35
15	0.242	−211.46
16	0.300	−231.71

different excitations. It is desirable to compute most, or all, of the 2^M cases to determine the best choice in terms of element embedded impedance.

When the phase of the pattern is not important, a power pattern may be synthesized; this may allow additional degrees of freedom. Now the iterative procedure operates on the magnitude squared of the unit circle product (Orchard et al., 1985). Another option pairs roots to produce a real array distribution (Ares et al., 1994, 1996).

When additional constraints are placed on the pattern, such as null positions, the array equations are overdetermined. These can be solved in a least mean squares sense by the singular value decomposition method (Ares et al., 1997).

3.10 THINNED ARRAYS

3.10.1 Probabilistic Design

Large arrays with half-wave spacing contain many elements, and may be unacceptable. Also, owing to the associated cost, weight, etc., mutual coupling may seriously degrade scanning performance. Since half-power beamwidth is approximately proportional to λ/L, array length L cannot be reduced without increasing beamwidth. In some cases, array size is dictated by beamwidth, but the directivity of a filled array is not needed. Thus a thinned array can offer essentially the same beamwidth with less directivity and fewer elements. Directivity is approximately equal to the number of elements, N. Mutual coupling effects are also significantly reduced (Agrawal and Lo, 1972). The average sidelobe level (power) for a linear array is $1/N$. Regular thinning produces grating lobes, but these can be partially suppressed by randomizing[1] the element spacings. There is then the question of how to select the "random" or nonuniform spacings. Most of the analytical work started with a desired pattern expressed as a Fourier series, then devised various ways of fitting the array sum to the Fourier series. Sandler (1960) expands in terms of a set of uniformly spaced arrays; Harrington (1961) considers an array with small departures from uniform; Unz (1962) expands each array term in a Bessel function series; Ishimaru (1962) expands in a delta-function series. All of these approaches suffer from a common difficulty: in order to expand the desired pattern into a Fourier or any other series, the pattern must be known in detail. If only amplitude is known, the problem is much more difficult; ideally pattern amplitude and phase are given. Rarely, however, is phase specified, and for narrow-beam, low-sidelobe arrays, the amplitude cannot even be specified. Rather the sidelobe envelope is usually specified, or the envelope below which all sidelobes must lie is specified. No Fourier series can be fitted to a sidelobe envelope as the series represents the actual lobes. Thus, analytical efforts have been found to be of little use.

Computer selection of element spacings has also been pursued. However, the problem is enormously difficult as a simple calculation will show. Take an array 100 wavelengths long, and allow possible element positions every eighth wavelength. Now take 40 elements, a 20% filled situation. There are

$$\binom{800}{40} \simeq 10^{70} \tag{3.80}$$

possible patterns to be evaluated if a direct search is used. Dynamic programming was tried as a search speedup, but is not applicable since the optimum position of an element depends upon the positions of elements on both sides of the element. Other search speedups have also proved unsatisfactory. Optimization algorithms using conjugate gradient methods, such as Fletcher–Powell or Fletcher–Reeves (in most computer scientific subroutine libraries), will yield an optimum thinned array, say minimum sidelobe envelope for a given array length, but are exceedingly slow. Most of the random-array computer studies have been on small arrays (King et al., 1960; Andreasen, 1962; Redlich, 1973; Steinberg, 1976). Use of such arrays is limited by the type of sidelobe structure that usually appears. High sidelobes occur at random angles, with a wide range of sidelobe heights in between these high lobes. Thus, the computer design usually attempts to keep all sidelobes below a fixed value. Algorithmic comparisons have been made by Steinberg (1976).

A major advance in understanding the behavior of random thinned arrays was made by Lo (1964). He assumed that isotropic elements with unit excitation were placed randomly, giving a probability density function. From this the probability of a narrow-beam pattern with all sidelobes below a given level was

---

[1] If the array spacings are selected by a deterministic algorithm, the array is aperiodic instead of random.

calculated by approximating the autocorrelation function by a rectangle of width $2\lambda/L$. However, the accuracy is poor for probability values that are small, or for small arrays. A different method was used by Agrawal and Lo (1972) wherein the probability of the pattern function crossing the specified sidelobe level was examined. The resulting integral equation was solved numerically by dividing the interval into many short segments, with the probability of a crossing assumed constant over a segment. The resulting product in the limit gives a closed-form probability result that compares very well with Monte Carlo simulations of arrays as short as 5 wavelengths, for low as well as high probabilities. The Agrawal and Lo result is in terms of the array length L/λ, and a combined parameter α which is number of elements N times power sidelobe level:

$$\alpha = \frac{N}{\text{SLR}}. \tag{3.81}$$

Here the sidelobe ratio SLR is in power. The result for probability of sidelobes below 1/SLR is

$$P = [1 - \exp(-\alpha)] \exp\left(-\frac{2L}{\lambda}\sqrt{\frac{\pi\alpha}{3}} \exp(-\alpha)\right). \tag{3.82}$$

Thus sidelobe level and number of elements can be directly traded. Figure 3.38 shows P versus α for $L/\lambda = 10, 30, 100, 300, 1000$. For example, an SLR of 20 dB is achieved with an 80% probability for $L = 1000\lambda$ with $N = 1028$; or an average element spacing of one wavelength. For large N the formula may be simplified (Lo, 1964):

$$P = [1 - \exp(-\alpha)]^{6L/\lambda}. \tag{3.83}$$

This gives a reasonably close fit for $P > 0.5$ and $L/\lambda > 100$. Figure 3.39 shows the approximate relationship for $P = 0.8, 0.09$, and 0.99 versus L/λ. The quantity plotted is $10\log N - \text{SLR}$ and again these results are only valid for large arrays. All of the results are for an unsymmetrical array. If the array is symmetric about its center, some degrees of freedom are lost, and the sidelobes are higher.

In using the curves of Fig. 3.39, care must be taken that the resulting N is not greater than that of a filled array, $2L/\lambda$; for example, a filled array with $L = 100\lambda$ had $N = 200$ (half-wave spacing), which gives $\alpha = 0.44$ for $\text{SLR} = 13.26\,\text{dB}$. From the graph the probability is near unity, which means that a 100λ aperture can achieve an SLR to 13.26 dB or better. However, for $L = 30\lambda$, $N = 60$, the value of α for the SLR is 2.83, which from the curve gives a probability near zero. This means that a thinned 30λ aperture will probably have sidelobes worse than -13 dB. Thus, for each aperture length there is a sidelobe ratio that is difficult to exceed, with longer apertures yielding better SLR capability. For example, for 80% probability, Table 3.16 gives these values. Thus, for reasonably low sidelobes, very large arrays are required

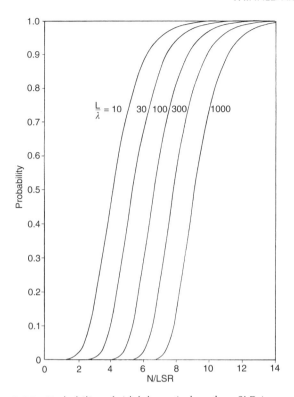

Figure 3.38 Probability of sidelobe ratio less than SLR (power).

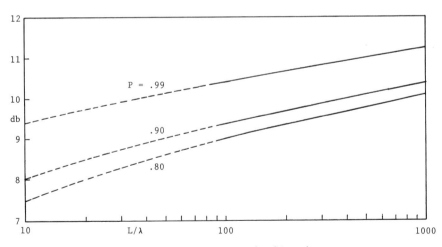

Figure 3.39 $10 \log N - $ SLR for thinned arrays.

to achieve a significant thinning. But because these very large arrays have very many sidelobes, only a small fraction of these sidelobes need to be at the envelope level to greatly reduce directivity and beam efficiency. This is expected, since directivity is roughly proportional to the number of elements.

TABLE 3.16 Upper Limit of Sidelobe Ratio

L/λ	SLR (dB)
10	5.7
30	9.6
100	14.1
300	18.2
1000	22.9
3000	27.2

Beam efficiency is important for radiometer and power-transfer situations. A tapered sidelobe envelope will reduce the sidelobe power and increase beam efficiency. This is especially important for planar arrays since there are so many more sidelobes at large angles from broadside.

The analysis and results above are for thinned arrays of isotropic elements. Half-wave elements will tend to reduce far-out sidelobes in one plane owing to the element pattern. Often it is advantageous to use elements larger than half-wave, and these further reduce far-out sidelobes. Steinberg (1976) has shown that the effect of element directivity is to reduce the effective array length used in Eqn. (3.82). His result is

$$\frac{L_{\text{eff}}}{\lambda} = \frac{L/\lambda}{\sqrt{\alpha\, W/\lambda}}. \tag{3.84}$$

This is only valid for $W/\lambda > 1/2$, owing to approximations used in evaluating an integral. An example will show the effect of element directivity. Take $L/\lambda = 1000$, $P = 80\%$, and SLR $= 15$ dB. Then a thinned array of isotropic elements needs 325 elements. Keeping the SLR and the number of elements fixed, consider half-wave dipole-type elements ($W/\lambda = 1/2$) and 2λ subarrays. Table 3.17 gives the probabilities of sidelobe ratio of 15 dB or greater. As expected then, directive elements reduce the sidelobes.

An interesting concept in thinned arrays uses fractal theory to randomize element placement (Kim and Jaggard, 1986). A three-element subarray with regular 0.8λ spacing is the "generator," while the fractal random process (the "initiator") locates the subarrays at pseudo-random locations. An example of a 360λ array with 180 elements shows a highest sidelobe of -12.5 dB, but the

TABLE 3.17 Thinned Array with Directive Elements $L = 1000\lambda$, $P = 0.8$, SLR $= 15$ dB

W/λ	P	Filling Factor
Isotropic	0.8	0.163
0.5	0.87	0.163
2	0.97	0.651

sidelobe envelope is roughly constant at −20 dB. In contrast, if the elements were uniformly spaced, the first sidelobe would be −13.2 dB and the envelope would taper down with angle off broadside to a value below −60 dB at 90 deg, but the beamwidth would be four times larger.

3.10.2 Space Tapering

A special kind of thinned array uses variable element spacing to produce an equivalent amplitude taper. The goal is to produce a sidelobe envelope that tapers down (Willey, 1962). As in all thinned arrays, the number of array elements must be large. It is particularly attractive in a distributed amplifier array where each element has its own transmitter/receiver module. Since it is much easier to make all modules alike, space tapering allows lower sidelobes to be realized, with all elements equally excited. Experience has shown, however, that the sidelobes do not fall off quite as fast as predicted by theory. Also, an occasional high sidelobe is observed. It may be necessary to make adjustments to the spacing or to the filling ratio.

Some calculations show the degree of thinning versus sidelobe level. The Taylor one-parameter distribution is used; see Section 3.3.1. The number of elements relative to the $\lambda/2$ filled array number N_0 is

$$N = \frac{N_0}{I_0(\pi B)} \int_0^1 I_0\left(\pi B \sqrt{1-p^2}\right) dp. \tag{3.85}$$

This can be integrated (using 11.4.10 and 10.2.13 of Abramowitz and Stegen, 1970) to give

$$N = \frac{N_0 \sinh \pi B}{\pi B \ I_0(\pi B)}. \tag{3.86}$$

Table 3.18 gives the values and also shows the relative beamwidths. Even for SLR = 40 dB, the thinning is less than half. So although space tapering does allow simulation of tapered amplitude distributions, the percentage of elements

TABLE 3.18 Spaced Tapered Linear Array

SLR (dB)	N/N_0	θ_3/θ_{30}
13.26	1	1
15	0.9133	1.042
20	0.7562	1.156
25	0.6647	1.260
30	0.6030	1.355
35	0.5577	1.443
40	0.5223	1.524

3.10.3 Minimum Redundancy Arrays

For some applications, such as aperture synthesis in radio astronomy, the spatial frequency spectrum is of use, and it is important to consider the simplest arrays that provide all spatial frequencies up to an upper limit. These arrays are called minimum redundancy arrays. The redundancy R is defined (Moffet, 1968) as

$$R = \frac{N(N-1)}{2 \sum \text{spacings}}, \qquad (3.87)$$

where the array has N elements, and all multiples of unit spacing $< N_s$ are present. For very large arrays the lower limit on redundance is 4/3 (Bedrosian, 1986). An example for a linear array, from Moffet, is sketched in Fig. 3.40. In the figure are seven elements, with spacings 1,3,6,2,3,2. The spatial frequency coverage is up to 17, and the redundancy is 1.24. Finding such arrays for large N is an interesting problem in number theory.

Figure 3.40 Seven-element minimum redundancy array.

ACKNOWLEDGMENT

Photograph courtesy of Dr. J. B. L. Rao.

REFERENCES

Abramowitz, M. and Stegun, L., *Handbook of Mathematical Functions*, NBS, 1970.

Agrawal, V. D. and Lo, Y. T., "Mutual Coupling in Phased Arrays of Randomly Spaced Antennas," *Trans. IEEE*, Vol. AP-20, May 1972, pp. 288–295.

Andreasen, M. G., "Linear Arrays with Variable Interelement Spacings," *Trans. IEEE*, Vol. AP-10, 1962, pp. 137–143.

Ares, F., Elliott, R. S., and Moreno, E., "Synthesis of Shaped Line-Source Antenna Beams Using Pure Real Distributions," *Electron. Lett.*, Vol. 30, No. 4, 1994, pp. 280–281.

REFERENCES

Ares, F., Rengarajan, S. R., and Moreno, E., "Remarks on Comparison between Real and Power Optimisation Methods for Array Synthesis of Antennas," *Electron. Lett.*, Vol. 32, No. 15, 1996, pp. 1338–1339.

Ares, F. et al., "Extension of Orchard's Pattern Synthesis Technique for Overdetermined Systems," *Electromagnetics*, Vol. 17, Jan.-Feb. 1997, pp. 15–23.

Bandler, J. W. and Charalambous, C., "Theory of Generalized Least Pth Approximation," *Trans. IEEE*, Vol. CT-19, 1972, pp. 287–289.

Barbiere, D., "A Method for Calculating the Current Distribution of Tschebyscheff Arrays," *Proc. IRE*, Vol. 40, 1952, pp. 78–82.

Bayliss, E. T., "Design of Monopulse Antenna Difference Patterns with Low Side Lobes," *BSTJ*, Vol. 47, May-June 1968, pp. 623–650.

Bedrosian, S. D., "Nonuniform Linear Arrays: Graph-Theoretic Approach to Minimum Redundancy," *Proc. IEEE*, Vol. 74, July 1986, pp. 1040–1043.

Bickmore, R. W. and Spellmire, R. J., "A Two-Parameter Family of Line Sources," Rep. TM 595, Hughes Aircraft Co., Culver City, Calif., 1956.

Blackman, R. B. and Tukey, J. W., *Measurement of Power Spectra*, Dover, 1958.

Brown, J. L., "A Simplified Derivation of the Fourier Coefficients for Chebyshev Patterns," *Proc. IEE*, Vol. 105C, 1957, pp. 167–168.

Brown, J. L., "On the Determination of Excitation Coefficients for a Tchebycheff Pattern," *Trans. IRE*, Vol. AP-10, 1962, pp. 215–216.

Brown, L. B. and Scharp, G. A., "Tschebyscheff Antenna Distribution, Beamwidth, and Gain Tables," NOLC Rep. 383, Feb. 1958.

Dolph, C. L., "A Current Distribution for Broadside Arrays which Optimizes the Relationship between Beam Width and Side-Lobe Level," *Proc. IRE*, Vol. 34, 1946, pp. 335–348.

Drane, C. J., "Derivation of Excitation Coefficients for Chebyshev Arrays," *Proc. IEE*, Vol. 110, 1963, pp. 1755–1758.

Drane, C. J., "Dolph–Chebyshev Excitation Coefficient Approximation," *Trans. IEEE*, Vol. AP-12, 1964, pp. 781–782.

Drane, C. J., "Useful Approximations for the Directivity and Beamwidth of Large Scanning Dolph–Chebyshev Arrays," *Proc. IEEE*, Vol. 56, 1968, pp. 1779–1787.

DuHamel, R. H., "Optimum Patterns for Endfire Arrays," *Proc. IRE*, Vol. 41, 1953, pp. 652–659.

Elliott, R. S., "Design of Line-Source Antennas for Sum Patterns with Sidelobes of Individually Arbitrary Heights," *Trans. IEEE*, Vol. AP-24, 1976, pp. 76–83.

Elliott, R. S., "On Discretizing Continuous Aperture Distributions," *Trans. IEEE*, Vol. AP-25, 1977, pp. 617–621.

Elliott, R. S., "Improved Pattern Synthesis for Equispaced Linear Arrays," *Alta Frequenza*, Vol. 21, Nov.-Dec. 1982, pp. 296–300.

Elliott, R. S. and Stern, G. J., "A New Technique for Shaped Beam Synthesis of Equispaced Arrays," *Trans. IEEE*, Vol. AP-32, Oct. 1984, pp. 1129–1133.

Fletcher, R., "A New Approach to Variable Metric Algorithms," *Computer J.*, Vol. 13, 1970, pp. 317–322.

Fletcher, R. and Powell, M. J. D., "A Rapidly Convergent Descent Method for Minimization," *Computer J.*, Vol. 6, 1963, pp. 163–168.

Fletcher, R. and Reeves, C. M., "Function Minimization by Conjugate Gradients," *Computer J.*, Vol. 7, 1964, pp. 149–154.

Friis, H. T. and Lewis, W. D., "Radar Antennas", *BSTJ*, Vol. 26, 1947, pp. 219–317.

Gill, P. E. and Murray, W., "Quasi-Newton Methods for Unconstrained Optimization," *J. Inst. Math. Appl.*, Vol. 9, 1972, pp. 91–108.

Hansen, R. C., *Microwave Scanning Antennas*, Vol. 1, Academic Press, 1964 [Peninsula Publishing, 1985], Chapter 1.

Hansen, R. C., "Aperture Efficiency of Villeneuve n̄ Arrays," *Trans. IEEE*, Vol. AP-33, June 1985, pp. 666–669.

Harrington, R. F., "Sidelobe Reduction by Nonuniform Element Spacing," *Trans. IRE*, Vol. AP-9, 1961, pp. 187–192.

Ishimaru, A., "Theory of Unequally-Spaced Arrays," *Trans. IEEE*, Vol. AP-10, 1962, pp. 691–702.

Jones, W. R. and DuFort, E. C., "On the Design of Optimum Dual-Series Feed Networks," *Trans. IEEE*, Vol. MTT-19, May 1971, pp. 451–458.

Kim, Y. and Jaggard, D. L., "The Fractal Random Array," *Proc. IEEE*, Vol. 74, Sept. 1986, pp. 1278–1280.

King, D. D., Packard, R. F., and Thomas, R. K., "Unequally-Spaced Broad-Band Antenna Arrays," *Trans. IEEE*, Vol. AP-8, 1960, pp. 380–384.

Kinsey, R. R., Tandem Series-Feed System for Array Antennas, US Patent 8509577, April 1970.

Kinsey, R. R., "The AN/TPS-59 Antenna Row-Board Design," *IEEE APS Symp. Dig.*, 1974, pp. 413–416.

Kirkpatrick, G. M., "Aperture Illuminations for Radar Angle-of-Arrival Measurements," *Trans. IRE*, Vol. ANE-9, Sept. 1953, pp. 20–27.

Levy, R., "Generalized Rational Function Approximation in Finite Intervals Using Zolotarev Functions," *Trans. IEEE*, Vol. MTT-18, Dec. 1970, pp. 1052–1064.

Levy, R., "Characteristics and Element Values of Equally Terminated Achieser–Zolotarev Quasi-Low-Pass Filters," *Trans. IEEE*, Vol. CT-18, Sept. 1971, pp. 538–544.

Lo, Y. T., "A Mathematical Theory of Antenna Arrays with Randomly Spaced Elements," *Trans. IEEE.*, Vol. AP-12, 1964, pp. 257–268.

Lopez, A. R., "Monopulse Networks for Series Feeding an Array Antenna," *Trans. IEEE*, Vol. AP-15, July 1986, pp. 436–440.

McNamara, D. A., "Optimum Monopulse Linear Array Excitations Using Zolotarev Polynomials," *Electron. Lett.*, Vol. 21, 1985, pp. 681–682.

McNamara, D. A., "Direct Synthesis of Optimum Difference Patterns for Discrete Linear Arrays Using Zolotarev Distributions," *Proc. IEE*, Vol. MAP-140, Pt. H, Dec. 1993, pp. 495–500.

McNamara, D. A., "Performance of Zolotarev and Modified-Zolotarev Difference Pattern Array Distributions," *Proc. IEE*, Vol. MAP-141, Feb. 1994, pp. 37–44.

Moffet, A. T., "Minimum-Redundancy Linear Arrays," *Trans. IEEE.*, Vol. AP-16, Mar. 1968, 172–175.

Orchard, H. J., Elliott, R. S., and Stern, G. J., "Optimising the Synthesis of Shaped Beam Antenna Patterns," *Proc. IEE*, Vol. 132, Pt. H, Feb. 1985, pp. 63–67.

Perini, J. and Idselis, M. H., "Radiation Pattern Synthesis for Broadcast Antennas," *Trans. IEEE*, Vol. BC-18, 1972, pp. 53–62.

Powers, E. J., "Utilization of the Lambda Functions in the Analysis and Synthesis of Monopulse Antenna Difference Patterns," *Trans. IEEE*, Vol. AP-15, Nov. 1967, pp. 771–777.

Rao, J. B. L. et al., "One-Dimensional Scanning X-Band Phased Array for an Engagement Radar," Naval Research Lab., NRL/MR/5310-96-9853, June 11, 1996.

Redlich, R. W., "Iterative Least-Squares Synthesis of Nonuniformly Spaced Linear Arrays," *Trans. IEEE*, Vol. AP-21, 1973, pp. 106–108.

Riblet, H. J., "Discussion on 'A Current Distribution for Broadside Arrays which Optimizes the Relationship between Beam Width and Side-Lobe Level'," *Proc. IRE*, Vol. 35, 1947, pp. 489–492.

Rothman, M., "Table of $\int I_0(x)\,dx$ for 0(0.1)20(1)25," *Q. J. Mech. Appl. Math.*, Vol. 2, 1949, pp. 212–217.

Salzer, H. E., "Calculating Fourier Coefficients for Chebyshev Patterns," *Proc. IEEE*, Vol. 63, 1975, pp. 195–197.

Sandler, S. S., "Some Equivalence between Equally and Unequally Spaced Arrays," *Trans. IEEE*, Vol. AP-8, 1960, pp. 496–500.

Schelkunoff, S. A., "A Mathematical Theory of Linear Arrays," *BSTJ*, Vol. 22, 1943, pp. 80–107.

Stark, P. A., *Introduction to Numerical Methods*, Macmillan, 1970.

Stegen, R. J., "Excitation Coefficients and Beamwidths of Tschebyscheff Arrays," *Proc. IRE*, Vol. 41, 1953, pp. 1671–1674.

Steinberg, B. D., *Principles of Aperture and Array System Design*, Wiley, 1976.

Taylor, T. T., "One-Parameter Family of Line Sources Producing Modified $\sin \pi u/\pi u$ Patterns," Rep. TM 324, Hughes Aircraft Co., Culver City, Calif., 1953.

Taylor, T. T., "Design of Line-Source Antennas for Narrow Beamwidth and Low Sidelobes," *Trans. IRE*, Vol. AP-3, 1955, pp. 16–28.

Temes, G. S. and Zai, D. Y. F., "Least Pth Approximation," *Trans. IEEE*, Vol. CT-16, 1969, pp. 235–237.

Unz, H. "Nonuniform Arrays with Spacings Larger Than One Wavelength," *Trans. IEEE*, Vol. AP-10, 1962, pp. 647–648.

Van Der Maas, G. J., "A Simplified Calculation for Dolph–Tschebycheff Arrays," *J. Appl. Phys.*, Vol. 25, 1954, pp. 121–124.

Villeneuve, A. T., "Taylor Patterns for Discrete Arrays," *Trans. IEEE*, Vol. AP-32, Oct. 1984, pp. 1089–1093.

Willey, R. E., "Space Tapering of Linear and Planar Arrays," *Trans. IEEE*, Vol. AP-10, 1962, pp. 369–376.

Woodward, P. M., "A Method of Calculating the Field over a Plane Aperture Required to Produce a Given Polar Diagram," *J. IEE*, Vol. 93, Part IIIA, 1947, pp. 1554–1558.

Woodward, P. M. and Lawson, J. D., "The Theoretical Precision with Which an Arbitrary Radiation Pattern May Be Obtained from a Source of Finite Size," *J. IEE*, Vol. 95, Part III, 1948, pp. 363–370.

CHAPTER FOUR

Planar and Circular Array Pattern Synthesis

4.1 CIRCULAR PLANAR ARRAYS

4.1.1 Flat Plane Slot Arrays

Many aircraft radars are now equipped with flat plane slot arrays instead of dish antennas, where the outline is approximately circular. These are usually fixed narrow-beam antennas with controlled sidelobes. Pattern synthesis typically uses the product of two linear syntheses; one along the slot "sticks," and a second across the sticks. These are often designed as Taylor $\bar{n}$ patterns, as described in Chapter 3. Since the sticks near the edge are shorter, with fewer slots, than those at the center, the overall pattern will deviate somewhat from the product of the two orthogonal patterns. In particular, some adjustments may be necessary to produce the desired sidelobe envelopes in the principal planes. As these arrays are often divided into quadrants for azimuth and elevation monopulse, a compromise may be advisable between the sum pattern sidelobes, and the difference pattern sidelobes and slope; see Chapter 3. An example of the design, using the zero adjustment methods described in Chapter 3, of a planar array of inclined slots, providing a cosecant type pattern, with a Taylor $\bar{n}$ pattern in the cross plane, is given by Erlinger and Orlow (1984).

Design of the array elements, to include mutual coupling, is more complex in that the entire array must be encompassed in the synthesis process. That synthesis is described in Chapter 6, and it can be extended to the flat plane array by using double sums for mutual coupling to include all slots. The iterative technique again converges rapidly. A typical flat plane slot array is shown in Fig. 4.1. See also Figs. 6.30 and 6.31.

Use of principal plane canonical patterns simplifies the synthesis process, but lower sidelobes can be obtained through synthesis of a rotationally symmetric pattern, as discussed in the next section.

Figure 4.1 Flat plane array. (Courtesy Hughes Aircraft Co.)

4.1.2 Hansen One-Parameter Pattern

A symmetric circular one-parameter distribution and pattern analogous to the Taylor one-parameter distribution was developed by Hansen (1976). Starting as before with the uniform amplitude pattern, which is $2J_1(\pi u)/\pi u$, the close-in zeros are shifted to suppress the close-in sidelobes. The single parameter is H, and the space factor (pattern) is

$$F(u) = \frac{2J_1\left(\pi\sqrt{u^2 - H^2}\right)}{\pi\sqrt{u^2 - H^2}}, \qquad u \geq H;$$
$$F(u) = \frac{2I_1\left(\pi\sqrt{H^2 - u^2}\right)}{\pi\sqrt{H^2 - u^2}}, \qquad u \leq H. \qquad (4.1)$$

J_1 and I_1 are the Bessel function and modified Bessel function of the first kind and order one. In exactly the same manner as for the Taylor pattern, the expression for $u \geq H$ forms the sidelobe region, and part of the main beam, while the expression for $u \leq H$ forms the remainder of the main beam. The transition occurs at $u = H$, and at this point the formula is 17.57 dB above the first sidelobe. From this point to the beam peak is $2I_1(\pi H)/\pi H$, so the sidelobe ratio is

$$\text{SLR} = 17.57\,\text{dB} + 20\log\frac{2I_1(\pi H)}{\pi H}. \tag{4.2}$$

The one-parameter pattern is easy to use, as both J_1 and I_1 are represented by absolutely convergent series and are quickly generated by common computer subroutines. Figure 4.2 shows a typical one-parameter pattern, for 25 dB SLR. Table 4.1 gives H and the normalized half-beamwidth u_3. Values up to 50 dB SLR are given, as the antenna art allows this performance. The 3 dB beamwidth is a solution of Eqn (43):

$$\frac{I_1(\pi H)}{2\pi H} = \frac{J_1\left(\pi\sqrt{u_3^2 - H^2}\right)}{\pi\sqrt{u_3^2 - H^2}}. \tag{4.3}$$

The aperture distribution is

$$g(p) = I_0\left(\pi H\sqrt{1-p^2}\right), \tag{4.4}$$

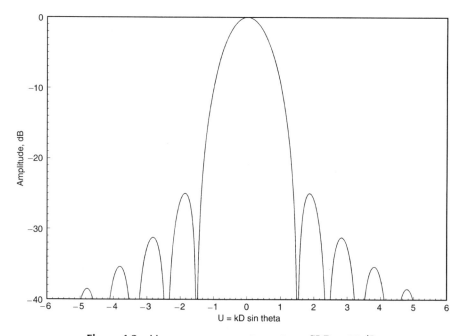

Figure 4.2 Hansen one-parameter pattern, SLR = 25 dB.

CIRCULAR PLANAR ARRAYS   109

TABLE 4.1 Characteristics of Hansen One-Parameter Distribution

SLR (dB)	H	u_3 (rad)	η_t	ηu_3	Edge Taper (dB)
17.57	0	0.5145	1	0.5145	0
20	0.4872	0.5393	0.9786	0.5278	4.5
25	0.8899	0.5869	0.8711	0.5113	12.4
30	1.1977	0.6304	0.7595	0.4788	19.3
35	1.4708	0.6701	0.6683	0.4478	25.8
40	1.7254	0.7070	0.5964	0.4216	32.0
45	1.9681	0.7413	0.5390	0.3996	38.0
50	2.2026	0.7737	0.4923	0.3809	43.9

where p is the radial coordinate, zero at the center, and unity at the edge. I_0 is a zero-order modified Bessel function. In Fig. 4.3 is given the beamwidth times aperture diameter D divided by wavelength, and it may be seen that the beamwidth is nearly independent of D/λ, depending primarily on SLR. The dots will be discussed in the next section. Figure 4.4 shows the (half-) aperture amplitude distribution for several values of SLR. Edge values are given in Table 4.1. The aperture excitation efficiency η_t is found by integrating the distribution

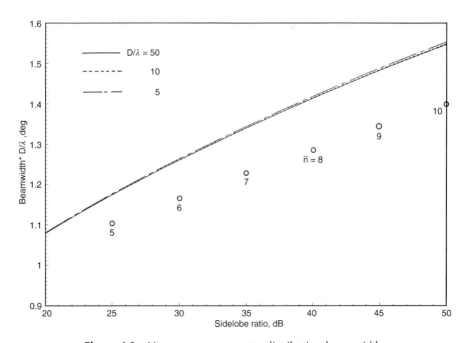

Figure 4.3 Hansen one-parameter distribution beamwidth.

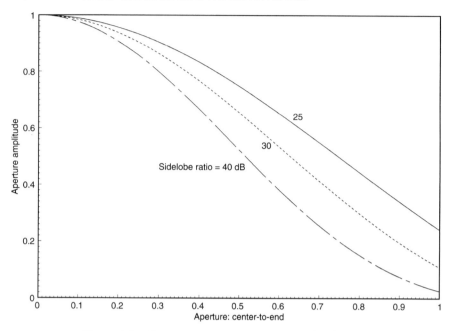

Figure 4.4 Hansen one-parameter aperture distribution.

$$\eta_t = \frac{2\left(\int_0^\pi g(p)p\,dp\right)^2}{\pi^2 \int_0^\pi g^2(p)p\,dp}$$

$$= \frac{4I_1^2(\pi H)}{\pi^2 H^2[I_0^2(\pi H) - I_1^2(\pi H)]}. \qquad (4.5)$$

This result for efficiency is readily calculated, and is shown in Table 4.1. Table 4.1 also gives the product of efficiency and normalized beamwidth. In comparison with the circular Taylor n̄ distribution, the Hansen distribution is slightly less efficient, but it is smoother and more robust.

Beam efficiency η_b, the fraction of power in the main beam, might require integration over the sidelobes, a tedious task. However, it can be written in terms of aperture efficiency η_t, which can be easily calculated from the aperture distribution. The aperture efficiency is

$$\frac{1}{\eta_t} = \frac{\pi^2 D^2}{2\lambda^2} \int_0^{\pi/2} \frac{E^2}{E_0^2} \sin\theta\,d\theta, \qquad (4.6)$$

while beam efficiency is

$$\eta_b = \frac{\int_0^{\theta_N} \frac{E^2}{E_0^2} \sin\theta \, d\theta}{\int_0^{\pi/2} \frac{E^2}{E_0^2} \sin\theta \, d\theta}. \tag{4.7}$$

Here θ_N is the main beam null angle. Now the beam efficiency can be expressed by a simple integral over the main beam:

$$\eta_b = \frac{\pi^2 D^2 \eta_t}{2\lambda^2} \int_0^{\theta_N} \frac{E^2}{E_0^2} \sin\theta \, d\theta. \tag{4.8}$$

Results are shown in Fig. 4.5. For diameters of 10 wavelengths or more, η_b is nearly independent of D/λ. Because of the low sidelobes at all azimuth angles, the beam efficiency is much higher than that for a linear distribution.

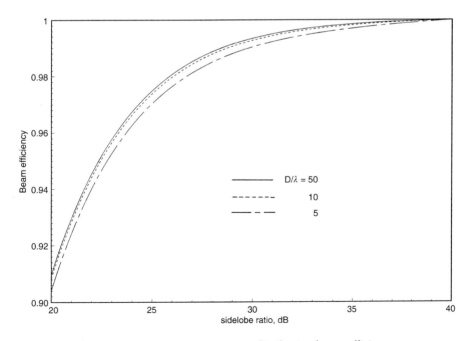

Figure 4.5 Hansen one-parameter distribution beam efficiency.

4.1.3 Taylor Circular n̄ Pattern

The circular Taylor n̄ distribution offers a slightly higher efficiency, through a slightly slower decay of the first n̄ sidelobes, exactly analogously to the linear Taylor n̄ distribution. Again the aperture distribution is symmetric. The uniform $2J_1(\pi u)/\pi u$ pattern is the starting point; n̄ zeros each side of the main beam are adjusted to provide roughly equal levels at the specified SLR. Farther-out zeros are those of the uniform pattern. As before, a dilation factor σ provides a transition between the n̄ zeros, and those adjacent. The sidelobe ratio is related to the general parameter A:

$$\text{SLR} = \cosh A. \tag{4.9}$$

The Taylor circular n̄ space factor (pattern) is given by (Taylor, 1960)

$$F(u) = \frac{2J_1(\pi u)}{\pi u} \prod_{n=1}^{\bar{n}-1} \frac{1 - u^2/u_n^2}{1 - u^2/\mu_n^2}. \tag{4.10}$$

The new zeros are

$$u_n = \pm \sigma \sqrt{A^2 + (n - \tfrac{1}{2})^2} \tag{4.11}$$

and the dilation factor is

$$\sigma = \frac{\mu_{\bar{n}}}{\sqrt{A^2 + (\bar{n} - \tfrac{1}{2})^2}}. \tag{4.12}$$

The zeros of $J_1(\pi u)$ are μ_n. A typical n̄ pattern is shown in Fig. 4.6. Beamwidth is the ideal line source beamwidth u_3 times σ; Table 4.2 gives the parameter A, half-beamwidth u_3, and σ for various SLR and n̄. Hansen (1960) gives tables of the aperture distribution.

The aperture distribution becomes (Taylor, 1960)

$$g(p) = \frac{2}{\pi^2} \sum_{n=0}^{\bar{n}-1} \frac{F_n J_0(\pi p \mu_n)}{J_0^2(\pi \mu_n)}. \tag{4.13}$$

Note that the radial variable for circular Taylor n̄ is $p = 2\pi\rho/D$. The coefficients F_m are

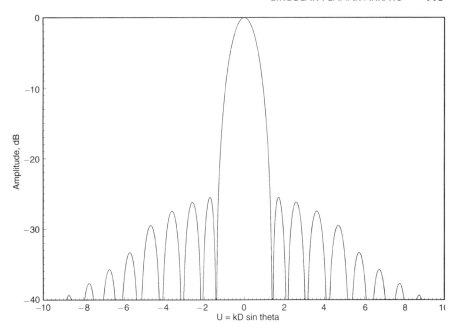

Figure 4.6 Taylor circular $\bar{n}$ Pattern, SLR = 25 dB, $\bar{n} = 5$.

TABLE 4.2 Taylor Circular $\bar{n}$ Pattern Characteristics

SLR (dB)	A	u_3	σ							
			$\bar{n}=3$	$\bar{n}=4$	$\bar{n}=5$	$\bar{n}=6$	$\bar{n}=7$	$\bar{n}=8$	$\bar{n}=9$	$\bar{n}=10$
20	0.9528	0.4465	1.2104	1.1692	1.1398	1.1186	1.1028	1.0906	1.0810	1.0732
25	1.1366	0.4890	1.1792	1.1525	1.1296	1.1118	1.0979	1.0870	1.0782	1.0708
30	1.3200	0.5284	1.1455	1.1338	1.1180	1.1039	1.0923	1.0827	1.0749	1.0683
35	1.5032	0.5653		1.1134	1.1050	1.0951	1.0859	1.0779	1.0711	1.0653
40	1.6865	0.6000		1.0916	1.0910	1.0854	1.0789	1.0726	1.0670	1.0620
45	1.8697	0.6327			1.0759	1.0748	1.0711	1.0667	1.0624	1.0583
50	2.0530	0.6639			1.0636	1.0628	1.0604	1.0573	1.0542	

$$F_m = -J_0(\pi\mu_m)\frac{\prod_{n=1}^{\bar{n}-1}\left(1 - \frac{\mu_m^2}{u_n^2}\right)}{\prod_{\substack{n=1\\n\neq m}}^{\bar{n}-1}\left(1 - \frac{\mu_m^2}{\mu_n^2}\right)}, \qquad m > 0 \tag{4.14}$$

$$F_0 = 1$$

Representative aperture distributions (half only) are shown in Figs. 4.7 and 4.8. Too high a value of $\bar{n}$ causes instability in the null spacings, reflected in a

114 PLANAR AND CIRCULAR ARRAY PATTERN SYNTHESIS

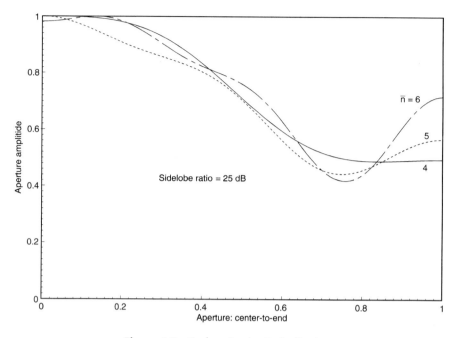

Figure 4.7 Taylor circular $\bar{n}$ distribution.

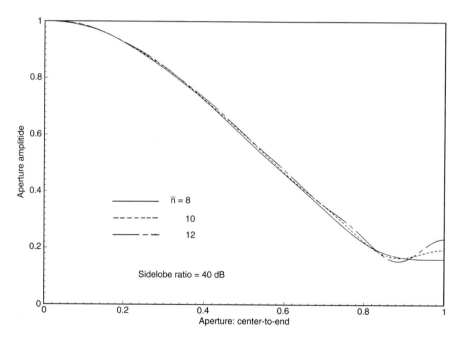

Figure 4.8 Taylor circular $\bar{n}$ distribution.

nonmonotonic distribution. As larger values of $\bar{n}$ give higher efficiency, for each SLR there is an optimum trade on $\bar{n}$.

Aperture efficiency is obtained from

$$\eta_t = \frac{2\left(\int_0^\pi g(p)p\, dp\right)^2}{\int_0^\pi g^2(p)p\, dp} \tag{4.15}$$

Using the above $g(p)$, the terms in the numerator for $m > 0$ contain the factor $J_0(\pi\mu_m)$, which is zero. Only the $m = 0$ term remains. In the denominator, the integration of the product of the two series produces an integral of two orthogonal Bessel functions; only a single series remains:

$$\frac{1}{\eta_t} = 1 + \sum_{n=1}^{\bar{n}-1} \frac{F_n^2}{J_0^2(\pi\mu_n)} \tag{4.16}$$

Excitation efficiencies for several SLR and $\bar{n}$ are given in Table 4.3, from Rudduck et al. (1971). Beam efficiency again is from an integral over the main beam. Table 4.4 gives this versus SLR.

TABLE 4.3 Taylor Circular $\bar{n}$ Efficiency

SLR (dB)	$\bar{n} = 4$	$\bar{n} = 5$	$\bar{n} = 6$	$\bar{n} = 8$	$\bar{n} = 10$
20	0.9723	0.9356	0.8808	0.7506	0.6238
25	0.9324	0.9404	0.9379	0.9064	0.8526
30	0.8482	0.8623	0.8735	0.8838	0.8804
35	0.7708	0.7779	0.7880	0.8048	0.8153
40	0.7056	0.7063	0.7119	0.7252	0.7365
45		0.6484	0.6494	0.6575	0.6663
50			0.5986	0.6016	0.6076

Courtesy Rudduck, R. C. et al, "Directive Gain of Circular Taylor Patterns," *Radio Sci.*, Vol. 6, Dec. 1971, pp. 1117–1121.

TABLE 4.4 Taylor Circular $\bar{n}$ Beam Efficiency

SLR (dB)	$\bar{n}$	η_b
20	4	0.8271
25	5	0.9142
30	6	0.9621
35	7	0.9844
40	8	0.9938
45	9	0.9976
50	10	0.9991

116 PLANAR AND CIRCULAR ARRAY PATTERN SYNTHESIS

4.1.4 Circular Bayliss Difference Pattern

Before discussing the Bayliss $\bar{n}$ difference pattern, it is useful to consider a uniform circular aperture of radius a with the two halves out of phase by π. The ϕ integral for the pattern function reduces to

$$\cos \phi \, H_0(k\rho \sin \theta \cos \phi), \qquad (4.17)$$

where H_0 is the zero-order Struve function (Hansen, 1995). The difference pattern in the $\phi = 0$ plane is then given by

$$F(\theta, \phi) = \cos \phi \int_0^a H_0(k\rho \sin \theta \cos \phi) \rho \, d\rho. \qquad (4.18)$$

The Struve function closely resembles the J_1 Bessel function, as seen in Fig. 4.9, although there appears to be no similarity for higher orders. Figure 4.10 gives a pattern produced by numerical integration, with the amplitude such that the corresponding sum pattern peak is unity. The difference pattern peaks occur at $u_0 = \pm 2.653$, and their level is -3.19 dB. First sidelobes are at $u = \pm 9.775$, with level -14.90 dB, or 11.72 dB below the difference peak.

The Bayliss difference pattern for a line source is discussed in Section 3.7.2. Bayliss (1968) developed a similar $\bar{n}$ difference pattern for a circular aperture. The starting pattern is the derivative of the "ideal" Taylor sum pattern

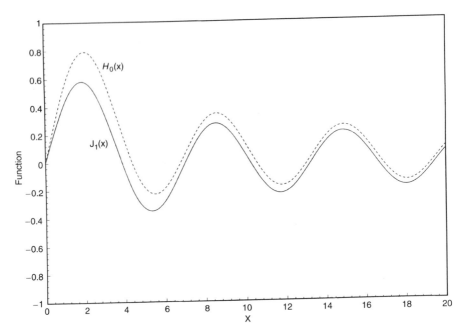

Figure 4.9 J_1 Bessel and H_0 Struve functions.

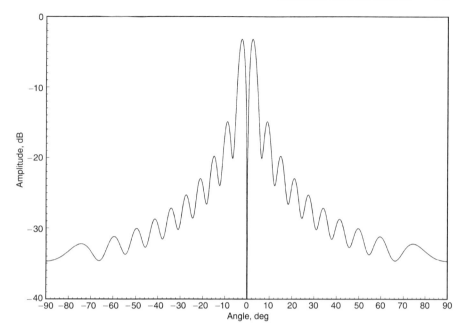

Figure 4.10 Uniform circular aperture difference pattern, $D = 20\lambda$

$\cos(\pi\sqrt{u^2 - A^2})$, as the uniform circular aperture difference pattern does not exist in usable form. The derivative, with coefficients deleted, is

$$f_{\text{ideal}}(u) = u \sin\left(\pi\sqrt{u^2 - A^2}\right). \tag{4.19}$$

The $\bar{n}$ difference pattern is achieved by replacing $\bar{n}$ of the zeros of the ideal pattern by shifted zeros z_n:

$$f(u) = f_{\text{ideal}}(u) \prod_{n=1}^{\bar{n}} \frac{z_n^2 - u^2}{\mu_n^2 - u^2}. \tag{4.20}$$

The new zeros are z_n; these replace the ideal pattern zeros μ_n. The space factor can be written in product form on the zeros:

$$f(u) = u \cos\phi \prod_{n=1}^{\bar{n}-1}\left(1 - \frac{u^2}{\sigma^2 z_n^2}\right) \prod_{n=\bar{n}}^{\infty}\left(1 - \frac{u^2}{\mu_n^2}\right). \tag{4.21}$$

As the zeros of the second product are those of $J_1'(\pi\mu_n)$, that product can be augmented with the missing zeros, and replaced by $J_1'(\pi u)$, with the result

$$f(u) = u\cos\phi\, J_1'(\pi u) \frac{\prod_{n=1}^{\bar{n}-1}\left(1 - \frac{u^2}{\sigma^2 z_n^2}\right)}{\prod_{n=0}^{\bar{n}-1}\left(1 - \frac{u^2}{\mu_n^2}\right)}. \quad (4.22)$$

An alternate version sums over $\bar{n}$ terms:

$$f(u) = u\cos\phi\, J_1'(\pi u) \sum_{n=0}^{\bar{n}-1} \frac{B_n J_1(\pi \mu_n)}{\mu_n^2 - u^2} \quad (4.23)$$

In this difference pattern result, the zeros for $n > \bar{n}$ are given by $z_n = \pm\sqrt{A^2 + n^2}$. Because the ideal difference pattern had close-in sidelobes of irregular levels, Bayliss adjusted these $\bar{n}$ zeros to obtain a monotonic sidelobe envelope. This was done numerically, with fourth-order polynomials fitted to the results. The parameter A and the position of the difference pattern peak were also fitted. These parameters are the same for the linear Bayliss pattern, and are given in Table 3.11. The polynomial coefficients are given in Table 3.12. Figure 4.11 shows a circular Bayliss pattern for SLR = 25 dB, and $\bar{n} = 5$.

The aperture distribution comes from the Bessel series:

$$g(p) = \cos\phi \sum_{n=0}^{\bar{n}-1} B_n J_0(\mu_n p), \quad (4.24)$$

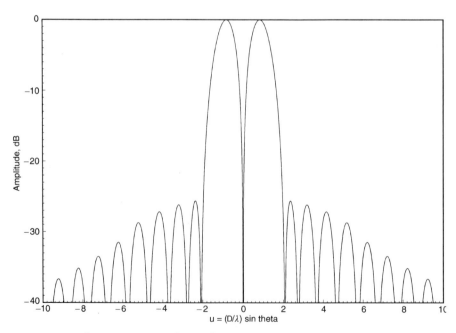

Figure 4.11 Circular Bayliss pattern, SLR = 25 dB, $\bar{n} = 5$.

where $p = \pi\rho$, with ρ the aperture radius from zero to one. The B_n coefficients are

$$B_m = \frac{2\mu_m^2 \prod_{n=1}^{\bar{n}-1}\left(1 - \frac{\mu_m^2}{\sigma^2 z_n^2}\right)}{J_1(\pi\mu_m) \prod_{\substack{n=0 \\ n \neq m}}^{\bar{n}-1}\left(1 - \frac{\mu_m^2}{\mu_n^2}\right)}, \quad m = 0, 1, 2, \ldots, \bar{n} = 1; \quad (4.25)$$

$$B_0 = 0, \quad m \geq \bar{n}.$$

The dilation factor σ is given by

$$\sigma = \frac{\mu_{\bar{n}}}{\sqrt{A^2 + \bar{n}^2}}. \quad (4.26)$$

Figure 4.12 shows the aperture distribution for the case of Fig. 4.10. As expected, larger values of $\bar{n}$ for a given SLR produce a rise at the aperture edge, which can be higher than the center value. Figure 4.13 from a Bayliss report shows a quadrant of the two-dimensional distribution.

Aperture efficiency is defined as the directivity at the difference pattern peak, divided by the uniform aperture sum pattern directivity. Owing to the orthogonality of the aperture series, the directivity integral is readily evaluated, giving

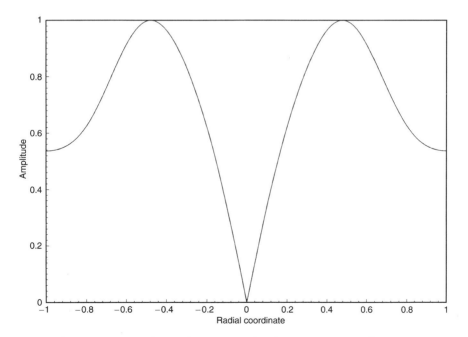

Figure 4.12 Circular Bayliss aperture distribution, SLR = 25 dB, $\bar{n} = 5$.

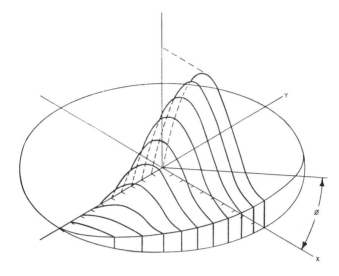

Figure 4.13 Circular Bayliss distribution, SLR = 35 dB, $\bar{n} = 10$. (Courtesy Bayliss, E. T., "Monopulse Difference Patterns with Low Sidelobes and Large Angle Sensitivity," Tech. Memo. 66-4131-3, Bell Labs., Dec. 1966.)

$$\frac{1}{\eta_t} = \frac{\pi^4}{8}\sum_{n=0}^{\bar{n}-1}|B_n|^2\left(1 - \frac{1}{\pi^2\mu_n^2}\right)J_1^2(\pi\mu_n). \qquad (4.27)$$

A second important parameter is slope, here normalized to that of the maximum slope aperture. This reference pattern is $J_2(\pi u)/\pi u$, and its slope is $(0.5/\pi)\sqrt{G_0}$ where G_0 is the directivity. For the Bayliss pattern the normalized slope becomes

$$S = \sqrt{\eta_t}\sum_{n=0}^{\bar{n}-1}\frac{B_n J_1(\pi\mu_n)}{\mu_n^2}. \qquad (4.28)$$

Table 4.5 gives efficiency and slope for typical combinations of SLR and $\bar{n}$.

TABLE 4.5 Circular Bayliss Efficiency and Slope

SLR (dB)	$\bar{n}$	η_t	Normalized Slope
25	5	0.4931	0.8754
30	6	0.4593	0.8156
35	8	0.4285	0.7638
40	10	0.3988	0.7144
45	12	0.3717	0.6691
50	14	0.3475	0.6286

4.1.5 Difference Pattern Optimization

The use of subarrays, where each can be fed with a different amplitude (and phase) for sum and difference patterns, was discussed in Chapter 3 for linear arrays. The same principle was applied to circular planar arrays by Josefsson et al. (1977). The array is divided into subarrays, with quadrantal symmetry. The shape of each subarray and the number of elements will generally be different; see Fig. 4.14. Symmetric pairs of subarrays are then connected to hybrid junctions; these hybrids are connected to the sum and to the two difference combiner networks; see Fig. 4.15. As in the linear array case, a stair-step approximation is provided to the three desired pattern distributions. Excellent results have been obtained with only a small number of subarrays per quadrant. This configuration is especially attractive with printed circuit antennas, as the implementation is natural for realization in stripline.

4.2 NONCIRCULAR APERTURES

4.2.1 Two-Dimensional Optimization

When the aperture (array) is not circular, or when x and y distributions are not separable, recourse must be had to numerical techniques, such as constrained optimization. Unfortunately, a relatively simple technique (dynamic programming) is not applicable, as it requires the function ahead of the change to be independent of the change. But with all arrays, changes to one zero or to one excitation coefficient affect the entire pattern. Optimization programs such as

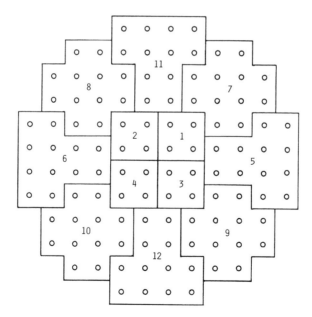

Figure 4.14 Planar array subarrays.

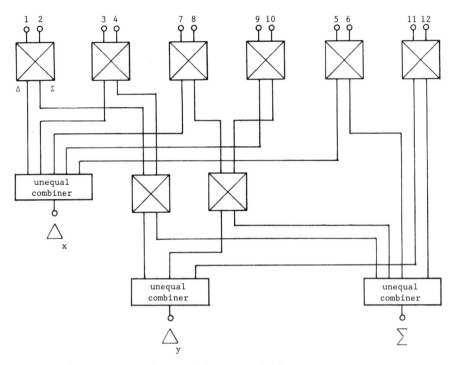

Figure 4.15 Feed network for sum and difference optimization.

conjugate gradient are suitable provided the objective function can be well specified, but slow. Since sidelobe positions are not predictable, maximizing directivity subject to the requirement that all sidelobes be below a given envelope is difficult. Either a sidelobe location routine that searches for each and every sidelobe peak at each function (array pattern) evaluation is used, or each pattern must be digitized sufficiently finely to accurately locate all sidelobe peaks. Sidelobe searching algorithms only work well on canonical patterns, with regularly spaced, well-behaved sidelobes. Numerically optimized patterns, and measured patterns, typically have a nonmonotonic sidelobe envelope. Further, many "split" sidelobes occur that are much narrower. The typical fine structure of sidelobes defeats most algorithms, even for one-dimensional patterns. Fine sampling with filtering to find the envelope is feasible, but for two-dimensional patterns it can be extraordinarily slow. Since an optimization may require hundreds of function evaluations, the computational cost may well be excessive.

For patterns where the array is small or modest in size, and the pattern features are coarse, optimization of directivity, slope, etc., with constraints on sidelobe envelope, bandwidth, tolerances, etc., may be feasible; see Perini and Idselis (1972), for example. Several classes of optimization algorithms exist. First are gradient schemes. Simple gradient and conjugate gradient schemes tend to be fast, but they often stop at a local maximum, or hang up at a saddle point. More sophisticated schemes avoid both these problems and will usually find the global maximum (Fletcher, 1986). Some of these are those of Fletcher and Powell (1963), Fletcher and Reeves (1964), and Gill and Murray (1972). An algorithm due to Hersey et al. (1972) allows bounds to be placed over regions, such as sidelobe regions; a fine discretization provides a REMEZ exchange minimax optimization (Mucci et al., 1975). A review of unconstrained optimization methods is given by Nocedal (1991). Search techniques are generally inefficient, except for simplex methods. Simplex compares function values, forming a sequence of simplexes; each simplex is a set of points with a vertex where the function value is maximum. This vertex is reflected in the centroid of the other vertices, forming a new simplex. The value at this new vertex is evaluated, and the process is repeated. Special rules re-start the process during stagnation. In the author's modest experience an excellent algorithm is the Fletcher variable metric (Fletcher, 1970), available in Fortran from Harwell, both in analytical and numerical derivative form.

Because the sidelobe positions change as the excitation changes, an iterative procedure is usually used. Here the sidelobe positions for an appropriate canonical pattern are used in the first iteration. At each iteration the pattern zeros are determined, with sidelobe peaks assumed midway between.

Synthesis with sidelobe constraints, or synthesis to minimize the oscillations in a shaped beam pattern, often minimize the sum of squares of the differences. However, use of least squares in the optimized function gives fits that are less good at the interval edges. A better scheme uses a least pth optimization, where the pth root of the sum of differences, each to the pth power, is minimized (Bandler and Charalambous, 1972; Temes and Zai, 1969). A value of $p = 10$ gives superior fits, while very large p values approach minimax results. However, the latter require careful floating-point number handling.

4.2.2 Ring Sidelobe Synthesis

For rectangular apertures, a sidelobe envelope that decays with θ is often desirable, analogous to the Hansen circular one-parameter pattern, or to the Taylor circular $\bar{n}$ pattern. Such a synthesis can be produced by the brute-force computer techniques of the previous section, or by the methods of Section 3.9.2, where the zeros of a (two-dimensional) polynomial are iteratively adjusted. This latter approach is far better.

A more analytical approach utilizes the Baklanov (1966) transformation, which converts the two-dimensional problem to a one-dimensional problem. Tseng and Cheng (1968) applied this transformation to a square array providing Chebyshev type ring sidelobes. Kim and Elliott (1988) extended this to an adjustable topology of ring sidelobes. For a square array of $N \times N$ elements, let the pattern be

$$F(u, v) = 4 \sum_{n=1}^{N/2} \sum_{m=1}^{N/2} A_{nm} \cos\left[(2n-1)u\right] \cos\left[(2m-1)v\right]. \tag{4.29}$$

The Baklanov transformation is

$$w = w_0 \cos u \cos v. \tag{4.30}$$

Now assume the desired ring sidelobe pattern is given by a polynomial of order $N - 1$ in the new variable w:

$$f_{N-1}(w) = \sum_{l=1}^{N/2} B_{2l-1} w^{2l-1}. \tag{4.31}$$

Through use of trig expansions the array excitation coefficients A_{nm} are found from the polynomial coefficients B_l using

$$A_{nm} = \sum^{N/2} \frac{B_{2l-1}}{2^{2l-1}} \binom{2l-1}{2l-n} \binom{2l-1}{2l-m} \left[\frac{w_0}{2}\right]^{2l-1} \tag{4.32}$$

The sum starts at the larger of n and m. The parameter w_0 is related to the sidelobe envelope, and is determined iteratively. The synthesis procedure, to determine the B_l coefficients, can utilize either the zero adjusting process or the Orchard process of Section 3.9.2. In any case, the polynomial zeros must be determined so that sidelobe (ring) locations can be calculated. This procedure is repeated each iteration. Fortunately, fewer than 10 iterations are usually required. Further use of a projection algorithm is described by Kim (1988). An example is shown in Fig. 4.16, which is a contour pattern for a 10×10

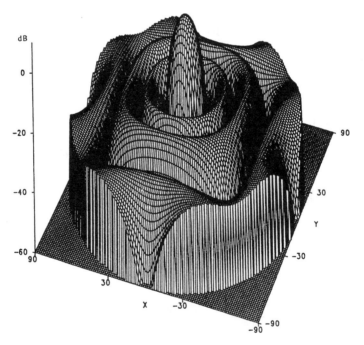

Figure 4.16 Baklanov ring sidelobe pattern. (Courtesy Kim, Y. U. and Elliott, R. S., "Extensions of the Tseng-Cheng Pattern Synthesis Technique," *J. Electromagnetic Waves Appl*, Vol. 2, No. 3/4, 1988, pp. 255–268.)

square array with first sidelobe at $-30\,\text{dB}$ (Kim, 1990). The sidelobe rings are not circular, but fit the aperture geometry. Such a synthesis provides higher gain for the same sidelobe level than a design based on the product of two linear distributions.

Acknowledgment

Photograph courtesy of Dr. Winifred Kummer.

REFERENCES

Baklanov, Y. V., "Chebyshev Distribution of Current for a Planar Array of Radiators," *Radio Eng. and Electron. Phys.*, Vol. 11, 1966, pp. 640–642.

Bandler, J. W. and Charalambous, C., "Theory of Generalized Least Pth Approximation," *Trans. IEEE*, Vol. CT-19, May 1972, pp. 287–289.

Bayliss, E. T. "Design of Monopulse Antenna Difference Patterns with Low Sidelobes," *Bell Systems Tech. J.*, Vol. 47, No. 5, May–June 1968, pp. 623–650.

Erlinger, J. J. and Orlow, J. R., "Waveguide Slot Array with $\text{Csc}^2\theta \cos\theta$ Pattern," *Proceedings 1984 Antenna Applications Symposium*, RADC-TR-85-14, Vol. 1, pp. 83–112, AD-A153 257.

Fletcher, R., "A New Approach to Variable Metric Algorithms," *Computer J.*, Vol. 13, Aug. 1970, pp. 317–322.

Fletcher, R., *Practical Methods of Optimization*, Wiley, 1986.

Fletcher, R. and Powell, M. J. D., "A Rapidly Convergent Descent Method for Minimization," *Computer J.*, Vol. 6, 1963, pp. 163–168.

Fletcher, R. and Reeves, C. M., "Function Minimization by Conjugate Gradients," *Computer J.*, Vol. 7, 1964, pp. 149–154.

Gill, P. E. and Murray, W., "Quasi-Newton Methods for Unconstrained Optimization," *J. Inst. Math. Appl.*, Vol. 9, 1972, pp. 91–108.

Hansen, R. C., "Tables of Taylor Distributions for Circular Aperture Antennas," *Trans. IEEE*, Vol. AP-8, Jan. 1960, pp. 23–26.

Hansen, R. C., "A One-Parameter Circular Aperture Distribution with Narrow Beamwidth and Low Sidelobes," *Trans. IEEE*, Vol. AP-24, July 1976, pp. 477–480.

Hansen, R. C., "Struve Functions for Circular Aperture Patterns," *Microwave and Opt. Technol. Lett.*, Vol. 10, Sept. 1995, pp. 6–7.

Hersey, H. S. Tufts, D. W. and Lewis, J. T., "'Interactive Minimax Design of Linear-Phase Nonrecursive Digital Filters Subject to Upper and Lower Function Constraints," *Trans. IEEE*, Vol. AU-20, June 1972, pp. 171–173.

Josefsson, L. Moeschlin, L. and Sohtell, V., "A Monopulse Flat Plate Antenna for Missile Seeker," presented at the Military Electronics Defense Expo, Weisbaden, West Germany, Sept. 1977.

Kim, Y. U. "Peak Directivity Optimization under Side Lobe Level Constraints in Antenna Arrays," *Electromagnetics*, Vol. 8, No. 1, 1988, pp. 51–70.

Kim, Y. U. "A Pattern Synthesis Technique for Planar Arrays with Elements Excited In-Phase," *J. Electromagnetic Waves Appl.*, Vol. 4, No. 9, 1990, pp. 829–845.

Kim. Y. U. and Elliott, R. S., "Extensions of the Tseng-Cheng Pattern Synthesis Technique," *J. Electromagnetic Waves Appl*, Vol. 2, No. 3/4, 1988, pp. 255–268.

Mucci, R. A. Tufts, D. W. and Lewis, J. T., "Beam Pattern Synthesis for Line Arrays Subject to Upper and Lower Constraining Bounds," *Trans. IEEE*, Vol. AP-23, Sept. 1975, pp. 732–734.

Nocedal, J. "Theory of Algorithms for Unconstrained Optimization," in *Acta Numerica*, Cambridge University Press, 1991, pp. 199–242.

Perini, J. and Idselis, M. H., "Radiation Pattern Synthesis for Broadcast Antennas," *Trans. IEEE*, Vol. BC-18, Sept. 1972, pp. 53–62.

Rudduck, R. C. et al., "Directive Gain of Circular Taylor Patterns," *Radio Sci.*, Vol. 6, Dec. 1971, pp. 1117–1121.

Taylor, T. T. "Design of Circular Apertures for Narrow Beamwidths and Low Sidelobes," *Trans. IRE*, Vol. AP-8, Jan. 1960, pp. 17–22.

Temes, G. S. and Zai, D. Y. F. "Least Pth Approximation," *Trans. IEEE*, Vol. CT-16, May 1969, pp. 235–237.

Tseng, F.-I. and Cheng, D. K. "Optimum Scannable Planar Arrays with an Invariant Sidelobe Level," *Proc. IEEE*, Vol. 56, Vol. 11, Nov. 1968, pp. 1771–1778.

CHAPTER FIVE

Array Elements

Fixed beam broadside arrays may employ low- or moderate-gain elements, but most arrays employ low-gain elements owing to the effects of grating and quantization lobes (see Chapter 3). This chapter is concerned with these low-gain elements. Moderate-gain elements, such as the spiral, helix, log-periodic, Yagi–Uda, horn, and backfire, generally do not have unique array properties, and are covered in numerous other books.

5.1 DIPOLES

5.1.1 Thin Dipoles

The dipole, with length L approximately a half-wavelength, is widely used because of its simple construction and good performance. Dipoles are made of two collinear and contiguous metallic rods or tubes with the feed between; or of conical conductors, typically hollow; or of strips or triangles printed on a thin dielectric support. Microstrip dipoles are discussed later. A strip dipole, where strip thickness is small compared to strip width, is equivalent to a cylindrical dipole of radius a equal to one-fourth the strip width w (Lo, 1953). A half-wave dipole has a pattern symmetric about the dipole axis; with θ measured from the axis it is

$$f(\theta) = \frac{\cos(\pi/2 \cos\theta)}{\sin\theta}. \tag{5.1}$$

The half-power beamwidth is 78.1 deg, and the directivity is $1.64 = 2.15$ dB. As the dipole length shortens, the pattern approaches that of a short dipole, $\sin\theta$, with half-power beamwidth of 90 deg and directivity of $1.5 = 1.76$ dB. For lengths longer than a half-wavelength, the pattern sharpens, then breaks up. At $L = 0.625\lambda$, the main lobe is broadside, and the two sidelobes are small. But at $L = 0.75\lambda$, the sidelobes (at 45 deg) are larger than the main beam. Finally, the full-wave dipole with sinusoidal current distribution has a null at broadside. Pattern bandwidth then goes from short dipole length to length of roughly 0.6λ. The impedance bandwidth is usually limiting. It was shown by Abraham in 1898 that a vanishingly thin dipole has a sinusoidal current distribution. The Carter zero-order impedance theory discussed in Chapter 7 is adequate for mutual-impedance calculations, and for self-impedance of very thin ($L/a > 1000$) dipoles. Full-wave dipoles, according to this theory, have zero feed current, so a higher-order theory is needed for them. However, owing to their high input impedance and multilobe pattern, full-wave dipoles are rarely used. For antennas in the vicinity of half-wave in length, the King second-order theory (King, 1956) agrees well with experiments. Unfortunately, no satisfactory simple function fit to the variation of impedance versus frequency has been possible.

Practical dipole bandwidths can range up to roughly 30–40%; the open sleeve dipole discussed below provides an octave of impedance bandwidth (73%). Figures 5.1 and 5.2 give thin dipole input resistance and reactance versus dipole length, for three values of L/a. Capacitance between the dipole arms produces resonance below half-wave length, with fatter resonant dipoles being shorter than thin resonant dipoles. Figure 5.3 shows dipole shortening versus dipole radius. Moment method calculations also agree well with experiments (Hansen, 1990), and can give good results for very fat dipoles. The designer should start with the simple Carter results, then refine as needed with moment method, and of course, validate with measurements. The advent of network analyzers has greatly simplified the task of antenna impedance measurement.

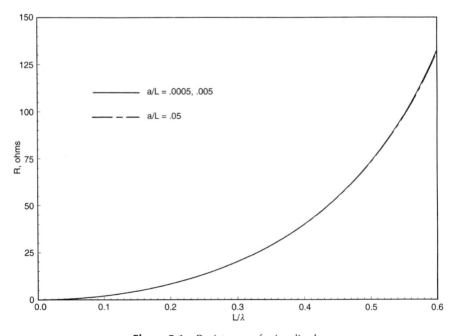

Figure 5.1 Resistance of wire dipole.

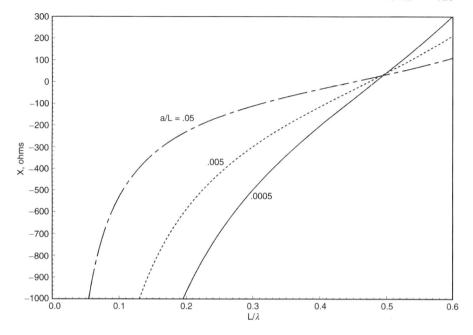

Figure 5.2 Reactance of wire dipole.

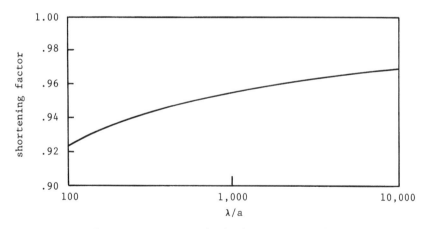

Figure 5.3 Resonant dipole shortening vs. radius.

Arrays with dipoles parallel to a ground or back screen are commonly used for beam angles around broadside. Such dipoles are fed by baluns, to convert the balanced two-wire dipole feed to a coax feed. Baluns then are normal to the screen, and utilize the roughly λ/4 spacing between dipole and screen. A simple but often used balun is the split tube balun of Figure 5.4, where the split is λ/4 long. Wider bandwidths are provided by coaxial type I or III baluns[1], where a

---

[1] Defined circa World War II by Nelson and Stavis (1947).

130 ARRAY ELEMENTS

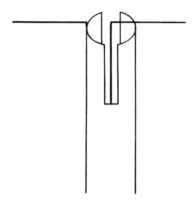

Figure 5.4 Split tube balun.

coax feed line is enclosed by a cylinder connected at the screen. In the type III balun, a dummy outer conductor (to the coax) is added to form a two-wire line inside the outer cylinder; see sketch in Fig. 5.5. Circular or adjustable polarization is provided by crossed dipoles connected to two sets of balun posts at right angles to each other. A simple way of producing circular polarization uses crossed dipoles connected to a single balun. One dipole is larger and thinner, the other shorter and fatter, so that the respective phases are ±45 deg. Bandwidth is significantly reduced, as is expected.

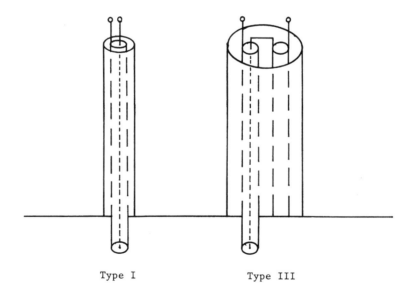

Figure 5.5 Coaxial baluns. (Courtesy Nelson, J. A. and Stavis, G., "Impedance Matching, Transformers and Baluns," in *Very High-Frequency Techniques*, Vol. 1, H. J. Reich, Ed., Radio Research Lab., McGraw-Hill, 1947.)

A dipole a distance *h* above a ground plane is equivalent to the dipole and its image (at a distance twice that of the dipole above the plane): thus this "two-element array" provides directivity. The pattern, but not including the dipole pattern, is $\sin(kh\cos\theta)$. Input impedance, as shown in Fig. 5.6, oscillates with *h*, but for small *h* becomes small. As the spacing decreases the pattern approaches $\cos\theta$. Directivity increases slowly as *h* decreases, as observed by Brown (1937). However, since the image current is reversed, as *h* decreases the heat loss increases. Thus the gain actually peaks. Figure 5.7 shows directivity versus h/λ, and also gain for several loss factors. The loss resistance shown is the equivalent resistance at the terminals. This is related to the surface resistance R_s by multiplying the latter by the metallic path length and dividing by the path width. For a half-wave dipole the path length is 1/2, while the width is the dipole circumference:

$$R_L = \frac{R_s L}{4\pi a} = \frac{R_s L}{\pi w}. \tag{5.2}$$

For example, a printed circuit dipole at 5000 MHz made of copper, and with a length/width ratio of 50, has a terminal resistance of 0.3 ohm, a small but nontrivial value. There is then a spacing between dipole and ground screen that maximizes the gain.

Direct (metallic) coupling of a dipole to a transmission line is not necessary; electromagnetic coupling may be used (Forbes, 1960). Here the shorted dipole is placed above a two-wire feed line at an angle to the line.

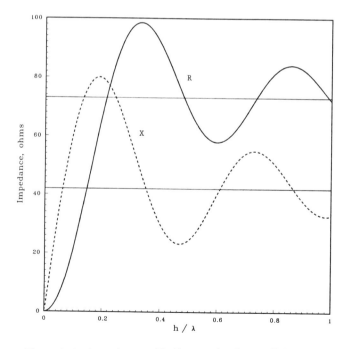

Figure 5.6 Impedance of half-wave dipole parallel over screen.

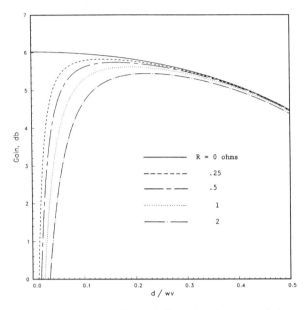

Figure 5.7 Gain of out of phase dipole pair with loss.

Dipoles with cross sections other than circular can, if the length/diameter is roughly ≥ 10, be equated to a cylindrical dipole. The most important is the thin strip dipole. Lo (1953) found the equivalent radius of polygons: for a strip dipole it is 1/4 the width; for a square conductor it is 0.59 times the side. An L shaped conductor has equivalent radius of 0.2 times the sum of the arm widths (Wolff, 1967). An elliptical conductor has equivalent radius half the sum of the major and minor radii (Balanis, 1982); also given there is a curve for a rectangular conductor.

Folded dipoles are used mostly in VHF–UHF arrays, as their primary advantage is the ability to match typical two-wire transmission lines. They are no more broadband than a fat dipole occupying the same volume. The impedance over a wide range of lengths, as seen in Fig. 5.8, has resonances at dipole length roughly $\lambda/3$ and $2\lambda/3$. In between, at length $\lambda/2$, is the normal operating point. The impedance is given by

$$Z = \frac{2(1+a)^2 Z_T Z_D}{(1+a^2)Z_D + 2Z_T}, \tag{5.3}$$

where Z_D is the dipole impedance and Z_T is the shorted stub line impedance: $Z_T = jZ_0 \tan(\pi L/\lambda)$, with L the dipole length. The impedance can be adjusted by changing the ratio of the radius of the fed arms ρ_1 to the radius of the coupled arm ρ_2, and the center-center spacing d. The impedance transformation ratio is $(1+a)^2$, which is shown in Fig. 5.9 versus spacing for several radius ratios. Figure 5.10 gives radius ratio versus spacing for several transformation ratios (Hansen, 1982).

DIPOLES   133

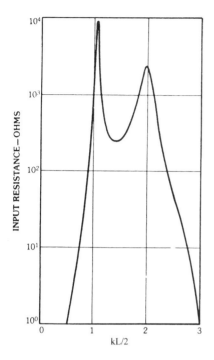
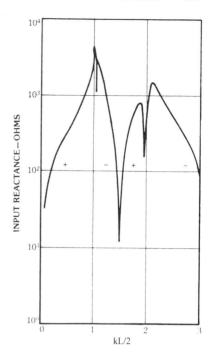

Figure 5.8 Folded dipole impedance, $d = 12.5a$, $a = 0.0005\lambda$. (Courtesy Rispin, L. W. and Chang, D. C., "Wire and Loop Antennas," in *Antenna Handbook—Theory, Applications, and Design*, Y. T. Lo and S. W. Lee, Eds., Van Nostrand Reinhold, 1988, Chapter 7.)

5.1.2 Bow-Tie and Open Sleeve Dipoles

For very fat configurations a bow-tie shape has demonstrated excellent bandwidth. Only numerical methods such as the moment method have proved useful in analyzing such fat dipoles. Figures 5.11 and 5.12 give impedance design data for bow-tie monopoles versus electrical half-length in degrees as calculated by Butler et al. (1979). The conical monopole, or biconical dipole, has slightly better bandwidth than its flat counterparts. An FDTD (finite difference time domain) analysis has been given by Maloney, Smith, and Scott (1990).

Sleeve dipoles (monopoles), wherein a cylindrical sleeve surrounds the central portion of the dipole, were developed during World War II at the Harvard Radio Research Laboratory (Bock et al., 1947). A significant improvement was made by Bolljahn (1950) in opening the sleeve, allowing a more compact and versatile structure. Reports and theses were produced by H. B. Barkley, John Taylor, A. W. Walters, and others; King and Wong (1972) and Wong and King (1973) optimize impedance over an octave, while Wunsch (1988) presents a numerical analysis. The sleeve is open as it does not surround the dipole but consists only of two tubes or plates on opposite sides of the dipole. Figure 5.13 sketches a linearly polarized open sleeve dipole. Crossed dipoles with crossed sleeves for circular polarization (CP) can also be constructed. The crossed

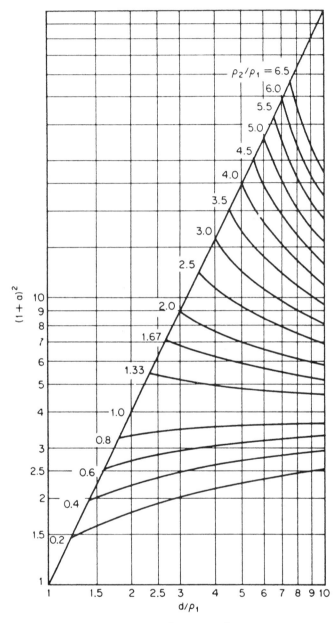

Figure 5.9 Impedance transformation.

DIPOLES 135

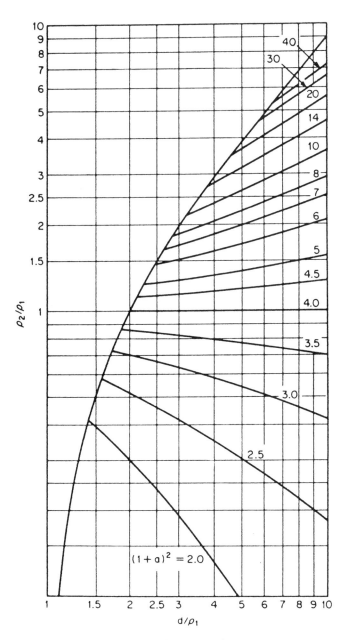

Figure 5.10 Impedance transformation.

136 ARRAY ELEMENTS

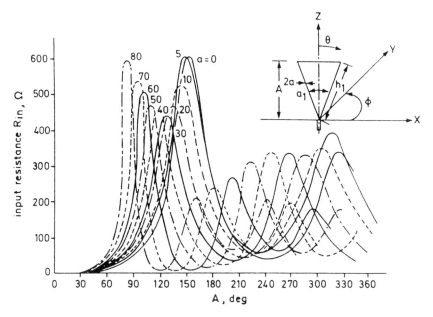

Figure 5.11 Bowtie monopole resistance. (Courtesy Butler, C. M. et al., "Characteristics of a Wire Biconical Antenna," *Microwave J.*, Vol. 22, Sept. 1979, pp. 37–40.)

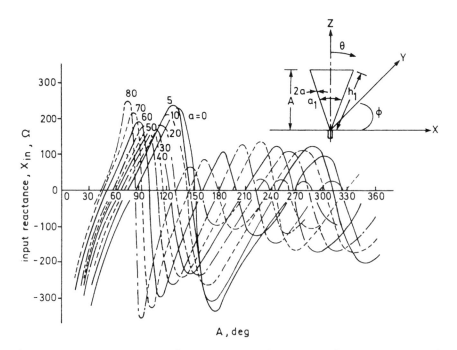

Figure 5.12 Bowtie monopole reactance. (Courtesy Butler, C. M. et al., "Characteristics of a Wire Biconical Antenna," *Microwave J.*, Vol. 22, Sept. 1979, pp. 37–40.)

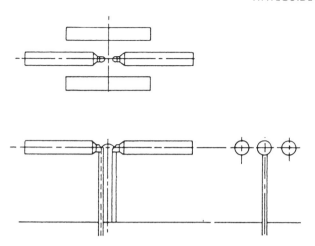

Figure 5.13 Open sleeve dipole. (Courtesy King, H. E. and Wong, J. L., "An Experimental Study of a Balun-Fed Open-Sleeve Dipole in Front of a Metallic Reflector," *Trans. IEEE,* Vol. AP-20, 1972, pp. 201–204.)

sleeves on each side can be replaced by a metallic circular or square disk. Bandwidth, for VSWR < 2 can be an octave. These dipoles have been arrayed over a ground plane with excellent results. The open sleeve dipole appears to be the low-gain element with the widest bandwidth, excluding of course resistively loaded elements where bandwidth is traded for efficiency. Again, moment methods—particularly patch versions—are the appropriate design tool.

5.2 WAVEGUIDE SLOTS

Waveguide slots were invented by Watson (1946, 1947) in 1943 at McGill University (Montreal, Canada), with related work by Stevenson, Cullen, and others in Britain. After World War II, slot work shifted to the Hughes Aircraft Microwave Laboratory at Culver City, California, under the leadership of Les Van Atta. Contributors there included Jim Ajioka, Al Clavin, Bob Elliott, Frank Goebbels, Les Gustafson, Ken Kelly, Lou Kurtz, Bernie Maxum, Joe Spradley, Lou Stark, Bob Stegen, George Stern, Ray Tang, and Nick Yaru.

A waveguide slot is simply a narrow slot cut into the broad wall or edge of a rectangular waveguide. Slots can be accurately milled, especially with numerically controlled machines, and the waveguide provides a linear feed that is low-loss. The precise control of aperture distribution afforded by slot arrays has led to their replacing reflector antennas in many missile and aircraft radar systems, and their wide use in many other applications. A narrow slot along the center line of the waveguide broad wall does not radiate; to produce radiation, the slot must be displaced toward the edge or rotated about the centerline. Similarly, a slot in the narrow wall does not radiate if it is normal to the

138 ARRAY ELEMENTS

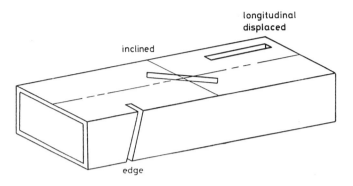

Figure 5.14 Waveguide slot types.

edge; rotating the slot couples it to the waveguide mode. Since an edge slot must usually be "wrapped around" to get resonant length, displacing such a slot toward the center-line is impractical. Figure 5.14 shows the three most important types. The displaced broad wall slot is often called a shunt slot as its equivalent circuit is a shunt admittance across the feed line. The rotated series slot, centered on the broad wall, is a series slot, as its equivalent circuit has a series impedance. And the edge slot is a shunt-type slot. The edge slot and displaced broadwall slot are most often used, with polarization often the deciding factor. In an edge slot the E polarization is along the guide axis, while for displaced slots it is across the axis. Pattern behavior of waveguide slots is close to that of slots in an infinite ground plane, except at angles near grazing, where edge effects are important.

Resonant arrays (Chapter 6) require in-phase elements at half-guide-wavelength spacing. To cancel the guide phase advance, every other slot is placed on the opposite side of the center line (for longitudinal slots), or the rotation angle is reversed (for edge slots). Now the array pattern is composed of an array factor with double spacing, and an element consisting of a pair of slots. For longitudinal slots the subarray (dual slot element) lobes can be controlled through element spacing. With edge slots, however, cross-polarized lobes are produced at certain angles. These lobes can be reduced by replacing each slot by a closely spaced pair of slots with smaller inclination angle.

5.2.1 Broad Wall Longitudinal Slots

Admittance (or impedance) of a waveguide slot consists of an external contribution which can be computed from the equivalent slot in ground plane or Babinet equivalent dipole, and an internal contribution due to energy storage in evanescent modes in the guide around the slot. Since the internal contribution is reactive, the conductance of a resonant slot can be found approximately from the Babinet dipole and from the coupling to the TE_{01} mode. Stevenson (1948) developed the formula

$$\frac{G}{Y} = \frac{480(a/b)}{\pi R_0(\beta/k)} \sin^2 \frac{\pi x}{a} \cos^2 \frac{\pi \beta}{2k} \tag{5.4}$$

where R_0 is the dipole resistance, x is the slot offset, a and b are the guide width and height, and β is the guide wavenumber. This formula is satisfactory for slot conductance, but there is no simple formulation for slot susceptance. A variational formulation of the susceptance problem was made by Oliner (1957a,b). He obtained closed-form simplifications which are of use, but not sufficiently accurate for array design. With the availability of powerful computers, a more complete evaluation of variational forms can be made. Yee (1974) developed such a formulation, and results from it are used below.

A different type of solution was developed by Khac and Carson (1973) and Khac (1974), who wrote coupled integral equations representing external and internal electric fields. The coupled integral equations are then solved by the moment method, using pulse expansion functions and delta testing functions. In both, wall thickness is included via Oliner's method of coupling external and internal fields by a waveguide transmission line, where the waveguide cross section is the slot and the waveguide length is the wall thickness. Satisfactory agreement with measured data has been realized with both approaches. Further improvement has been made by Elliott and his students, using piecewise sinusoidal expansion and testing functions, a Galerkin stationary formulation (Park, Stern, and Elliott, 1983).

Using the variational method, admittance of slots in WR-90 waveguide was calculated: resonant slot length versus slot offset and resonant conductance. Offset x is conveniently normalized to guide width a, while resonant length l_r is normalized to free-space wavelength λ_0. Figure 5.15 shows curves for $a = 0.9$, wall thickness $= 0.05$ and slot width $= 1/16$, all in inches. Frequencies of 9.375 and 10 GHz are shown, with $b = 0.4$ and 0.2 inches. Standard WR-90 has 0.4 height, but the half-height guide is of considerable interest for receiving arrays. Resonant conductance is shown in Fig. 5.16 for one case; it matches well the Stevenson formula given earlier.

Next the calculated admittance is normalized by resonant conductance G_r, with slot length normalized by resonant length. These universal curves were developed by Kaminow and Stegen (1954) on the basis of careful measurements. Figure 5.17 gives these data for WR-90 at 9375 MHz, again for two slot offsets. A plotting error in the original curves has been corrected. The curves are universal in the following sense. To first order the variation of Y/G_r with $1/l_r$ and with x/a is independent of β/k, that is, of frequency. Thus measurement or calculation of data for a given waveguide size at one frequency is adequate. For small offsets (0.05 and below), the normalized admittance variation is also independent of a/b. However, for larger offsets, data should be obtained for the exact a/b. "Universal" curves are shown in Fig. 5.18 for half-height WR-90 guide at 9375 MHz; two slot offsets are shown.

The variational or moment method calculation of admittance is much too slow to use with iterative methods such as those described in Chapter 6; polynomial fits are more suitable. However, the universal curves are difficult to fit with one expression. Dividing up the curve allows a more accurate fit, but care

140 ARRAY ELEMENTS

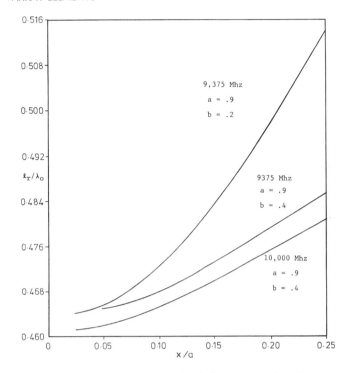

Figure 5.15 Longitudinal slot resonant length.

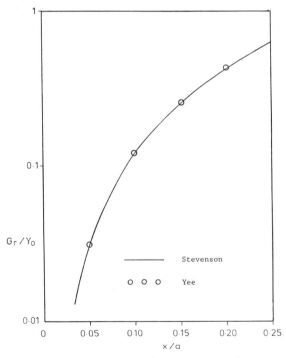

Figure 5.16 Longitudinal slot resonant conductance.

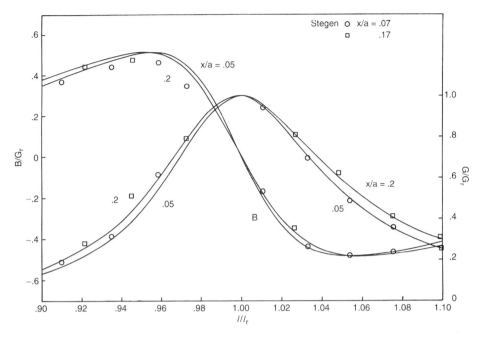

Figure 5.17 Longitudinal slot admittance, 9375 MHz, $a = 0.9$, $b = 0.4$. (Courtesy Stegen, R. J., "Longitudinal Shunt Slot Characteristics," Tech. Memo. 261, Hughes Aircraft Co., Nov. 1951.)

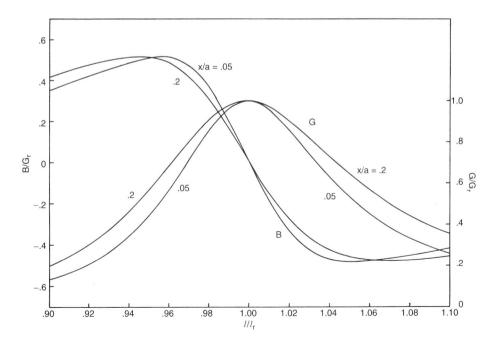

Figure 5.18 Longitudinal slot admittance, 9375 MHz, $a = 0.9$, $b = 0.2$.

must be taken to avoid slope discontinuities at the joins as these affect the gradient methods used for array design. An alternative method uses impedance rather than admittance. When slot impedance is normalized to resonant resistance, the curves are as shown in Figs. 5.19 and 5.20. Resistance is a straight line (to first order), with the slope independent of a/h. There is a slope change with β/k as shown. The reactance curve is a slightly curved line, which again is independent of β/k, and for small offsets is independent of a/b also. For larger offsets the reactance curves depend on both x/a and a/b. Thus the impedance curves are just as "universal" as the admittance curves, and are much easier to fit with polynomials. If data are not available for a particular size and wall thickness of waveguide to be used in an array, it is necessary either to calculate or to measure sufficient data to plot universal curves. The impedance curves are preferable as they are slowly varying.

For most waveguide slots an assumed sinusoidal field distribution is satisfactory. However, dielectric filled guides induce a slot that is better approximated by a half-cosinusoidal distribution (Elliott, 1983). Unfortunately, this means that mutual admittance requires numerical integration over both slots, which greatly impedes the synthesis of slot array patterns. A simple relationship between the two mutual admittances has been given by Malherbe and Davidson (1984). Dielectric filling that occupies only part of the waveguide width can help suppress grating lobes; a moment method analysis of a transverse slot is given by Joubert (1995).

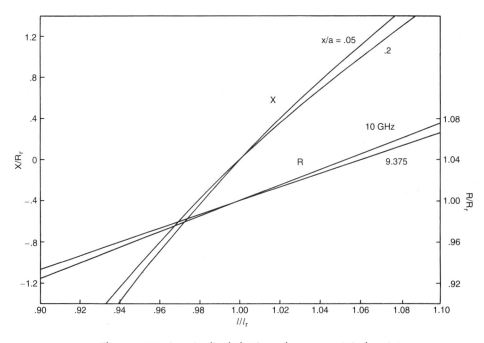

Figure 5.19 Longitudinal slot impedance, $a = 0.9$, $b = 0.4$.

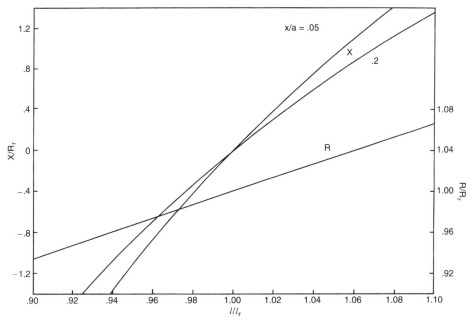

Figure 5.20 Longitudinal slot impedance, $a = 0.9$, $b = 0.2$.

It is convenient to make waveguide slots with round ends, to accommodate milling machines. To first order, such a slot is equivalent to a rectangular slot of the same width and same area (Oliner, 1957a,b). For more accurate results, a moment method analysis can be used (Sangster and McCormick, 1987).

Longitudinal displaced broadwall slots are commonly used owing to the relative simplicity of design and manufacture. A more general slot is the compound slot, which is both displaced and rotated. The compound slot offers a wider range of amplitude and phase couplings. Design data are given by Rengarajan (1989).

5.2.2 Edge Slots

Edge slots are difficult to analyze owing to their wrap-around nature. For the same reason, the wall thickness has a significant effect on admittance. For a reduced-height guide, where the wrap-around is severe, even the pattern is difficult to calculate. At present, no theories for edge slots have appeared, using either variational methods or moment methods. Array design is based on measurements, and, even here, edge slots are difficult. Because of the strong mutual coupling between edge slots (as compared with displaced broad-wall slots) incremental conductance is usually measured. That is, a series of resonant slots is measured; then one slot is taped up and the remaining slots are measured again. The resulting incremental conductance (Watson, 1946) is that of a resonant slot in the presence of mutual coupling. Trial and error is necessary to find the resonant length for a set of slots for a given angle, however. For inclination angles below 15 deg, the resonant conductance developed by Stevenson (Watson, 1946) varies as $\sin^2 \theta$:

$$\frac{G}{Y_0} = \frac{480 a/b}{\pi R_0 (2a/\lambda)^4 \beta/k} \sin^2 \theta. \qquad (5.5)$$

This and the incremental values are shown in Fig. 5.21 for WR-90 at 9375 MHz, for a 1/16 inch wide slot. Both appropriately follow a $\sin^2 \theta$ behavior, especially for small angles, with the incremental value larger than the single slot conductance. Resonant lengths will be different in an array, of course; so extensive measurements are necessary to develop sufficient data for design. Although many industrial organizations have done this, the available edge slot data base is meager. A finite element analysis, where the slot region is filled with many cylinders of length equal to the wall thickness and of triangular cross section, has been given by Jan et al. (1996). This paper includes limited measured data.

The wrap-around nature of edge slots can be avoided by using an H slot (Chignell and Roberts, 1978), where a slot normal to the guide edge is augmented by slots at each end that are parallel to the edges. Thus the slot looks like an H. By making the outside arms asymmetric, the coupling can be varied; in fact circular polarization can be achieved (Hill, 1980). A moment method analysis of H slots is given by Yee and Stellitano (1992).

Other types of slots such as probe coupled, iris excited, and crossed are discussed by Oliner and Malech (1966).

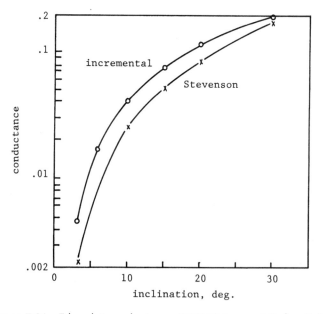

Figure 5.21 Edge slot conductance, 9375 MHz, $a = 0.0$, $b = 0.4$.

5.2.3 Stripline Slots

Slots in stripline are typically narrow rectangular resonant slots cut into the outer conducting plate of the stripline, with the strip center conductor oriented normal to the slot and offset from the center, and with slot length adjusted for resonance. Conductance can then be calculated (Breithaupt, 1968). For example, a metal-clad dielectric sheet would be etched to produce the slot radiators, then a double-clad dielectric sheet would be etched on one side to produce the feed network. These two would be fastened together to make the stripline slot array. Although the external (radiation) behavior of a stripline slot is similar to that of a slot in an infinite ground plane, the internal admittance is strongly affected by the closely spaced ground plane underneath. The net result is to severely restrict the impedance bandwidth.

Another stripline slot configuration uses boxed stripline (stripline with shorting edges parallel to the strip conductor). Each array stick looks like waveguide (see Fig. 5.22), except that the slots are all on the centerline (Park and Elliott, 1981). The strip conductor is symmetrically angled below each slot, with the coupling proportional to the strip inclination angle. Snaking the strip conductor with the slots along the centerline eliminates unbalance and loading problems experienced in the earlier work of Strumwasser (circa 1952)[2], where the strip was straight and the slots angled. The slots here are one wavelength apart (in the stripline medium), so no phase reversals are needed. Mutual coupling is accommodated in the design by techniques described in Chapter 7. Transverse displaced slots have also been used (Sangster and Smith, 1994).

5.2.4 Open End Waveguides

Waveguide radiators are different from slots and dipoles in that evanescent modes are important. Indeed, these modes can be controlled to minimize scan effects. Because of this these elements are discussed in Chapter 7 on Mutual Coupling.

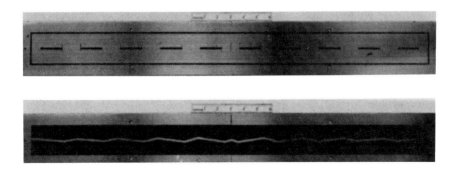

Figure 5.22 Boxed stripline array. (Courtesy Hughes Aircraft Co.)

---

[2]Unpublished Hughes report.

5.3 TEM HORNS

5.3.1 Development of TEM Horns

Two conductors that flare out, either continuously or in steps, with a TEM mode between them constitute a TEM horn. Unlike a waveguide horn, where a feed mode cutoff exists, the TEM horn can exceed an octave in bandwidth. Low-frequency impedance performance is controlled primarily by the horn mouth. All of the various types of TEM horns discussed below are derived from the biconical antenna, as sketched by Schelkunoff (1943) in Fig. 5.23. World War II work curved the vee dipole, and rotated it, as seen in Fig. 5.24 (Alford, 1947), but details have not been found. A two-dimensional tapered TEM horn was built of metal strips by Sengupta and Ferris (1971); see Fig. 5.25. Three-dimensional TEM horns flare both in separation and in arm width,

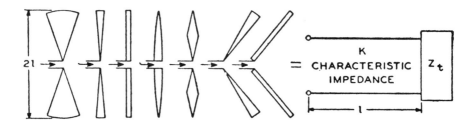

Figure 5.23 Transition from biconical to transmission line. (Courtesy Schelkunoff, S. A., *Electromagnetic Waves*, Van Nostrand, 1943, Section 11.6.)

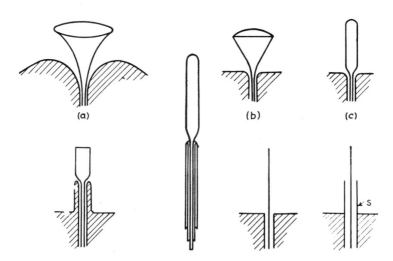

Figure 5.24 Sleeve dipole prototypes. (Courtesy Alford, A., "Broad-Band Antennas," in *Very High-Frequency Techniques*, Vol. 1, H. J. Reich, Ed., McGraw-Hill, 1947, Chapter 1.)

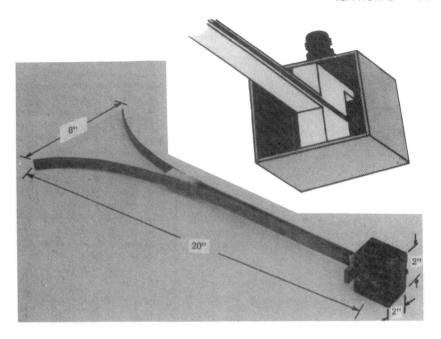

Figure 5.25 Rudimentary TEM horn. (Courtesy Sengupta, D. L. and Ferris, J. E., "Rudimentary Horn Antenna," *Trans. IEEE*, Vol. AP-19, Jan. 1971, pp. 124–126.)

from throat to mouth, providing better impedance performance over a wide bandwidth (Evans and Kong, 1983). A sketch of a three-dimensional TEM horn is shown in Fig. 5.26. Kerr (1973) used a heavily ridged waveguide horn, where the ridge opening formed a roughly exponential taper from throat to mouth. He then found that the side walls could be removed, making a TEM horn. Top and bottom walls are also not important. The ridges alone constitute a good horn; the next step was to extend the conductor on each side, and to fabricate the horn by etching away a horn shaped opening in a plated dielectric sheet. These configurations were disclosed at the 1979 European Microwave Conference by Prasad and Mahapatra (1983) and by Gibson (1979). The latter carried the whimsical name "Vivaldi Antenna," but this term is deprecated as it has no content.

Several variants exist of these printed TEM horns. The metal comprising the sides of the horn may extend a short distance, or may connect to that of adjacent horns. The horn conductors may be thin, or even petal shaped (Gazit, 1988; Fourikis et al. 1993). The horn opening may be curved and roughly exponential; it may contain a fixed horn angle (like a vee dipole), or it may be stepped, as in a stepped waveguide horn. The constant angle and the stepped versions have been called "linear tapered," and "constant width" slot antennas (Yngvesson et al., 1985, 1989; Janaswamy et al., 1986; Schaubert, 1993). This nomenclature is poor, as "slot antenna" implies radiation through the slot, while these antennas are all endfire TEM horn antennas. With all of

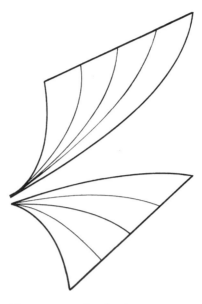

Figure 5.26 Sketch of TEM horn.

these printed antennas it is often convenient to use stripline, microstrip, slot line, coplanar guide, etc., for feeding the horn. In all cases there is much art in devising broadband feeds that transition from the transmission line to the horn. There are many types of printed circuit baluns; a microstrip balun is shown in Fig. 5.27 (Simons et al., 1995).

In arraying printed TEM horns it is important to distinguish between horns with finite width conductors and those where the conductors are part of a conducting sheet. The latter case allows currents produced by excitation of one horn to couple directly to adjacent horns. This may or may not be advantageous in impedance matching of large arrays.

5.3.2 Analysis and Design of Horns

Single TEM horns can be analyzed by surface patch moment methods (Janaswamy, 1989), in which the dielectric sheet is incorporated through perturbation theory. Gridded space methods are also applicable; Thiele and Taflove (1994) apply FDTD to isolated TEM horns, and to small arrays thereof. Results are good, but the computational process is not cost effective. Less powerful approaches approximate the field inside the horn by replacing the horn contour by many small steps, using slotline equations for each step, and finally equating power flow between steps. The resulting aperture distribution is used with a half-plane Green's function formulation to get patterns (Eleftheriades et al. 1991; Janaswamy and Schaubert, 1987). Results are not as good as those from mode matching at small steps in waveguide horns, but are better for horn mouths several wavelengths or more. Because of the many

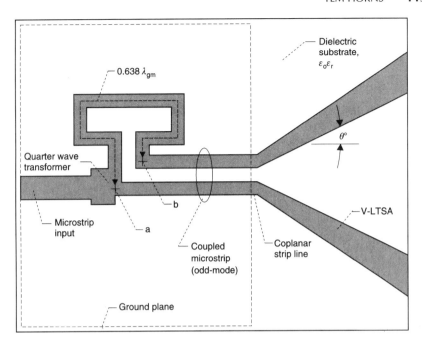

Figure 5.27 Microstrip balun. (Courtesy Simons, R. N. et al., "Integrated Uniplanar Transition for Linearly Tapered Slot Antenna," *Trans. IEEE*, Vol. AP-43, Sept. 1995, pp. 998–1002.)

variables—mouth width, horn length, horn conductor geometry, dielectric sheet (if used) dielectric constant and thickness, and type of feed—it is not feasible to give design data, especially for the relatively small horns that are useful in arrays. Moment method analysis and design is recommended, starting with no dielectric and an ideal feed, then adding these items. Users should be cautioned that two-dimensional horns may have high cross polarization in the diagonal plane (Kim and Yngvesson, 1990). Polarization agility is obtained simply with two horns at right angles, or, in a unique arrangement, with four horns with a circular cavity underneath. Here the horns form a four-pointed star, with the horn mouths facing inward (Povinelli, 1987).

5.3.3 TEM Horn Arrays

The behavior of an array of TEM horns depends upon whether the individual horns are connected only at the feeds, or whether direct current paths exist from each horn to the next. When the horn openings are etched out of a conducting layer, current paths exist from throat to mouth of each horn and to throat to mouth of adjacent horns. As a result, the embedded impedance versus frequency for a fixed beam array, or the *scan impedance* for a scanning array, will be different. For all but small arrays moment methods are the best approach. Of course much can be learned from the behavior and design trades of infinite arrays, and in the spectral (Floquet) domain the number of equations to be solved is modest. Cooley et al. (1991) have given a moment method analysis of an infinite array of linearly tapered TEM horns; Schaubert et al. (1994) extend this to curved TEM horns. These analyses are able to predict important phenomena such as blind angles in E-plane scan. Both linear polarizations can be provided by "egg-crating" two printed circuit boards together (Hunt and Ventresca, 1977). Another configuration uses printed horns for one polarization, and a waveguide mode between the printed planes for the other polarization (Monser, 1981, 1984). In all configurations the horn axes are in the main beam direction, as the horn lengths are greater than half-wave, thus not allowing the axes to be in the array plane. An extensive discussion of TEM horn arrays is given in Chapter 7.

5.3.4 Millimeter Wave Antennas

In recent years much effort has been expended on combining solid-state circuit devices with antennas, primarily vee strip dipoles and TEM horns (Kotthaus and Vowinkel, 1989). At millimeter wavelengths the power source or preamplifier/mixer can be produced on the same substrate as the antenna. Rutledge et al., 1978, 1983) give reviews of these. Many of the papers are concerned primarily with the solid-state devices rather than with the antenna, and are not discussed here. A variant on the printed circuit flat TEM horn is the square linear taper horn with dipole feed at the mouth opening; the horn walls are etched and metallized on a thick substrate, as in the feed (Guo et al., 1991; Rebeiz et al., 1990; Rebeiz and Rutledge, 1992).

5.4 MICROSTRIP PATCHES AND DIPOLES

A microstrip patch or dipole is a resonant metallic shape in the top microstrip conductor. Patches were invented by Deschamps (see Deschamps and Sichak, 1953), probably based on the earlier partial sleeve (shorted quarter-wave patch) work at the University of Illinois Antenna Laboratory. Further development was by Munson (1974) and others. Because of the extensive literature on microstrip antennas, only a cursory coverage is given here. Books include *Microstrip Antennas* (Bahl and Bhartia, 1980), *Microstrip Antenna Theory and Design* (James et al., 1981), *Handbook of Antenna Design* (2 volumes) (James and Hall, 1989), *Millimeter-Wave Microstrip and Printed Circuit Antennas* (Bhartia et al., 1991), *Broadband Patch Antennas* (Zurcher and Gardiol, 1995) and *Advances in Microstrip and Printed Antennas* (Lee & Chen, 1997). A useful review of microstrip design and fabrication techniques has been given by Gardiol (1988).

Patches can be of many shapes, but most are rectangular (nearly square) or circular. A patch may be fed by an in-plane microstrip feed line, as sketched in Fig. 5.28, or by a coaxial cable connected to the underside of the ground plane, with the center conductor extending up to the patch. A typical inset microstrip line feed for matching is shown in Fig. 5.29. Patches are typically square or

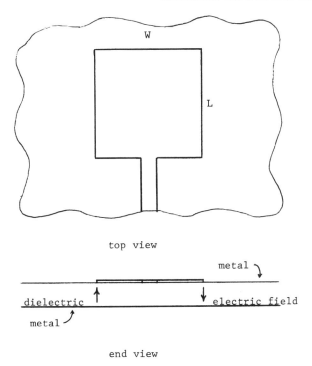

Figure 5.28 Microstrip patch geometry.

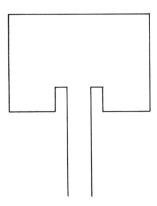

Figure 5.29 Inset feed impedance matching.

circular, and may have electromagnetically coupled portions; the last are discussed later. Square and circular annular slots in stripline are also used. Polarization and pattern for patches are similar to those of half-wave dipoles or slots. The zero-order pattern for a square patch is given by

$$E_x = \cos\theta \, \text{sinc}\, \frac{kLv}{2} \cos\frac{kWu}{2}, \tag{5.6}$$

where $u = \sin\theta\cos\phi$ and $v = \sin\theta\sin\phi$. For a circular patch of radius a, with $\beta = ka\sin\theta$, the zero-order pattern is

$$E_x = \cos\theta[J_0(\beta) - \cos(2\phi)J_2(\beta)]. \tag{5.7}$$

A typical measured pattern is shown in Fig. 5.30 (Kerr, 1979). Cross polarization (XP) occurs in diagonal planes, with maximum XP levels as shown in Table 5.1 (Hansen, 1987). Note that the formulas in the reference are correct except that the square patch XP coefficient should be $2uw$ instead of $4uw$. However, the tables there are incorrect.

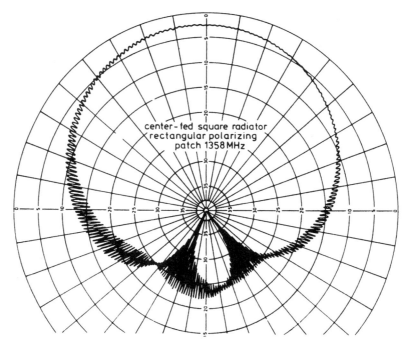

Figure 5.30 Typical patch axial ratio pattern.
(Courtesy Kerr, J. L., "Microstrip Antenna Developments," *Proceedings Printed Circuit Antenna Technology Workshop,"* New Mexico State University, Las Cruces, N.M., Oct. 1979.)

TABLE 5.1 Maximum Cross Polarization Levels

	Square Patch		Circular Patch	
	$\theta_{max,XP}$ (deg)	Max. XP (dB)	$\theta_{max,XP}$ (deg)	Max. XP (dB)
1	51.51	−17.93	51.96	−17.37
2	54.74	−23.54	53.40	−22.59
3	54.74	−26.82	53.86	−25.84
5	54.74	−31.07	54.22	−30.36
10	54.74	−36.95	54.48	−35.91

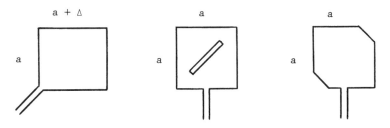

Figure 5.31 CP patches. (Courtesy Kerr, J. L., "Microstrip Antenna Developments," *Proceedings Printed Circuit Antenna Technology Workshop*, New Mexico State University, Las Cruces, N.M., Oct. 1979.)

Circular polarization (CP) is achieved simply by using two feeds at 90 deg, with a 90 deg hybrid connected to the feed lines. If the hybrid is relatively wideband, the CP bandwidth will be the same as that for linear polarization. A simpler but slightly narrower-band version replaces the hybrid by a power divider, with an extra $\lambda/4$ line length in one feed. Single feeds can also be used to produce CP: the principle is to modify the patch so that the x-directed resonance is below and the y-directed resonance is above the operating band. When the phases differ by 90 deg, CP results. However, the bandwidth over which CP occurs is significantly reduced from the intrinsic patch bandwidth. For example, the patch is rectangular with one side shorter and one side longer than a square patch. A circular patch can be modified to an elliptical shape. Simple formulas relating dimensions to bandwidth were derived by Lo and Richards (1981). Other schemes for exciting CP are sketched in Fig. 5.31.

Microstrip dipoles, where the patch width is small, giving a strip dipole, have been investigated extensively by Alexopoulos and colleagues (Uzunoglu et al., 1979). However, for thin substrates the bandwidth tends to be much smaller than that of a patch on the same substrate (Pozar, 1983).

5.4.1 Transmission Line Model

The square patch acts as a half-wave resonant transmission line, with the end gaps producing the radiated field, as sketched in Fig. 5.32. Simplest of the analytical models is the transmission line model (Sengupta, 1984), where an effective dielectric constant is used, and where the ends are terminated with a radiation conductance and susceptance. The effective dielectric constant varies with the ratio of patch thickness to width (Hammerstad, 1975):

$$\varepsilon_e = \tfrac{1}{2}(\varepsilon_r + 1) + \frac{\varepsilon_r - 1}{2\sqrt{1 + 12t/W}}. \tag{5.8}$$

A more complex formulation has been developed by Hammerstad and Jensen (1980), but the simpler form may be adequate. The radiation admittance implies a radiation Q, which is (Vandesande et al., 1979)

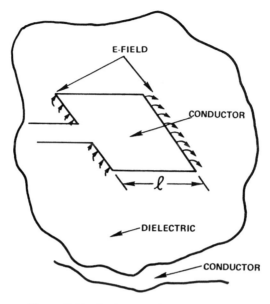

Figure 5.32 Radiating edge slots of patch.

$$Q_{rad} = \frac{\sqrt{\varepsilon_e}\lambda_0}{4t} - \frac{\varepsilon_e \Delta L}{t}, \tag{5.9}$$

where ΔL is the change in effective length produced by field fringing (Hammerstad, 1975):

$$\frac{\Delta L}{t} = 0.412 \frac{(\varepsilon_e + 0.300)(W/t + 0.262)}{(\varepsilon_e - 0.258)(W/t + 0.813)}. \tag{5.10}$$

A more accurate, and again more complex, formula is given by Kirschning et al. (1981). An advantage of the transmission line approach is that the feed point impedance can easily be computed. Since the bandwidth for a matched antenna of small bandwidth is $1/(\sqrt{2}Q)$ for VSWR ≤ 2 the bandwidth is approximately

$$\text{BW} \simeq \frac{4t}{\sqrt{2\varepsilon_r}\lambda_0}. \tag{5.11}$$

That is, it is proportional to thickness in free space wavelengths, and inversely to square root of dielectric constant. Figure 5.33 shows bandwidth for VSWR < 2 for dielectric constants of 2.5, 5.0, and 10.0. Note that this approximate bandwidth does not include conduction and dielectric losses (which increase bandwidth). Surface wave effects are also not included, but these are usually important only for large arrays. Bandwidth for any other VSWR is given by

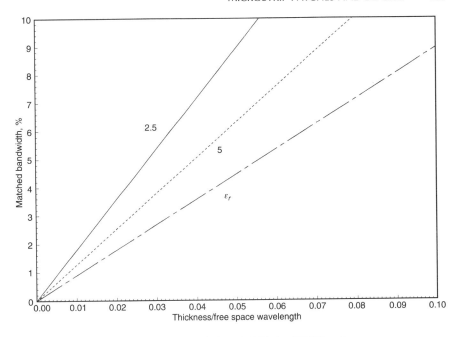

Figure 5.33 Patch bandwidth for VSWR = 2.

$$\text{BW} = \text{BW}_2 \frac{\sqrt{2}(\text{VSWR} - 1)}{\sqrt{\text{VSWR}}} \tag{5.12}$$

For half-power, VSWR = 5.828, and the bandwidth is $2\sqrt{2}$ times larger than the VSWR = 2 bandwidth.

5.4.2 Cavity and Other Models

The cavity model, developed by Lo and colleagues (Lo et al., 1979) at the University of Illinois-Urbana, and by Carver and colleagues (Carver and Mink, 1981) at New Mexico State University, represents the field between the patch and the ground plane as a series of rectangular (or circular) cavity modes. These modes are then matched to the gap admittance at the patch edges. Almost all of the radiated energy is from the dominant TM_{01} mode; from this a radiation Q_{rad} can be calculated. All modes contribute to the stored energy. Using the dominant mode current and field, the conduction and dielectric losses and Q values are obtained. Finally, the overall Q is simply $1/Q = 1/Q_{\text{rad}} + 1/Q_{\text{cond}} + 1/Q_{\text{diel}}$. Bandwidth is found from Q as before, and the input impedance is calculated from the Q and the stored and radiated power. Although the cavity model is more accurate than the transmission line model in predicting resonant frequency, a small error or shift usually appears.

Moment methods have also been used to determine the currents on the conducting surfaces. Newman and Tulyathan (1981) used quadrilateral expansion and testing functions in the spatial domain. This avoids the difficulties associated with calculation of the spectral domain Green's function; the spectral domain is more appropriate for infinite array analysis via Floquet's theorem. Since that first paper, many have appeared; other types of subsectional expansion and test functions have been used, as have entire expansion and test functions. The latter are usually the cavity modes.

FDTD methods have also been used to analyze patch elements (Wu et al., 1992; Luebbers and Langdon, 1996). However, the cavity and moment method approaches are usually satisfactory, and are simpler.

5.4.3 Parasitic Patch Antennas

Bandwidth can be increased by adding one or more resonant circuits. Fano (1950), in two classic papers, derived limits on bandwidth improvement due to use of additional resonant circuits, under the assumptions of narrow bandwidth and lossless networks. His results were recast into filter theory format by Matthaei, Young, and Jones (1964), and their formulation has been used for calculations. Table 5.2 gives the bandwidth multiplication factor for VSWR of 2 and 5.828 (half-power). Actual matching devices will have loss, and this will increase the bandwidth at the expense of efficiency. Note that for the commonly used VSWR = 2, an infinite number of matching circuits will increase bandwidth 3.8 times.

There are several clever ways of incorporating matching circuits into patch antennas. Perhaps the simplest is the parasitic patch, where a parasitic patch is supported above the driven patch by a (usually) low-dielectric-constant material, such as foam or hexcel (see sketch, Fig. 5.34). The parasite length and width, and the spacing and dielectric constant add four more design variables. As a result these antennas were difficult to design until moment methods were applied. Kastner et al. (1988) use spectral domain sampling and FFT. Barlatey et al. (1990) use entire domain expansion and test functions (cavity modes) in the spatial domain. Tulintseff et al. (1991) use entire domain expansion and test functions plus a polynomial for edge currents, in the spectral domain. These powerful numerical simulation techniques have allowed parasitic patch antennas to be designed, with roughly double the bandwidth of the driven patch alone.

TABLE 5.2 Maximum Bandwidth Improvement Factors

Number of Additional Matching Circuits	VSWR = 2	Half-power
1	2.3094	2.0301
2	2.8596	2.4563
3	3.1435	2.6772
4	3.3115	2.8083
∞	3.8128	3.2049

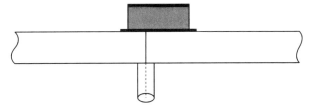

Figure 5.34 Probe fed patch with parasite.

Another way of incorporating a matching section is through electromagnetic coupling. Oltman (1977) reported a strip dipole electromagnetically coupled to an open ended microstrip line; an electromagnetically coupled patch was analyzed by Zhang et al. (1985). A modification utilized coupling through an aperture in either stripline or microstrip ground plane (Pozar, 1985). Sullivan and Schaubert (1986), using moment method, analyzed an open ended microstrip line electromagnetically coupled to a narrow slot (across the line), which in turn is electromagnetically coupled to the patch; see sketch in Fig. 5.35. Reciprocity can be used to separate the analysis into two coupled problems: one for the antenna, and a second for the microstrip line (Pozar, 1986). Careful analyses have been done by Ittipiboon et al. (1991) and Yang and Shafai (1995). Again, with careful design the bandwidth can be at least doubled.

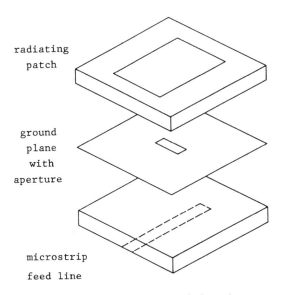

Figure 5.35 Aperture coupled patch.

ACKNOWLEDGMENT

Photographs courtesy of Dr. P. K. Park and Dr. Joe Ferris.

REFERENCES

Alford, A., "Broad-Band Antennas," in *Very High-Frequency Techniques*, Vol. 1, H. J. Reich, Ed., McGraw-Hill, 1947, Chapter 1.

Bahl, I. J. and Bhartia, P., "Microstrip Antenna Arrays," in *Microstrip Antennas*, Artech House, 1980, Chapter 7.

Balanis, C. A., *Antenna Theory—Analysis and Design*, Harper & Row, 1982.

Barlatey, L., Mosig, J. R., and Sphicopoulos, T., "Analysis of Stacked Microstrip Patches with a Mixed Potential Integral Equation," *Trans. IEEE*, Vol. AP-38, May 1990, pp. 608–615.

Bhartia, P., Rao, K. V. S., and Tomar, R. S., *Millimeter-Wave Microstrip and Printed Circuit Antennas*, Artech House, 1991, Chapters 5 and 7.

Bock, E. L., Nelson, J. A., and Dorne, A., "Sleeve Antennas," in *VHF Techniques*, Vol. 1, H. J. Reich, Ed., Radio Research Lab., McGraw-Hill, 1947, Chapter 5.

Bolljahn, J. T., Broad Band Antenna, US Patent No. 2505751, May 2, 1950.

Breithaupt, R. W., "Conductance Data for Offset Series Slots in Stripline," *Trans. IEEE*, Vol. MTT-16, Nov. 1968, pp. 969–970.

Brown, G. H., "Directional Antennas," *Proc. IRE*, Vol. 25, No. 1, Jan. 1937.

Butler, C. M. et al., "Characteristics of a Wire Biconical Antenna," *Microwave J.*, Vol. 22, Sept. 1979, pp. 37–40.

Carver, K. R. and Mink, J. W., "Microstrip Antenna Technology," *Trans. IEEE*, Vol. AP-29, Jan. 1981, pp. 2–24.

Chignell, R. J. and Roberts, J., "Compact Resonant Slot for Waveguide Arrays," *Proc. IEE*, Vol. 125, 1978, pp. 1213–1216.

Cooley M. E. et al., "Radiation and Scattering Analysis of Infinite Arrays of Endfire Slot Antennas with a Ground Plane," *Trans. IEEE*, Vol. AP-39, Nov. 1991, pp. 1615–1625.

Deschamps, G. A. and Sichak, W., "Microstrip Microwave Antenna," *Proc. 1953 USAF Antenna R&D Symposium*, Allerton, Ill., Oct. 1953.

Eleftheriades G. V. et al., "Millimeter-Wave Integrated-Horn Antennas: Part I—Theory," *Trans. IEEE*, Vol. AP-39, Nov. 1991, pp. 1575–1581.

Elliott, R. S., "An Improved Design Procedure for Small Arrays of Shunt Slots," *Trans. IEEE*, Vol. AP-31, Jan. 1983, pp. 48–53.

Evans, S. and Kong, F. N., "TEM Horn Antenna: Input Reflection Characteristics in Transmission," *Proc. IEE*, Vol. 130H, Oct. 1983, pp. 403–409.

Fano, R. M., "Theoretical Limitations on the Broadband Matching of Arbitrary Impedances," *J. Franklin Inst.*, Vol. 249, Jan. 1950, pp. 57–83; Feb. 1950, pp. 139–154.

Forbes, G. R., "An Endfire Array Continuously Proximity-Coupled to a Two-Wire Line," *Trans. IEEE*, Vol. AP-8, Sept. 1960, pp. 518–519.

Fourikis, N., Lioutas, N., and Shuley, N. V., "Parametric Study of the Co- and Crosspolarisation Characteristics of Tapered Planar and Antipodal Slotline Antennas," *Proc. IEE*, Vol. 140H, Feb. 1993, pp. 17–22.

Gardiol, F. E., "Design and Layout of Microstrip Structures," *Proc. IEE*, Vol. 135H, June 1988, pp. 145–157.

Gazit, E., "Improved Design of the Vivaldi Antenna," *Proc. IEE*, Vol. 135H, Apr. 1988, pp. 89–92.

Gibson, P. J., "The Vivaldi Aerial," *9th European Microwave Conference*, 1979, pp. 101–105; also in *Advanced Antenna Technology*, P. H. B. Clarricoats, Ed., Microwave Exhibitions and Publisher, 1981, pp. 200–204.

Guo, Y., et al., "Aperture Efficiency of Integrated-Circuit Horn Antennas," *Microwave and Opt. Technol. Lett.*, Vol. 4, No. 1, Jan. 5, 1991, pp. 6–9.

Hammerstad, E., "Equations for Microstrip Circuit Design," *Proc. 5th European Microwave Conference*, 1975, pp. 268–272.

Hammerstad, E. and Jensen, O., "Accurate Models for Microstrip," *IEEE MTT Symposium*, 1980, pp. 407–409.

Hansen, R. C., "Folded and T-Match Dipole Transformation Ratio," *Trans. IEEE*, Vol. AP-30, Jan. 1982, pp. 161–162.

Hansen, R. C., "Cross Polarization of Microstrip Patch Antennas," *Trans. IEEE*, Vol. AP-35, June 1987, pp. 731–732.

Hansen, R. C., "Moment Methods in Antennas and Scattering," Artech House, 1990.

Hill, D. R., "Circularly Polarized Radiation from Narrow Wall Slots in Rectangular Waveguide," *Electron. Lett.*, Vol. 16, 1980, pp. 559–560.

Hunt, C. J. and Ventresca, P., "Large Bandwidth Antenna with Multipolarization Capability," *Proceedings Antenna Applications Symposium*, Allerton, Ill., April 1977.

Ittipiboon, A. et al., "A Modal Expansion Method of Analysis and Measurement on Aperture-Coupled Microstrip Antenna," *Trans. IEEE*, Vol. AP-39, Nov. 1991, pp. 1567–1574.

James, J. R. and Hall, P. S., Eds., *Handbook of Microstrip Antennas*, Vols. 1 and 2, IEE/Peter Peregrinus, 1989.

James, J. R., Hall, P. S., and Wood, C., *Microstrip Antenna Theory and Design*, IEE/Peregrinus, 1981, Chapter 6.

Jan C.-G. et al., "Analysis of Edge Slots in Rectangular Waveguide with Finite Waveguide Wall Thickness," *Trans. IEEE*, Vol. AP-44, Aug. 1996, pp. 1120–1126.

Janaswamy, R., "An Accurate Moment Method Model for the Tapered Slot Antenna," *Trans. IEEE*, Vol. AP-37, Dec. 1989, pp. 1523–1528.

Janaswamy, R. and Schaubert, D. H., "Analysis of the Tapered Slot Antenna," *Trans. IEEE*, Vol. AP-35, Sept. 1987, pp. 1058–1065.

Janaswamy, R. Schaubert, D. H., and Pozar, D. M., "Analysis of the Transverse Electromagnetic Mode Linearly Tapered Slot Antenna," *Radio Sci.*, Vol. 21, Sept.-Oct. 1986, pp. 797–804.

Joubert, J., "A Transverse Slot in the Broad Wall of Inhomogeneously Loaded Rectangular Waveguide for Array Applications," *IEEE Microwave and Guided Wave Lett.*, Vol. 5, Feb. 1995, pp. 37–39.

Kaminow, I. P. and Stegen, R. J., "Waveguide Slot Array Design," TM-348, Hughes Aircraft Co., Culver City, Calif., July 1954.

Kastner, R., Heyman, E., and Sabban, A., "Spectral Domain Iterative Analysis of Single- and Double-Layered Microstrip Antennas Using the Conjugate Gradient Algorithm," *Trans. IEEE*, Vol. AP-36, Sep. 1988, pp. 1204–1212.

Kerr, J. L., "Short Axial Length Broad-Band Horns," *Trans. IEEE*, Vol. AP-21, Sept. 1973, pp. 710–714.

Kerr, J. L., "Microstrip Antenna Developments," *Proceedings Printed Circuit Antenna Technology Workshop*, New Mexico State University, Las Cruces, N.M., Oct. 1979.

Khac, T. V., A Study of Some Slot Discontinuities in Rectangular Waveguides, PhD thesis, Monash University, Australia, 1974.

Khac, T. V. and Carson, C. T., "Impedance Properties of a Longitudinal Slot Antenna in the Broad Face of a Rectangular Waveguide," *Trans. IEEE*, Vol. AP-21, Sept. 1973, pp. 708–710.

Kim, Y.-S. and Yngvesson, K. S., "Characterization of Tapered Slot Antenna Feeds and Feed Arrays," *Trans. IEEE*, Vol. AP-38, Oct. 1990, pp. 1559–1564.

King, R W. P., *The Theory of Linear Antennas*, Harvard University Press, 1956.

King, H. E. and Wong, J. L., "An Experimental Study of a Balun-Fed Open-Sleeve Dipole in Front of a Metallic Reflector," *Trans. IEEE*, Vol. AP-20, March 1972, pp. 201–204.

Kirschning, M., Jansen, R. H., and Koster, N. H. L., "Accurate Model for Open End Effect of Microstrip Lines," *Electron. Lett.*, Vol. 17, Feb. 5, 1981, pp. 123–125.

Kotthaus, U. and Vowinkel, B., "Investigation of Planar for Submillimeter Receivers," *Trans. IEEE*, Vol. MTT-37, Feb. 1989, pp. 375–380.

Lee, K. F., and Chen, W., *Advances in Microstrip and Printed Antennas*, Wiley, 1997.

Lo, Y. T. "A Note on the Cylindrical Antenna of Noncircular Cross-Section," *J. Appl. Phys.*, Vol. 24, Oct. 1953, pp. 1338–1339.

Lo, Y. T. and Richards, W. F., "Perturbation Approach to Design of Circularly Polarised Microstrip Antennas," *Electron. Lett.*, Vol. 17, May 28, 1981, pp. 383–385.

Lo, Y. T., Solomon, D., and Richards, W. F., "Theory and Experiment on Microstrip Antennas," *Trans. IEEE*, Vol. AP-27, Mar. 1979, pp. 137–145.

Luebbers, R. J. and Langdon, H. S., "A Simple Feed Model that Reduces Time Steps Needed for FDTD Antenna and Microstrip Calculations," *Trans. IEEE*, Vol. AP-44, July 1996, pp. 1000–1005.

Malherbe, J. A. G. and Davidson, D. B., "Mutual Impedance for Half-Cosinusoid Slot Voltage Distribution: An Evaluation," *Trans. IEEE*, Vol. AP-32, Sept. 1984, pp. 990–991.

Maloney, J. G., Smith, G. S., and Scott, W. R., Jr., "Accurate Computation of the Radiation from Simple Antennas Using the Finite-Difference Time-Domain Method," *Trans. IEEE*, Vol. AP-38, July 1990, pp. 1059–1068.

Matthaei, G. L., Young, L., and Jones, E. M. T., *Microwave Filters, Impedance-Matching Networks, and Coupling Structures*, McGraw-Hill, 1964.

Monser, G., "New Advances in Wide Band Dual Polarized Antenna Elements for EW Applications," *Proc. Antenna Applications Symposium*, Allerton, Ill., Sept. 1981.

Monser, G., "Considerations for Extending the Bandwidth of Arrays Beyond Two Octaves," *Proc. Allerton Antenna Applications Symposium*, Allerton, Ill., Sept. 1984.

Munson, R. E., "Conformal Microstrip Antennas and Microstrip Phased Arrays," *Trans. IEEE*, Vol. AP-22, Jan. 1974, pp. 74–78.

Nelson, J. A. and Stavis, G., "Impedance Matching, Transformers and Baluns," in *Very High-Frequency Techniques*, Vol. 1, H. J. Reich, Ed., Radio Research Lab., McGraw-Hill, 1947, Chapter 3.

Newman, E. H. and Tulyathan, P., "Analysis of Microstrip Antennas Using Moment Methods," *Trans. IEEE*, Vol. AP-29, Jan. 1981, pp. 47–53.

Oliner, A. A., "The Impedance Properties of Narrow Radiating Slots in the Broad Face of Rectangular Waveguide," *Trans. IRE*, Vol. AP-5, Jan. 1957a, pp. 4–11.

Oliner, A. A., "The Impedance Properties of Narrow Radiating Slots in the Broad Face of Rectangular Wavegide—Part II: Comparison with Measurement," *Trans. IRE*, Vol. AP-5, Jan. 1957b, pp. 12–20.

Oliner, A. A., and Malech, R. G., "Radiating Elements and Mutual Coupling," in *Microwave Scanning Antennas*, Vol. 2, R. C. Hansen, Ed., Academic Press, 1966 [Peninsula Publishing, 1985], Chapter 2.

Oltman, H. G., "Electromagnetically Coupled Microstrip Dipole Antenna Elements," in *Proceedings 8th European Microwave Conference*, 1977, pp. 281–285.

Park, P. K. and Elliott, R. S., "Design of Collinear Longitudinal Slot Arrays Fed by Boxed Stripline," *Trans. IEEE*, Vol. AP-29, Jan. 1981, pp. 135–140.

Park, P. K., Stern, G. J., and Elliott, R. S., "An Improved Technique for the Evaluation of Transverse Slot Discontinuities in Rectangular Waveguide," *Trans. IEEE*, Vol. AP-31, Jan. 1983, pp. 148–154.

Povinelli, M. J., "A Planar Broad-Band Flared Microstrip Slot Antenna," *Trans. IEEE*, Vol. AP-35, Aug. 1987, pp. 968–972.

Pozar, D. M., "Considerations for Millimeter Wave Printed Antennas," *Trans. IEEE*, Vol. AP-31, Sept. 1983, pp. 740–747.

Pozar, D. M., "Microstrip Antenna Aperture-Coupled to a Microstripline," *Electron. Lett.*, Vol. 21, Jan. 17, 1985, pp. 49–50.

Pozar, D., "A Reciprocity Method of Analysis for Printed Slot and Slot-Coupled Microstrip Antennas," *Trans. IEEE*, Vol. AP-34, Dec. 1986, pp. 1439–1446.

Prasad, S. N. and Mahapatra, S., "A New MIC Slot-Line Aerial," *Trans. IEEE*, Vol. AP-31, May 1983, pp. 525–527.

Rebeiz, G. M. and Rutledge, D. B., "Integrated Horn Antennas for Millimeter-Wave Applications," *Ann. Telecommun.*, Vol. 47, 1992, pp. 38–48.

Rebeiz, G. M. et al., "Monolithic Millimeter-Wave Two-Dimensional Horn Imaging Arrays," *Trans. IEEE*, Vol. AP-38, Sept. 1990, pp. 1473–1482.

Rengarajan, S. R., "Compound Radiating Slots in a Broad Wall of a Rectangular Waveguide," *Trans. IEEE*, Vol. AP-37, Sept. 1989, pp. 1116–1123.

Rispin, L. W. and Chang, D. C., "Wire and Loop Antennas," in *Antenna Handbook—Theory, Applications, and Design*, Y. T. Lo and S. W. Lee, Eds., Van Nostrand Reinhold, 1988, Chapter 7.

Rutledge, D. B., Schwarz, S. E., and Adams, A. T., "Infrared and Submillimetre Antennas," *Infrared Phys.*, Vol. 18, 1978, pp. 713–729.

Rutledge, D. B., Neikirk, D. P., and Kasilingam, D. P., "Integrated-Circuit Antennas," in *Infrared and Millimeter Waves*, Vol. 10, Academic Press, 1983, Chapter 1.

Sangster, A. J. and McCormick, A. H. I., "Moment Method Applied to Round-Ended Slots," *Proc. IEE*, Vol. 134H, June 1987, pp. 310–314.

Sangster, A. J. and Smith, P., "Optimisation of Radiation Efficiency for a Transverse Ground-Plane Slot in Boxed-Stripline," *IEE Proc.—Microwave Antennas Prop.*, Vol. 141, Dec. 1994, pp. 509–516.

Schaubert, D. H., "Wide-Bandwidth Radiation from Arrays of Endfire Tapered Slot Antennas," in *Ultra-Wideband, Short-Pulse Electromagnetics*, H. L. Bertoni, L. Carin, and L. B. Felsen, Eds., Plenum, 1993, pp. 157–165.

Schaubert, D. H. et al., "Moment Method Analysis of Infinite Stripline-Fed Tapered Slot Antenna Arrays with a Ground Plane," *Trans. IEEE*, Vol. AP-42, Aug. 1994, pp. 1161–1166.

Schelkunoff, S. A., *Electromagnetic Waves*, Van Nostrand, 1943, Section 11.6.

Sengupta, D. L., "Transmission Line Model Analysis of Rectangular Patch Antennas," *Electromagnetics*, Vol. 4, 1984, pp. 355–376.

Sengupta, D. L. and Ferris, J. E., "Rudimentary Horn Antenna," *Trans. IEEE*, Vol. AP-19, Jan. 1971, pp. 124–126.

Simons, R. N. et al., "Integrated Uniplanar Transition for Linearly Tapered Slot Antenna," *Trans. IEEE*, Vol. AP-43, Sept. 1995, pp. 998–1002.

Stevenson, A. F., "Theory of Slots in Rectangular Waveguides," *J. Appl. Phys.*, Vol. 19, 1948, pp. 24–38.

Sullivan, P. L. and Schaubert, D. H., "Analysis of an Aperture Coupled Microstrip Antenna," *Trans. IEEE*, Vol. AP-34, Aug. 1986, pp. 977–984.

Thiele, E. and Taflove, A., "FD-TD Analysis of Vivaldi Flared Horn Antennas and Arrays," *Trans. IEEE*, Vol. AP-42, May 1994, pp. 633–641.

Tulintseff, A. N., Ali, S. M., and Kong, J. A., "Input Impedance of a Probe-Fed Stacked Circular Microstrip Antenna," *Trans. IEEE*, Vol. AP-39, Mar. 1991, pp. 381–390.

Uzunoglu, N. K., Alexopoulos, N. G., and Fikioris, J. G., "Radiation Properties of Microstrip Dipoles," *Trans. IEEE*, Vol. AP-27, Nov. 1979, pp. 853–858.

Vandesande, J., Pues, H., and Van de Capelle, A., "Calculation of the Bandwidth of Microstrip Resonator Antennas," *Proceedings 9th European Microwave Conference*, Brighton, Sept. 1979, pp. 116–119.

Watson, W. H., "Resonant Slots," *J. IEE*, Vol. 93, part IIIA, 1946, pp. 747–777.

Watson, W. H., *The Physical Principles of Waveguide Transmission and Antenna Systems*, Oxford University Press, 1947.

Wolff, E. A., *Antenna Analysis*, Wiley, 1967.

Wong, J. L. and King, H. E., "A Cavity-Backed Dipole Antenna with Wide-Bandwidth Characteristics," *Trans. IEEE*, Vol. AP-22, Sept. 1973, pp. 725–727.

Wu, C. et al., "Accurate Characterization of Planar Printed Antennas Using Finite-Difference Time-Domain Method," *Trans. IEEE*, Vol. AP-40, May 1992, pp. 526–534.

Wunsch, A. D., "Fourier Series Treatment of the Sleeve Monopole Antenna," *Proc. IEE*, Vol. 135H, Aug. 1988, pp. 217–225.

Yang, X. H. and Shafai, L., "Characteristics of Aperture Coupled Microstrip Antennas with Various Radiating Patches and Coupling Apertures," *Trans. IEEE*, Vol. AP-43, Jan. 1995, pp. 72–78.

Yee, H. Y., "Impedance of a Narrow Longitudinal Shunt Slot in a Slotted Waveguide Array," *Trans. IEEE*, Vol. AP-22, 1974, pp. 589–592.

Yee, H. Y. and Stellitano, P., "I-Slot Characteristics," *Trans. IEEE*, Vol. AP-40, Feb. 1992, pp. 224–228.

Yngvesson, K. S. et al., "Endfire Tapered Slot Antennas on Dielectric Substrates," *Trans. IEEE*, Vol. AP-33, Dec. 1985, pp. 1392–1400.

Yngvesson, K. S. et al., "The Tapered Slot Antenna—A New Integrated Element for Millimeter-Wave Applications," *Trans. IEEE*, Vol. MTT-37, Feb. 1989, pp. 365–374.

Zhang, Q., Fukuoka, Y., and Itoh, T., "Analysis of a Suspended Patch Antenna Excited by an Electromagnetically Coupled Inverted Microstrip Feed," *Trans. IEEE*, Vol. AP-33, Aug. 1985, pp. 895–899.

Zurcher, J.-F. and Gardiol, F. E., *Broadband Patch Antennas*, Artech House, 1995, Chapter 5.

CHAPTER SIX

Array Feeds

Fixed beam arrays are usually either linear arrays or assemblies of linear arrays to make a planar array. Thus the linear array is a basic building block. When the array elements are in series along a transmission line, the array is termed "series." Similarly, when the elements are in parallel with a feed line or network, the array is termed "shunt." A further division of series feeds is into resonant or standing wave feeds, and travelling wave feeds. Shunt feeds are generally of corporate type, or distributed. Each of these types is discussed below.

6.1 SERIES FEEDS

Waveguide slots lend themselves to series feeding, as they are spaced along the transmission feed line (waveguide). This is independent of whether the slot equivalent circuit is series or shunt; see Chapter 5. Dipoles are usually more suited to shunt feeds. Microstrip patches can be fed in series by microstrip lines connecting adjacent patches, but these arrays offer much less control over the array excitation. Since most series-fed arrays use waveguide slots, these are emphasized, both for resonant and for travelling wave arrays. Figure 6.1 shows an experimental single waveguide array with longitudinal slots.

6.1.1 Resonant Arrays

6.1.1.1 Impedance and Bandwidth. The most common resonant array is a waveguide with slots at half guide wavelength spacings, with the guide ends shorted, and with power coupled in (or out) along the guide. The end shorts allow standing waves to exist, and the $\lambda_g/2$ spacing allows all slot radiators to be in phase. Such a waveguide array is called "stick"; planar arrays are made up of closely packed sticks. Ideally, the sum of slot conductances should equal unity, and the sum of slot susceptances should equal zero, for shunt slots. The obvious limitation of resonant feeds is that they are resonant at only one frequency! Over a bandwidth the slot excitations

Figure 6.1 Longitudinal slot array. (Courtesy California State University Northridge.)

will change in amplitude and phase, thereby introducing pattern distortion and impedance mismatch. Three approaches to quantifying resonant array performance will be given. First, a simple end element equation; second, an analysis by Watson which is useful for uniformly excited (constant amplitude) arrays; and third, an exact analysis.

For a symmetric N-element array with element spacing d and propagation constant β, the last element is $\lambda/2$ out of phase when $N\pi d = N\pi + \pi$, or $f/f_0 = 1 + 1/N$. This gives a product of number of elements times percent bandwidth $NB = 114$. Unfortunately, this simple merit factor calculation is much too optimistic.

Watson (1947) set up a product of factors, where each factor represents a length of transmission line and a shunt load admittance Y_n. When multiple reflections between loads are neglected, the finite product can be approximated by a sum. His result, which is often quoted, is

$$\Gamma = \frac{2 - \sum_{n=1}^{N} Y_n - \exp(j2N\beta d) \sum_{n=1}^{N} Y_n \exp(-j2n\beta d)}{\left(2 + \sum_{n=1}^{N} Y_n\right) \exp(j2N\beta d) + \sum_{n=1}^{N} Y_n \exp(j2n\beta d)}. \quad (6.1)$$

At center frequency f_0, $\beta d = \pi$, and the formula reduces to a simple result:

$$\Gamma = \frac{1 - \sum_{n=1}^{N} Y_n}{1 + \sum_{n=1}^{N} Y_n}. \quad (6.2)$$

This is the genesis of the simple rule mentioned above, that the conductance sum be unity and that the susceptance sum be zero. For uniform excitation, $Y_n = 1/N$ and the series in Eqn. (6.1) can be summed exactly to give

$$\Gamma = \frac{1 - \exp[j(N-1)\beta d] \dfrac{\sin N\beta d}{N \sin \beta d}}{3 \exp j2N\beta d + \exp[j(N+1)\beta d] \dfrac{\sin N\beta d}{N \sin \beta d}}. \quad (6.3)$$

The $(\sin N\beta d)/(N \sin \beta d)$ factor is the pattern of an N-element array, and it produces the oscillations in VSWR and Γ as βd changes. These oscillations were shown by Fry and Goward (1950) who plotted Eqn. (6.3). The VSWR rises with increasing frequency, crossing a value of 2 at a normalized frequency below $1 + 1/2N$, and then oscillating about VSWR = 2. The merit factor is $NB = 89$ for $N = 20$. This value is more conservative than the previous simplistic merit factor of $NB = 114$.

A more accurate approach, which also gives results for tapered excitation, uses a cascade of A–B–C–D matrices (Altman, 1964). Each section of waveguide with a shunt admittance load (slot) is represented by a prototype section, as shown in Fig. 6.2. For brevity call $C = \cos \beta d$ and $S = \sin \beta d$. Then the section matrix (Altman, 1964) is

$$\begin{pmatrix} C & jS \\ jS & C \end{pmatrix} \begin{pmatrix} 1 & 0 \\ Y_n & 1 \end{pmatrix} = \begin{pmatrix} C + jY_n S & jS \\ Y_n C + jS & C \end{pmatrix}. \tag{6.4}$$

The matrix of the resonant array is then the product of the N section matrices, starting at one end, with a shorted quarter-wave section as the last matrix. A large Y_s simulates the short, and the overall matrix is

$$\begin{pmatrix} A & B \\ C & D \end{pmatrix} = \left[\prod_{n=1}^{N} \begin{pmatrix} C + jY_n S & jS \\ Y_n C + jS & C \end{pmatrix} \right] \begin{pmatrix} C_2 + jY_2 S_2 & jS_2 \\ Y_2 C_2 + jS_2 & C_2 \end{pmatrix}. \tag{6.5}$$

C is used both as a trig term and as part of A–B–C–D but the usage should be obvious. Subscript 2 is for the shorted section. Input impedance and reflection coefficient are readily obtained:

$$Z_{in} = \frac{A + B}{C + D}, \quad \Gamma = \frac{2}{A + B + C + D}. \tag{6.6}$$

Figure 6.3 shows VSWR versus normalized frequency for a uniform 20 element array. The curve is symmetric about f, so only half is shown. A 25 dB Taylor distribution array produces a similar curve with a damped oscillation. The parts of these curves below the first peak are plotted versus NB in Fig. 6.4. From these data the merit factor for uniform excitation is $NB = 66$ for VSWR = 2, and $NB = 50$ for VSWR = 2 for the 25 dB Taylor pattern.

An understanding of bandwidth limits is obtained from Smith chart plots. For uniform excitation, all B movements are equal, but the first few are

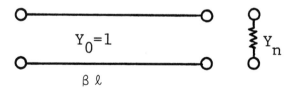

Figure 6.2 Prototype section.

SERIES FEEDS   **167**

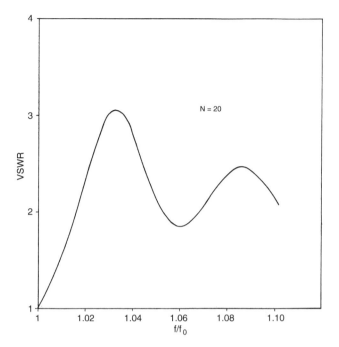

Figure 6.3 Uniform resonant array VSWR.

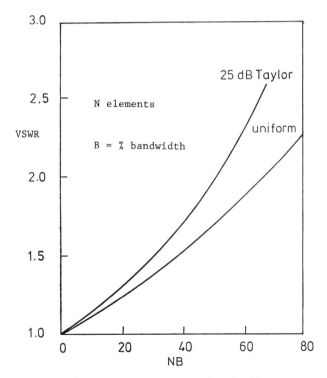

Figure 6.4 Resonant array bandwidth.

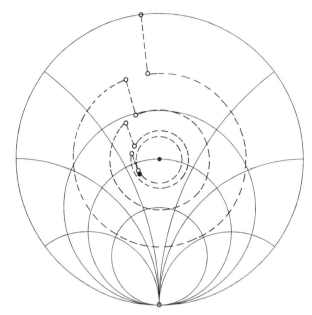

Figure 6.5 Resonant array impedance transformation, 4 elements, uniform, $f/f_0 = 0.96$.

larger and nearly radial on the Smith chart, as shown in Fig. 6.5. These move the impedance toward match. With a tapered excitation, Fig. 6.6, the first few are smaller owing to the taper, and are smaller on the chart. The last few are larger, and tend to bypass the match point. In both cases, large numbers of elements give a nearly completed rotation around the chart for each step. Location of the feed near strongly coupled elements exacerbates the reduced bandwidth.

In practice the bandwidth is smaller, owing to the change of slot admittance with frequency. For all types of waveguide slots the admittance change can be included, using moment methods, but the slot modeling is simpler and more accurate for broadwall slots than for edge slots. A further complication is beam splitting, which can occur when the two travelling waves that constitute the standing wave produce beams to the left and right of center (Kummer, 1966). This occurs at roughly twice the VSWR = 2 value of NB. Keeping VSWR ≤ 2 obviates these split beams.

6.1.1.2 Resonant Slot Array Design. Because waveguide slot arrays provide high efficiency with excellent pattern shaping and control, they deserve an exposition on their design. With the half guide wavelength spacing requirement, it is necessary to change the coupling sign for every other slot. With longitudinal slots the N slots alternate on each side of the broadwall centerline. For inclined broadwall slots and for inclined edge slots the inclination angle is simple reversed at the next slot. The result is broadside radiation in phase from all slots.

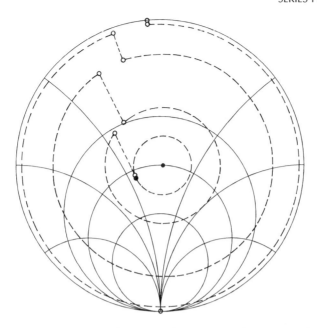

Figure 6.6 Resonant array impedance transformation, 4 elements, 25 dB, $f/f_0 = 0.96$.

The difficult task is to obtain the slot lengths and offsets, or lengths and angles, that will provide the desired array distribution, while satisfying the resonant array conditions. Were it not for mutual impedance this would be a simple task. However, mutual coupling is both ubiquitous and necessary, and must be included. Thus the array design task must encompass all slots simultaneously. Two relationships allow a solution to exist. First the sum of slot admittances (at resonance) must equal the specified input admittance, which is $1 + j0$ for a single stick, but will vary for sticks in a planar array. Second, the mode voltages of the slots must all be equal. In what follows, broadwall longitudinal slots are used as an example. Mutual impedance can be found by replacing the slots by Babinet equivalent strip dipoles, and these in turn can be replaced by cylindrical dipoles of diameter half the strip width (Tai, 1961). It is assumed here that the waveguide face is a flat ground plane. Carter's 1932 mutual impedance formulas can be used; a convenient FORTRAN formulation of these has been given by Hansen and Brunner (1979). For a dielectric-filled guide, a more accurate mutual impedance based on a cosinusoidal distribution along the slot can be used (Elliott, 1983); however, the mutual impedance calculation now requires numerical integration. Slot self-admittance includes the effects of energy storage inside the guide, and must be calculated more carefully or measured. These calculations are either of variational type (Yee, 1974) or of moment method type. The latter, initiated by Khac (1974), has been successfully exploited by Elliott and colleagues (Elliott, 1979; Elliott and Kurtz, 1978). See Chapter 5 for details.

Using the Stevenson slot result from Chapter 5, and the combined slot impedance Z_n, the slot admittance is

$$\frac{Y_n}{G_0} = \frac{480(a/b)C_n^2}{\pi(\beta/k)Z_n}. \tag{6.7}$$

The coefficient C_n is given in Chapter 5, and Z_n is the inverse of the sum of slot self admittance and slot mutual admittance. Using network equations, the combined slot impedance is given by

$$Z_n = \frac{480(a/b)C_n^2}{\pi(\beta/k)Y_n^0/G_0} + \sum_{m=1}^{N} \frac{V_m Z_{nm}}{V_n}. \tag{6.8}$$

The first term includes slot self-admittance Y_n^0, and the sum covers all the mutual terms, where Z_{nm} is the mutual impedance between the n and the m dipoles. Slot voltage is V_n. Given the slot lengths and offsets, Eqns. (6.7) and (6.8) can be used to obtain embedded slot admittances. Since the sum is constrained,

$$\sum_{n=1}^{N} \frac{Y_n}{G_0} = 1 + j0. \tag{6.9}$$

This complex equation gives two real equations. Now the mode voltages are equated. From Chapter 5, these are

$$v_n = j\frac{2\sqrt{2a/b}\,C_n V_n}{\pi(\beta/k)Y_n/G_0} \tag{6.10}$$

The result is a set of $2N - 2$ equations which, together with Eqn. (6.9), gives $2N$ equations for the $2N$ unknowns (slot length and offset). These nonlinear and transcendental equations can be readily solved via the Newton–Raphson method (Stark, 1970), where a gradient matrix is formed of derivatives of the equations, for each variable. A starting set of lengths and offsets must be provided. Through iteration, a solution is found that satisfies the input admittance and mode voltage equations. This technique generally converges in 6–10 iterations, and it has been used to design many linear and planar slot arrays, including those with shaped sidelobe envelopes (see Chapter 3). A sketch of a waveguide longitudinal slot array is shown in Fig. 6.7.

6.1.2 Travelling Wave Arrays

A travelling wave (TW) array, like a resonant array, has radiating elements dispersed along a transmission line. However, this array is fed at one end, with a load at the other end. As the excitation wave travels along, the array part of it is radiated, leaving only a small part at the end to go to the load. Thus elements

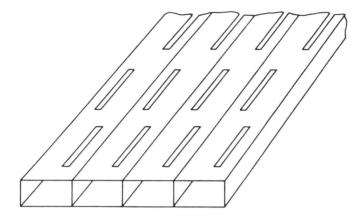

Figure 6.7 Longitudinal shunt slots.

near the feed must couple lightly to the transmission lines, while elements near the load must couple heavily. To avoid reflections in phase, the elements must not be spaced at half-wave intervals, so that the TW array is nonresonant. Because of the nonresonant spacing, the pattern main beam will move slightly with frequency. Element spacing can be less than, or greater than, half-wave; this is discussed further below. Element coupled susceptance should be zero or small, so that the pseudorandom addition of these results in a small value. Given a desired array excitation it is necessary to solve a transmission line problem to obtain element coupling values.

6.1.2.1 Frequency Squint and Single Beam Condition.

Both TEM and dispersive feed lines are used, and they may have phase velocity above (waveguide), equal to, or below (dielectric loaded coax) that of free space. For what range of spacings will only one beam exist? Two examples are representative. First is a TEM line, where the element couplings are π out of phase. The need for out-of-phase coupling will be explained later. With θ measured from broadside and d element spacing as usual, the interelement phase shift needed for a main beam at θ_0 is $kd \sin \theta_0$. Phase along the feed line is $kd + \pi$, where the π is the out-of-phase term. Thus the phase equation is

$$kd \sin \theta_0 + 2n\pi = kd - \pi. \tag{6.11}$$

Each $n \geq 0$ represents a real beam, with the nth beam starting at $d/\lambda = (n + \frac{1}{2})/2$. Figure 6.8 shows beam angles versus d/λ for the first three beams. No beam with complete addition exists for $d/\lambda < 0.25$, and for $d/\lambda = 0.75$ two beams exist. So the useful range is $0.25 < d/\lambda < 0.75$. Either forward squint ($d/\lambda > 0.5$) or backward squint ($d/\lambda < 0.5$) is possible. Beam-squint change with frequency is indicated by the slope of the $n = 0$ line. Spacing of 0.5λ represents a special case of broadside radiation. However,

172 ARRAY FEEDS

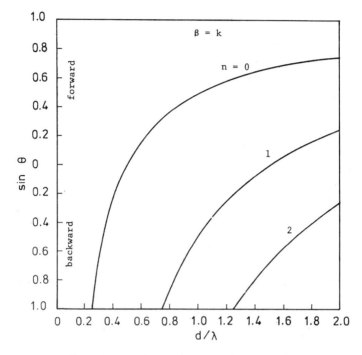

Figure 6.8 TW array beam angle vs. spacing.

this spacing allows all element admittances to add together, with the resulting narrowband performance of the standing-wave array of Section 6.1.1.

Now it will be evident why TW arrays with in-phase coupling are undesirable. The phase equation in this case is

$$kd \sin \theta_0 + 2n\pi = kd. \tag{6.12}$$

This allows a forward endfire main beam for $n = 0$ ($\sin \theta_0 = 1$) at all spacings. Although beams for $n = 1, 2$, etc. can be positioned at a desired angle, the endfire beam is always present.

The second example is that of a waveguide array with out-of-phase coupling. Again, the in-phase coupling allows a beam for all spacings, at $\sin \theta_0 = \beta/k$. For the out-of-phase case the phase equation is

$$kd \sin \theta_0 + 2n\pi = \beta d - \pi. \tag{6.13}$$

Now both plus and minus values of n give real beams. Figure 6.9 shows beam position versus d/λ for the first three beams. The graph is plotted for $\beta/k = 0.6$. For $\beta/k = 0.5$ both $n = 1$ modes start at $d/\lambda = 1$. Since waveguides are usually operated with higher β/k, the range of operations is

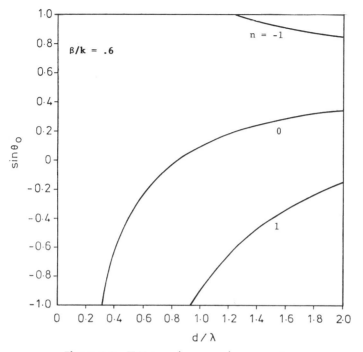

Figure 6.9 TW array beam angle vs. spacing.

$$\frac{0.5}{1+(\beta/k)} < \frac{d}{\lambda} < \frac{1.5}{1+(\beta/k)}, \quad \frac{\beta}{k} > 0.5. \tag{6.14}$$

For the example chosen, the $n = -1$ beam starts at $d/\lambda = 1.25$ where the $n = 1$ beam starts at $d/\lambda = 0.9375$. Not shown is the $n = 2$ beam starting at $d/\lambda = 1.563$. The figure gives a picture of the mode appearance but is not useful for design because it is for $\beta/k = 0.6$ only. Figure 6.10 plots beam angle versus spacing for various values of β/k: $\beta/k = 0.5(0.1)0.9$. Using this graph, the trade of β/k and d/λ to give a fixed beam angle may be made.

Change of beam angle with frequency is important, as can be seen from the slope of the curves. An explicit formula for slope is useful. Taking the derivative of $\sin\theta_0$ w.r.t. frequency, and multiplying to normalize the slope gives

$$f\frac{d\sin\theta_0}{df} = \frac{\lambda^2/\lambda_c^2}{\beta/k} + \frac{1}{2d/\lambda}. \tag{6.15}$$

This contains three interrelated variables, but fortunately it simplifies to

$$f\frac{d\sin\theta_0}{df} = \frac{1}{\beta/k} - \sin\theta_0. \tag{6.16}$$

Normalized slope is plotted versus $\sin\theta_0$ in Fig. 6.11. As expected, β/k nearer to unity gives lower slopes, as do larger values of $\sin\theta_0$. The curves stop at the

174 ARRAY FEEDS

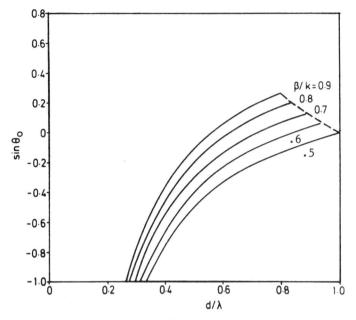

Figure 6.10 Single beam: angle vs. spacing.

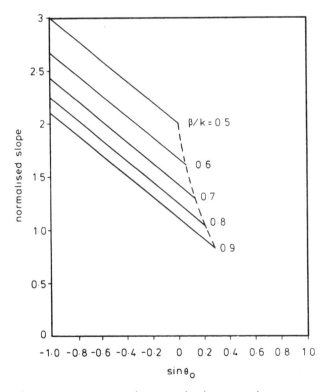

Figure 6.11 TW array beam angle change vs. frequency.

angle where a second beam emerges; thus these curves indicate the practical design range. The beam angle θ_0, waveguide β/k, element spacing, and normalized slope can all be traded for an optimum design, given a particular application.

6.1.2.2 Calculation of Element Conductance.
An exact calculation of the coupling in a travelling wave array would use A–B–C–D matrices to combine each segment of waveguide (between slots) and one slot admittance with other segments, finding the overall input admittance, as is done for resonant arrays. However, since TW arrays usually have many elements, each coupling is then small, and since the element admittances add in a pseudorandom manner owing to the irrational spacing, it is usually sufficient to assume that the array is matched at any point. At each element the power divides, with part producing the element excitation and the remainder going on to the next element after some waveguide attenuation. Let the desired array excitation power distribution be proportional to F_n, where $n = 1$ to N. Call the fraction of input power that is dissipated in the end load L,

$$L = \frac{P_l}{P_{in}}, \qquad (6.17)$$

and call the conductance distribution along the waveguide G_n. If waveguide loss is included, the factor s representing power loss between slots is appropriate:

$$s = \exp(-2\alpha d) \qquad (6.18)$$

Interelement spacing is d and the attentuation coefficient α can be found in standard texts on electromagnetic theory. Using the sketch of Fig. 6.12, the last coupling is written first, then the next, and so on.

$$G_N = \frac{F_N}{P_N} = \frac{F_N}{P_l}, \qquad G_{N-1} = \frac{F_{N-1}}{P_{N-1}} = \frac{F_{N-1}}{F_N/s + P_l/s}. \qquad (6.19)$$

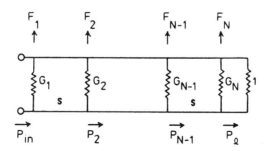

Figure 6.12 TW array circuit.

$$G_n = \frac{F_n}{\sum_{m=n+1}^{N} F_m s^{n-m} + P_l s^{n-N}},$$ (6.20)

$$G_1 = \frac{F_1}{P_{in} - F_1}.$$

The input power is the load power plus the sum of coupled powers, with each multiplied by the appropriate number of segment loss factor $1/s$:

$$P_{in} = \sum_{m=1}^{N} F_m s^{1-m} + P_l s^{1-N}.$$ (6.21)

The last two equations can be used to eliminate P_{in}, giving P_l:

$$P_l = \frac{L}{1 - Ls^{1-N}} \sum_{n=1}^{N} \frac{F_n}{s^n}.$$ (6.22)

Inserting this in the expression for G_n yields the conductance equation, after some rearrangement:

$$G_n = \frac{F_n/s^n}{\frac{1}{1 - Ls^{1-N}} \sum_{m=1}^{N} \frac{F_m}{s^m} - \sum_{m=1}^{n} \frac{F_m}{s^m}}.$$ (6.23)

For negigible loss, of course, $s = 1$. Loss generally needs to be included only when it is large, or when the loss between elements is comparable to the smallest conductance. Usually the F_n excitation coefficients are obtained from a distribution such as Taylor one-parameter or Taylor ñ, and the scale factor of the F_n will make $P_{in} = 1$. Note that at each slot the fraction of power coupled out is $G/(1 + G)$. The continuous version of Eqn. (9.169) is often seen (Kummer, 1966).

$$G(x) = \frac{F(x)}{\frac{1}{1 - L} \int_{-L/2}^{L/2} [F(y) + R(y)] \, dy - \int_{-L/2}^{x} [F(y) + R(y)] \, dy}.$$ (6.24)

Here $R(y)$ is the loss per unit length of guide. However, as might be expected, sampling the continuous results to get slot coupling is less accurate than using the sum formula (Eqn. 6.23). For small arrays, or for low-sidelobe arrays, it is important to use the more exact equation.

Figure 6.13 shows the conductance values for a 29-element uniformly excited array, for load power percentages of 5%, 10%, and 25%. It may be seen that the 5% and 10% curves go off scale, while the 25% curve peaks just

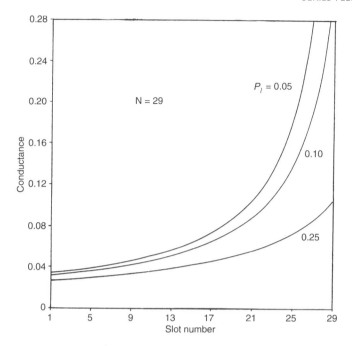

Figure 6.13 Uniform slot conductance.

above 0.1. It is always necessary to keep all conductances small; 0.1 is a good maximum. Not only are larger values often difficult to realize, but the impedance mismatch effect may no longer be negligible. Larger arrays, of course, will have lower conductances per slot. The curves for $N = 29$ may be converted to general-purpose curves (approximately) by multiplying values by $N/2 = 14.5$. Conductance values for different N are then obtained by dividing the new values by $N/2$. However, for precise results, a new calculation should be made after the number of slots is established based on the approximate scaling just described.

Figures 6.14 and 6.15 give conductances for the Taylor one-parameter distribution and for the Taylor $\bar{n}$ distribution ($\bar{n} = 5$). For both figures, $N = 29$ and SLR $= 25$ dB. Note that the heavy taper of the one-parameter distribution gives curves with a single peak, whereas the $\bar{n}$ distribution, which has a higher pedestal (see Chapter 3), tends to rise again near the load, especially for small load power. Since all the curves for 5% and 10% load power have peak above 0.1, it is in practice necessary either to increase load power, thereby reducing gain, or to increase the number of slots. Again, the curves can be scaled approximately using $N/2$.

6.1.2.3 TW Slot Array Design.
Without the effects of mutual coupling, the design of a travelling-wave slotted waveguide array is simple: a table of conductances is obtained (see previous section) and the universal slot curves of Chapter 5 are used to find length and position of each slot, to give the

178 ARRAY FEEDS

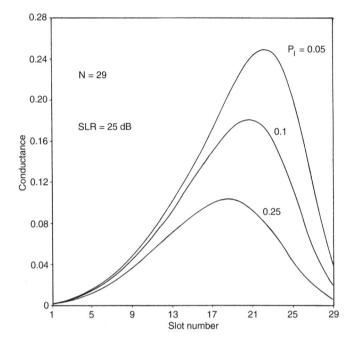

Figure 6.14 Taylor one-parameter slot conductance.

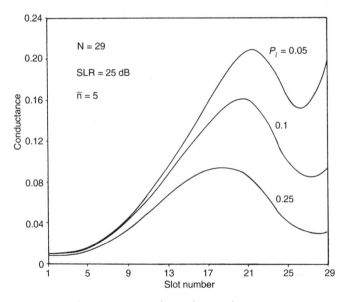

Figure 6.15 Taylor $\bar{n}$ slot conductance.

proper resonant conductances. With mutual coupling, the slots interact strongly so that it is necessary to consider the entire array at once. This has been successfully done for broadwall shunt (displaced) slots by Elliott (1979), and the procedure can be applied equally well for any array in which mutual coupling between elements can be computed accurately. A salient exception is the inclined-edge slot waveguide array, as the replacement of slots by Babinet equivalent dipoles for the computation of mutual impedance is not sufficiently accurate owing to the wrap-around of the slots.

The design procedure for travelling-wave arrays is similar to, but more complicated than, that for resonant arrays. As is the case there, the mode voltage at each slot is proportional to the slot voltage divided by the *scan impedance*:

$$v_n = \frac{2\sqrt{a/b}\, C_n V_n}{\pi(\beta/k) Y_a^n / G_0}. \tag{6.25}$$

Unlike the resonant array, the mode voltages are unequal, since the slots are spaced an irrational number of guide wavelengths apart. Notation is that of Section 6.1.1, except the mode voltages are subscripted here. In fact, the nth mode voltage is related to the $(n-1)$st mode voltage and $(n-1)$st admittance, where elements are numbered starting at the load. This relation is

$$\frac{v_n}{v_{n-1}} = \cos\beta d + j\frac{Y_{n-1}}{G_0}\sin\beta d. \tag{6.26}$$

Y_n is defined below, and d is the slot spacing along the guide. Combining with Eqn. (6.25), an equation is obtained for the ratio of two successive slot voltages:

$$\frac{V_n}{V_{n-1}} = \frac{C_{n-1} Y_{n-1}^A}{C_n Y_n^A}\left(\cos\beta d + j\frac{Y_{n-1}}{G_0}\sin\beta d\right). \tag{6.27}$$

Since the slot voltage is specified, if the fractional power into the load L, and all the slot lengths and offsets, are known, Eqn. (6.27) represents $2N-2$ equations that must be satisfied. That is, using the correct slot parameters gives the desired slot voltage ratios. A perturbation technique is then used with starting values of slot lengths and offsets. The N-element array has $2N$ unknowns; the remaining two equations come from the input admittance. Like the resonant array it should be $1+j0$. However, input admittance here is not just a simple sum of slot active admittances; rather, a transmission-line concatenation must be used. Calling Y_n the guide admittance seen at the nth slot (including the guide admittance beyond the slot and the slot active admittance), it becomes:

$$\frac{Y_n}{G_0} = \frac{Y_n^A}{G_0} + \frac{Y_{n-1}\cos\beta d + jG_0\sin\beta d}{G_0\cos\beta d + jY_{n-1}\sin\beta d}. \tag{6.28}$$

Starting at the slot nearest to the load where $Y_1 = Y_1^A + G_0$, Eqn. (6.28) is applied successively until the input admittance is obtained. This should be $1 + j0$ if the correct slot lengths and offsets were chosen. These two real equations, with the $2N - 2$ mode voltage equations, give the $2N$ real equations necessary to allow a definitive problem. These also can be solved iteratively using a Newton–Raphson solution. Figure 6.16 depicts a travelling wave patch array; note the change in patch sizes to accommodate coupling resistance of the form of Fig. 6.14. Figure 6.17 shows the AWACS array, in which the horizontal waveguide sticks are travelling wave arrays of edge slots. This is a very low sidelobe level array ($-50\,\text{dB}$).

The TW arrays just discussed have a beam angle that changes with frequency, owing to the series nature of the feed. This beam scan can be removed by use of an equal-path-length feed, as sketched in Fig. 6.18 (Cheston and

Figure 6.16 TW patch array.

Figure 6.17 Low-sidelobe edge slot array. (Courtesy Westinghouse Electric Corp.)

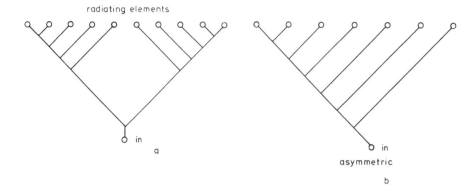

Figure 6.18 Equal-path-length feeds.

Frank, 1970). Unfortunately, much of the low-cost simplicity of the series TW feed is lost. The series feed slots in this scheme do not radiate directly, but each feeds a waveguide. Radiating elements are arranged at the waveguide end. The geometry is arranged so that the waveguide path length from the input port to each radiating element is the same; thus no change in beam position with frequency will occur. Another advantage is that the radiators experience the same amount of waveguide dispersion. The radiating waveguides should be along the squint angle so that the aperture is broadside. This specifies the angle between feed and radiating guides.

6.1.3 Frequency Scanning

Advantage can be taken of the beam squint with frequency by increasing the interelement path length such that a small frequency change scans the main beam. Such arrays are travelling wave arrays but are important enough to warrant a separate section. Assume that the radiating elements spaced d apart are connected by a serpentine (snake) feed of length s and wave number β. The wavefront is defined by

$$kd \sin \theta_m = \beta s - 2m\pi. \tag{6.29}$$

(See sketch in Fig. 6.19.) To avoid an endfire beam, $kd \geq \beta s - 2m\pi$. More than one main beam can be avoided (see Section 2.1) by keeping

$$\frac{d}{\lambda} < \frac{1}{1 + \sin \theta_m}. \tag{6.30}$$

The phase equation is often written as

$$\sin \theta_m = \frac{s\lambda}{d\lambda_g} - \frac{m\lambda}{d}. \tag{6.31}$$

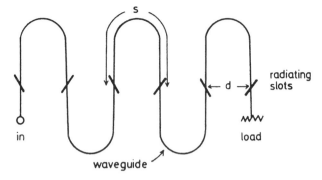

Figure 6.19 Serpentine feed geometry.

Clearly, larger s/λ, and correspondingly larger m, gives a faster change of beam angle with frequency. If the frequency scan passes through broadside there will be a significant change in impedance owing to the addition of all the element conductances. Element conductances, or couplings, are determined just as for a TW array. If the waveguide bends needed to make a snake are not well matched over the frequency band, it may be necessary to calculate the input admittance using the techniques developed for resonant arrays. The coupler reflections tend to produce a sidelobe at the conjugate beam direction; the level of this reflection sidelobe depends upon the coupling conductances, and is higher when the scan is closer to broadside.

In a straight TW feed, losses are usually not important, but the extra length of the sinuous feed often introduces significant loss. Maximum gain occurs when $\alpha s = 1$, where s is the total snake feed length (Begovich, 1966). The corresponding feed loss is 4.3 dB. Maximum radar target tracking accuracy, a combination of gain and beamwidth, occurs for $\alpha s = 2$, with 8.6 dB feed loss (Begovich, 1966).

Dispersion in the snake feed will distort short pulses if the pulse length is less than or comparable to the feed travel time. Detailed calculations have been made by Bailin (1956) and Tseng and Cheng (1964).

As an example, assume each serpentine loop to be five free-space wavelengths ($s = 5\lambda_0$), a waveguide wavenumber ratio of $\beta/k = 0.6$ at center frequency, and a slot spacing of half free-space wavelength. The beam angle is given by

$$\sin\theta_0 = 10\left[\sqrt{1 - \left(\frac{0.8f_0}{f}\right)^2} - 0.6\frac{f_0}{f}\right]. \tag{6.32}$$

This is shown in Fig. 6.20. A 10% bandwidth covers a 117 deg scan. A typical sinuous feed using sidewall couplers is shown in Fig. 6.21. Sinuous feeds have also been used to feed cylindrical reflectors. Another way of packaging the snake is in a helix. Here a helical waveguide with the outside wall open is machined from a solid block; a cover is later fastened around the open top

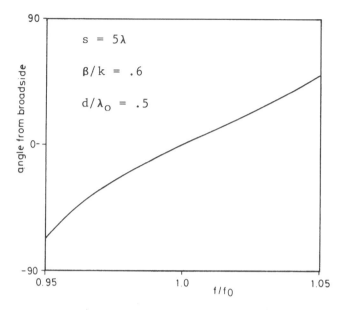

Figure 6.20 Frequency scan example.

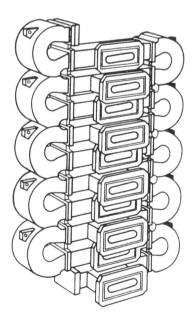

Figure 6.21 Serpentine feed. (Courtesy ITT Gilfillan.)

184 ARRAY FEEDS

(Croney and Foster, 1964). Higher dispersion, and hence a larger slope of degrees per megahertz, can be obtained from a dispersive, periodically loaded waveguide, analogous to the waveguide delay lines used in travelling-wave tubes (Hockham and Wolfson, 1978).

When the operating frequency is too near the resonant array frequency, the reflections from the sinuous feed bends and couplers tend to add to give a significant reflection. Kurtz and Gustafson (Gustafson, 1957) developed impedance matching nondirective couplers as seen from the feed. This allows scanning through broadside with small degradation. The coupler used a rectangular guide normal to the narrow wall of the feed guide, with a centered shunt slot in that wall. A resonant inductive iris was placed in the feed guide, opposite the coupling slot.

For low-sidelobe designs, an improved 4-port coupler uses a cross-guide coupler with two pairs of crossed coupling slots. A load for this directional coupler is placed at the end of the radiating waveguide stick (Ajioka, 1993). Figure 6.22 shows the coupling region for an S-band coupler. Phase shifts versus coupling are shown in Fig. 6.23 for that coupler, and a typical implementation is shown in Fig. 6.24 (Ajioka, 1993). An L-shaped coupling slot has been used by Park (1995). Figure 6.25 shows an airport surveillance radar antenna. Use of wideband directional couplers made out of aluminum billets by numerical controlled machining allows low sidelobes to be obtained.

6.1.4 Phaser Scanning

Linear series feeds may be scanned by use of phasers either in the feed line or adjacent to the elements; these are series phasers and parallel phasers, respectively. Figure 6.26 sketches these arrangements. With waveguide slot array sticks there is no room for phaser devices. Many other configurations do allow phasers to be incorporated, for example, the nonsquint feed of Fig. 6.18. An advantage of series phasers is that each phaser handles only $1/N$ of the transmitted power, and each element path has the phaser loss. A disadvantage is that the basic interelement phase shift $\phi = kd \sin \theta_0$ must be multiplied by the number of the element. As almost all phasers are now digitally controlled, this is a minor disadvantage. Modern digital controls are inexpensive. For parallel phasers the advantage is that all phasers have the same setting equal to ϕ. Disadvantages are that the first phaser must handle almost the entire power, and that the phaser losses are in series. In both topologies there are systematic errors due to impedance mismatch; these are treated later.

A word about phasers is appropriate, although a full discussion is outside the scope of this book. Phasers, often called phase shifters, are primarily of two basic types: switched line and ferrite. Switched line phasers use semiconductor switches to switch in or out lengths of transmission line, where the latter is usually coax, microstrip, or stripline. Discrete components may be used, but more often the switches, loads, and controls are in integrated circuit packages that are integrated into microstrip or stripline. MMIC (monolithic microwave integrated circuits) configurations are becoming more common, especially above 10 GHz. Of course, these phasers utilize time delay, but because the

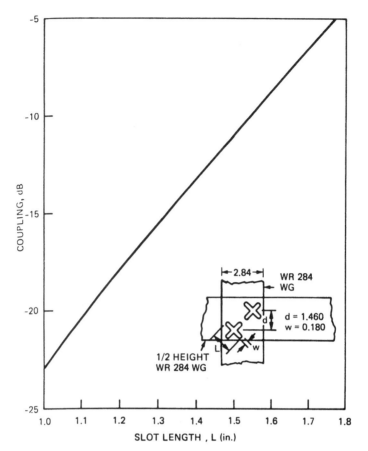

Figure 6.22 Cross slot coupling. (Courtesy Ajioka, J. S., "Frequency-Scan Antennas," in *Antenna Engineering Handbook*, R. C. Johnson, Ed., McGraw-Hill, 1993, Chapter 19.)

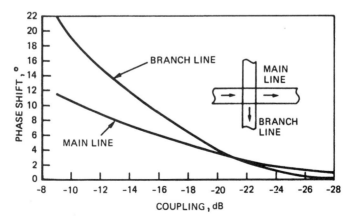

Figure 6.23 Cross slot coupler phase shift. (Courtesy Ajioka, J. S., "Frequency-Scan Antennas," in *Antenna Engineering Handbook*, R. C. Johnson, Ed., McGraw-Hill, 1993, Chapter 19.)

186   ARRAY FEEDS

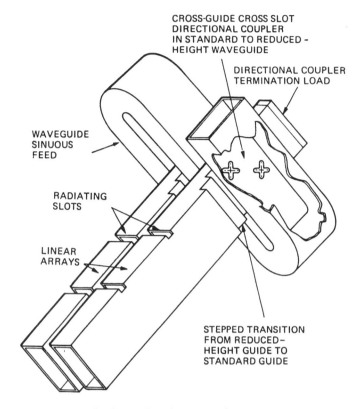

Figure 6.24 Serpentine feed on edge slot array. (Courtesy Ajioka, J. S., "Frequency-Scan Antennas," in *Antenna Engineering Handbook*, R. C. Johnson, Ed., McGraw-Hill, 1993, Chapter 19.)

largest bit is usually 180 deg, they are classed as phasers. Digital phasers are binary: a 5-bit phaser has phases of 180, 90, 46, 22.5, and 11.25 deg. The phaser design must keep the impedances matched as the bits switch, and in extreme applications must also match the losses. If amplifiers are incorporated to offset losses, the phasers become nonreciprocal.

Ferrite phasers are also of two types, and although these are intrinsically analog phasers, current practice has digitized the phase control. They are almost always embodied in waveguide. Toroidal ferrite phasers use one or several ferrite toroids of rectangular shape placed in a waveguide such that the axis of the toroid is along the guide axis. A drive wire threaded through the toroid provides either a positive or negative current pulse. The pulse drives the toroid to saturation, and the toroid is latched at a positive or negative remanent-induction point. The difference between the electrical lengths of the two states provides the phase shift. Typically several toroids are placed serially, with lengths chosen to provide the N bits of phase shift. Advantages and disadvantages are obvious: these phasers are nonreciprocal, and must be reset after each pulse for radar operation; drive power is used only when switching, so that power and heat dissipation are minimized.

SERIES FEEDS   187

Figure 6.25 Airport surveillance antenna. (Courtesy ITT Gilfillan.)

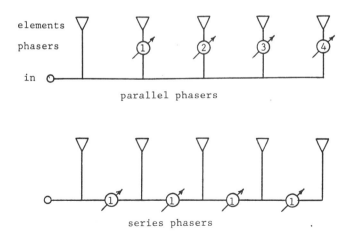

Figure 6.26 Linear series feeds.

188 ARRAY FEEDS

The Faraday rotation ferrite phaser uses a ferrite rod along the axis of the waveguide, with quarter-wave plates at each end to convert linear to circular polarization, and vice versa. In practice, the ferrite rod is plated with silver or gold to form a small ferrite waveguide. This ferrite guide is then placed inside the regular guide, with nonplated tapered transition sections on each end to match between the two guides. An external solenoidal coil around the waveguide provides the magnetization that controls the phase shift. Again, typical drive circuits are digitally controlled. These phasers are reciprocal, and are often used in high-power applications such as radars. Sophisticated drive compensation circuits have been developed to adjust the drive as the temperature of the ferrite increases; the drive power must be applied at all times.

Useful references on phasers include a special issue of *Transactions of the IEEE* on array control devices (Whicker, 1974), a chapter on phasers and delayers (Ince and Temme, 1969), and a two-volume monograph (Koul and Bhat, 1991).

6.2 SHUNT (PARALLEL) FEEDS

6.2.1 Corporate Feeds

Corporate feeds are common in arrays of dipoles, open-end guides, and patches, and are named after the structure of organization charts, where the feed divides into two or more paths, then each path divides, and so on. Such feeds are commonly binary, but sometimes the divider tree includes 3-way, or even 5-way dividers, depending upon the number of array elements. Figure 6.27 shows a simple binary parallel feed with phasers. As in the case of main line series feeds, the phase shift needed is progressively larger. Now, however, each path length has only one phaser. For wideband applications, where modulo 2π phase shift is inadequate, time delay units (delayers) can be used. In principle, a unit of delay is needed at each power division level. For wide-angle scanning, delayers of $\frac{1}{2}Nkd\sin\theta_0$ would be

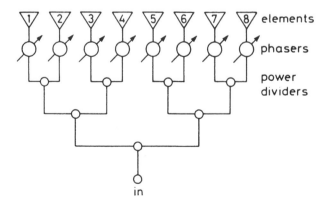

Figure 6.27 Parallel (corporate) feed (numbers are phase shift units).

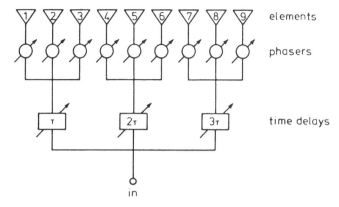

Figure 6.28 Phase-delay corporate feed.

placed after the first power division; four delayers with half that at the next level; and finally eight phasers at the element level, where all phase delays are less than 2π. A more practical compromise for large arrays is to have one or two levels of delayers in addition to the phasers at each element. Figure 6.28 shows such an example.

The critical component in the corporate feed is the power divider (combiner); bifurcated T waveguide or coaxial-line T junctions, or hybrid junctions, can be used. The former are simpler, but hybrid junctions reduce the effects of element impedance mismatch. Typical hybrids are the short-slot hybrid (Riblet, 1952; Levy, 1966) which use two guides with a contiguous narrow wall, the top-wall hybrid (Hadge, 1953), the magic tee (Young, 1947), and the ring hybrid (Tyrell, 1947). Power dividers and hybrids can also be implemented in stripline or microstrip (Shimizu and Jones, 1958; Levy, 1966; Young, 1972). A broadband TEM line coupler was developed by Schiffman (1958). The popular Wilkinson divider (Wilkinson, 1960) incorporates resistive loads to improve the match; it is commonly a stripline divider with the load across the separated lines. Another stripline divider is the Lange coupler (Lange, 1969), which is a 90 deg hybrid utilizing interdigitated coupled lines. However, many microstrip patch arrays use the simple split line reactive power divider.

6.2.2 Distributed Arrays

In a distributed array each element is connected to its own receiver/transmitter module. Such modules often contain duplexers, circulators, filters, preamps, power amplifiers, phasers, and control components. Sometimes the element is part of the module. Several advantages accrue. Many low-power semiconductor sources may be used instead of a single high-power tube source. Feed network loss, which reduces S/N, is minimized. Graceful degradation allows system operation with slight gain and sidelobe changes as modules fail. Mean time before failure (MTBF) is greatly increased (Hansen, 1961). Against these advantages is the higher cost. The nominal few hundred dollar module has been a long-sought goal, but nonrecurring engineering costs are high, especially for MMIC, so that a large number of production units is needed to make the modulus cost-effective. Solid-state and tube modules are discussed by Ostroff et al. (1985); the current art is reviewed by Cohen (1996).

A module may be utilized for each element, or for a line array in an array scanning in only one plane. Figure 6.29 shows a module that feeds six printed dipoles. Modules for planar arrays are discussed in Section 6.3.

Calculation of G/T for an active array is straightforward provided no resistances are used to provide amplitude taper; of course losses of all components, and active device gain and noise figure, must be known. Third-order intercept and the spur-free dynamic range (see Section 6.4.1) are also often important. When resistive tapers are used, care must be taken with G/T calculations (Lee, 1993). A comparison of active weighting (for amplitude taper) utilizing the gain of the MMIC preamplifiers, and passive weighting utilizing unequal combiners, has been given by Holzman and Agrawal (1996).

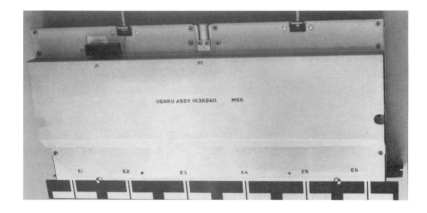

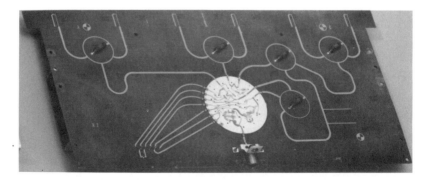

Figure 6.29 Subarray module. (Courtesy Hughes Aircraft Co.)

6.3 TWO-DIMENSIONAL FEEDS

6.3.1 Fixed Beam Arrays

Planar arrays may be composed of linear arrays, or may be an assemblage of individual elements, such as open waveguides. When the main beam direction is fixed, the constituent branch linear arrays and the main feed line that feeds the linear arrays can be travelling wave or resonant, or corporate. Thus, the flat plane array (FPA), which has replaced parabolic reflector antennas in many aircraft and missile radars, is comprised of branch "sticks," which are resonant arrays of longitudinal shunt slots in waveguide, with the waveguide sticks fed by a resonant main feed. Figure 6.30 shows a quadrant of the FPA, where each stick array is divided into subarrays, with five comprising the center stick. This is done to allow increased bandwidth since each resonant array now has fewer than 21 slots. The complete array has 1368 slots. Figure 6.31 shows the back, where the cross-guide feeds can be seen. A waveguide corporate feed excites the cross-guide feeds. A hybrid monopulse feed network is used to provide optimized sum and azimuth and elevation difference patterns. Figure 6.17 shows a very low (−50 dB) sidelobe edge slot waveguide array shaped to fit into a radome. Here the stick arrays are travelling wave arrays; since the array rotates in azimuth, the beam squint is easily accommodated. Achievement of such low

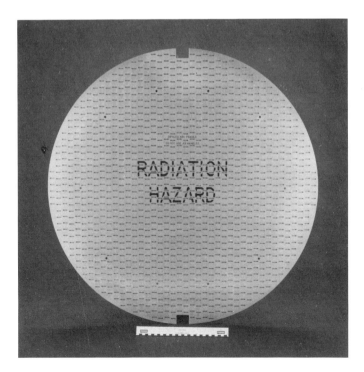

Figure 6.30   Flat plane slot array. (Courtesy Hughes Aircraft Co.)

192 ARRAY FEEDS

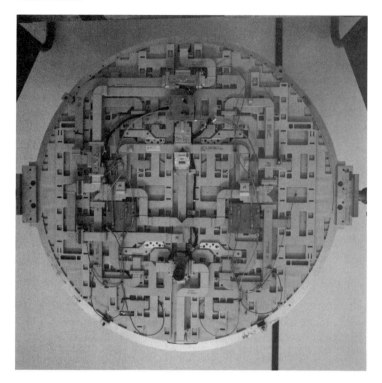

Figure 6.31 Optimized sum and difference feed network. (Courtesy Hughes Aircraft Co.)

sidelobe levels requires meticulous accommodation of edge effects and mutual coupling. Precise fabrication is essential for low-sidelobe antennas. Microstrip patch antennas with in-plane corporate feeds suffer from feed radiation and feed mutual coupling (Hall and Hall, 1988). A large C-band patch array using resonant subarray feeds is shown in Fig. 6.32 (Granholm et al., 1994).

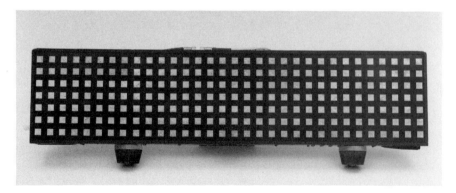

Figure 6.32 Dual polarization C-band SAR patch array. (Courtesy Technical University of Denmark.)

6.3.2 Sequential Excitation Arrays

A planar array of linearly polarized elements may be sequentially oriented to produce circular polarization (Huang, 1986, 1995; Hall and Hall, 1988). Figure 6.33 sketches a circular patch array where the patches are oriented and phased for circular polarization (CP). The basic feature is a 2×2 subarray with proper patch rotation and phasing. Advantages are better control of E- and H-plane beamwidths, and reduced mutual coupling (Huang, 1986). A major disadvantage is the large grating lobes in the diagonal planes. These are eliminated by using CP elements with sequential orientation and excitation. The same technique can be used with imperfect CP elements or feed errors to reduce grating lobes (Hall et al., 1989; Smith and Hall, 1994). Large subarrays, such as 4×4, may be used with new sequential rotation configurations to reduce sidelobes (Hall and Smith, 1994). Finally, linear polarized elements with sequential excitation can be arranged in rings (Iwasaki et al., 1995). A variation on linear polarization elements utilizes gridded patches, where each patch consists of a set of narrow parallel strips, with the strip directions rotated around the 2×2 subarray as before. Use of gridded patches reduces the mutual coupling between polarizations (Brachat and Baracco, 1995).

6.3.3 Electronic Scan in One Plane

Phase shift due to interelement line length and frequency, or produced by phasers, can provide scan in one plane. Most electronically scanned arrays are of this type, owing to lower losses and cost. Scan in a second plane is often provided by mechanical rotation. The phasers may be in series with the feed line, or in parallel at the elements; see the next section for details. Figure 6.34 shows part of a planar array of L-band patches; horizontal subarrays of 18 elements are fed as a resonant array, driven by a phaser module. Phasers provide elevation scanning, with the modules fed by a corporate feed (Elachi, 1988).

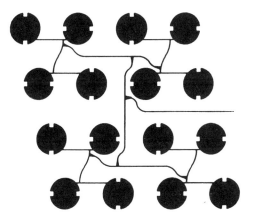

Figure 6.33 Sequentially rotated patch array.

194 ARRAY FEEDS

Figure 6.34 SIR-C subarray module. (Courtesy Ball Aerospace and Communications.)

Figure 6.25 is a photograph of an airport surveillance radar utilizing frequency scan (see next section) in elevation and rotation in azimuth. Edge slots are used.

6.3.4 Electronic Scan in Two Planes

Most two-dimensional (2-D) arrays have radiating elements arranged on regular rows and columns, and the feeds tend to follow the same arrangement. In the case of fixed beam planar waveguide slot arrays the elements are integral with the feed. As discussed in Section 6.1 and 6.2, the main and branch (secondary) feed components may be resonant, travelling wave, or corporate. Here a new criterion is used to demarcate feed topologies: the phaser locations in the feed network. Branch line topologies will be examined first.

Phasers can be located in series or in parallel with respect to the branch lines, as shown in Fig. 6.26. The branch feed can be resonant, travelling wave, or corporate. Of course, with a TW feed the phaser settings must accommodate the varying phase along a TW feed. The main feed, which feeds all the branch feeds, can also be series or parallel. Thus for three-dimensional (3-D) arrays many feed combinations are possible; however, some lend themselves to fabrication better than others.

Figure 6.35 shows a series–series feed arrangement, with both branch line and main line feeds of TW type. An advantage is that all branch phasers have the same setting $\theta = kd\sin\theta_0$, and all main phasers have the same setting $\Phi = kd\sin\phi_0$. Both phases are corrected for the TW feed line phase shifts. With all 2-D scan configurations it is possible to simplify the feed to 1-D scan. For example, the branch lines can be fixed beam, and the main line can be a TW frequency scan feed. Series branch lines can be resonant; these are often used when there is no branch line scan.

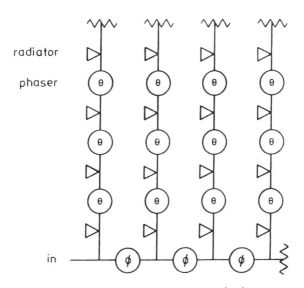

Figure 6.35 Series–series TW feed.

A parallel–series feed has a set of branch feed lines with parallel phasers, and a series phaser main feed line; see Fig. 6.36. Now the phase progression along the phasers of each branch feed are multiples of Φ. However, each parallel phaser handles equal power, and each element sees equal loss (for the branch feed). Again the branch lines and main line can be resonant or travelling wave.

A corporate-series feed has a corporate feed for each row of elements, with the corporate feeds fed from a series main line. Figure 6.37 sketches this topology. Again the main feed line can be of several types. Again the phaser values in a corporate feed are multiples; power is divided among the phasers.

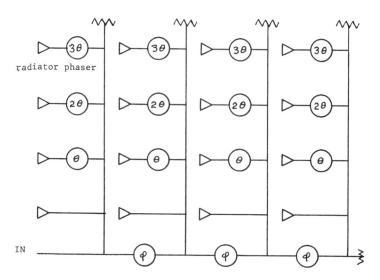

Figure 6.36 Parallel–series TW feed.

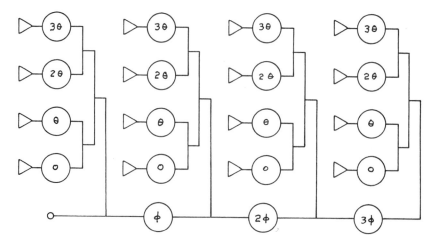

Figure 6.37 Corporate–series feed.

TWO-DIMENSIONAL FEEDS 197

All of these branch feed schemes can also be used with parallel or corporate main feeds. Figure 6.38 shows a series–parallel topology, with series phaser TW branch lines, and a resonant main feed with parallel main phasers. Main phasers again progress in phase, and all main phasers conduct the same power (for uniform array excitation). A variation on this feed topology is the 20 GHz array shown in Fig. 6.39 (Fitzsimmons et al., 1994). The rear block contains

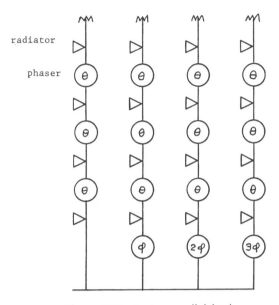

Figure 6.38 Series–parallel feed.

Figure 6.39 Hexagonal lattice array with modules. (Courtesy Boeing Defense and Space Group.)

198   ARRAY FEEDS

travelling wave waveguides with edge couplers to the circular guides. The center block contains phaser modules and dipole radiators; the right block is dielectric loaded guides (below air cutoff) and polarizers.

A series–corporate feed uses a corporate feed to excite the branch lines, as shown in Fig. 6.40. The principal advantage of the corporate feed is intrinsically wider bandwidth; there is no interbranch or interelement phase that varies with frequency. All corporate feed lines are of equal length. To utilize this wideband potential the phasers would be replaced by time delayers, to maintain proper path lengths at all frequencies of concern.

Finally, a full corporate feed encompasses both branch and main feeds; see Fig. 6.41. Now each phaser requires a different phase value, but the power is evenly divided. This type of feed is often used with microstrip patch arrays; see Fig. 6.42. As mentioned in the previous section, the improvement and size reduction of TR modules is continuing; Fig. 6.43 shows four generations of X-band modules that feed circular waveguide and TEM horn radiators. Arrays using TR modules are shown in Fig. 6.44; the top picture shows the MERA array, circa 1965, while the bottom picture shows the RASSR array, circa 1974.

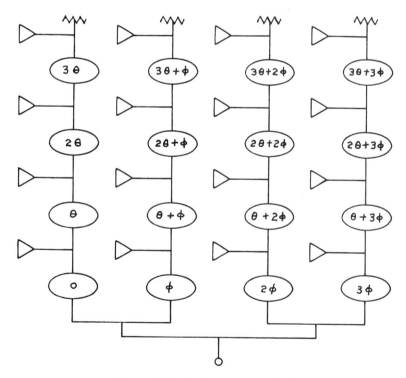

Figure 6.40   Series–corporate feed.

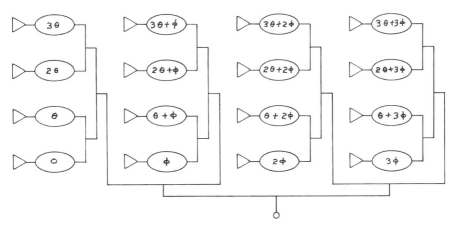

Figure 6.41 Full corporate feed.

Figure 6.42 Corporate fed patch array. (Courtesy Ball Aerospace and Communications.)

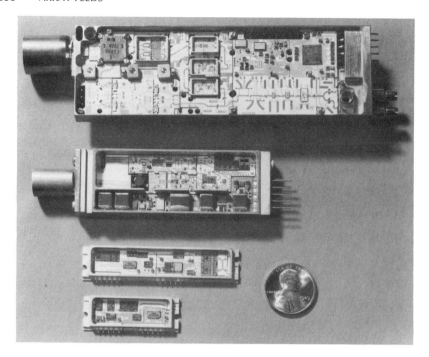

Figure 6.43 Four generations of modules. (Courtesy Texas Instruments.)

6.4 PHOTONIC FEED SYSTEMS

The potential advantages of optical hardware for the care and feeding of a phased array are twofold: to provide time delay without bulky, heavy, and lossy transmission lines; and to reduce weight, volume, and cost. This rapidly developing field is briefly examined here. Two books on the subject are worthwhile: those by Zmuda and Toughlian (1994), and Kumar (1996). Also, ARPA has sponsored an annual series of Symposia on Photonic Systems for Antenna Applications, available to U.S. Department of Defense contractors. Proceedings of these are listed below:

PSAA-1	Dec. 1990	AD-B174 896L & 850L
2	Dec. 1991	AD-B174 895L
3	Jan. 1993	AD-B177 110L
4	Jan. 1994	AD-B193 121L
5	Jan. 1995	AD-B205 913
6	Jan. 1996	AD-B218 390
7	Jan. 1997	AD-B223 415L

First, delay line options are explored, including switched delay, and dispersive delay. Then less conventional approaches are considered.

PHOTONIC FEED SYSTEMS 201

Figure 6.44 Array with TR modules. (Courtesy Texas Instruments.)

6.4.1 Fiber Optic Delay Feeds

6.4.1.1 Binary Delay Lines.
Photonic delay lines are used much as microwave delay lines are, with switches to select various delay segments. Of course, optical modulators and detectors are needed; these are discussed below. In the binary delay chain, semiconductor optical switches are used to switch in or out delay segments that are binary multiples of a minimum delay, say 1λ. An n-bit delay chain then provides 1λ, 2λ, 4λ, ..., $2^{n-1}\lambda$ delays. If phasers are not used for delays below 1λ, the photonic delay chain can include these also. Figure 6.45 shows a block diagram of a delay chain; these are sometimes called BIFODEL (binary fiber optic delay) (Goutzoulis and Davies, 1990). An efficient use of photonic delay is for subarrays only, where each antenna element is connected to a conventional phaser, with largest phase bit of 180 deg; each subarray is then connected to a photonic delay chain (Ng et al., 1991; Lee at al., 1995).

Problems encountered in the utilization of photonic delay are connected with uniformity, stability and drift, and IP3 (see below). Unfortunately a 1 dB optical nonuniformity produces a 2 dB microwave error. IP3 is the third-order intercept point, and it applies to the amplifiers used to offset link loss. The amplifier noise floor and the IP3 determine the spur-free dynamic range, an important parameter for many applications. If an array uses 5-bit phasers, the smallest phase bit is 11.25 deg, or $\lambda/32$. At the other extreme, the longest time delay needed to scan an array of length L to θ_0 is $L\sin\theta_0$. For example, a 100λ array scanned to 60 deg needs a longest delay of 87λ. With

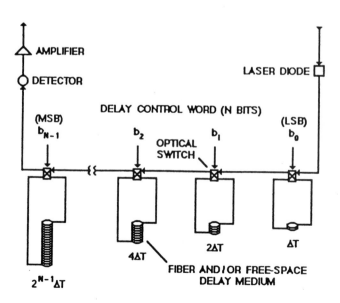

Figure 6.45 Switched binary fiber delay. (Courtesy Goutzoulis, A. P. and Davies, D. K., "Hardware-Compressive 2-D Fiber Optic Delay Line Architecture for Time Steering of Phased-Array Antennas," *Appl. Opt.*, Vol. 29, Dec. 20, 1990, pp. 5353–5359.)

binary steps there would be 7 bits of time delay. Maintaining delays of 64λ, 32λ, and 16λ, for example, to a small part of the least phaser bit $\lambda/32$, is a difficult and challenging task. If the allowable error is 2 deg, the time delay must be maintained to better than one part in 10 000. Photonic links are lossy, and the loss must be controlled and compensated so that the array amplitude taper is not unacceptably distorted.

Another type of delay chain uses two sets of switchable segments in series. This is called square root delay (Soref, 1984; Goutzoulis and Davies, 1990). It provides a set of delays, with each being larger by a fixed delay: delay $= N\tau$. The number of delay lines is $2\sqrt{N}$, while the number of switches is $2N - 1$. This sequence of equal step delays would be suitable to excite an array at a fixed scan angle, but the square root delay produces only one delay per switch setting. All the switch permutations would be used to produce the set of delays for the N array elements. Since there appears to be no way to latch (store) the delays, this type of delay chain is not useful for arrays.

The third-order intercept is obtained by sending two passband frequencies f_1 and f_2 through an amplifier. The third-order product is $2f_2 - f_1$, and IP3 is the power level where the linear (extrapolated) gain curve intersects the third-order product gain curve. The spur-free dynamic range is from the noise floor to the IP3 level; see Fig. 6.46. IP3 of a cascade of amplifiers is simply calculated:

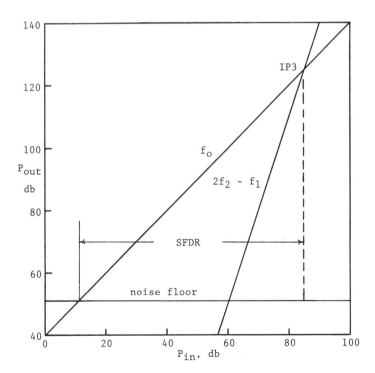

Figure 6.46 IP3 and spur-free dynamic range.

$$\frac{1}{IP3} = \frac{1}{IP3_1} + \frac{G_1}{IP3_2}, \quad (6.33)$$

where the two amplifiers have $IP3_1$ and $IP3_2$, with gains G_1 and G_2. This IP3 is referred to the input. Referred to the output, the IP3 is (Lee et al., 1995)

$$\frac{1}{IP3} = \frac{1}{IP3_2} + \frac{1}{G_2 \cdot IP3_1}. \quad (6.34)$$

Units of IP3 are $dBm/Hz^{2/3}$.

6.4.1.2 Acousto-Optical Switched Delay. An alternative to the use of switched series delays is an N-way "switch" that selects one of N delays. The optical switch is an acousto-optical Bragg cell, in which an applied microwave signal launches a travelling acoustic wave in the acousto-optic crystal. An incident optical beam is modulated by the phonon–phonon interactions, and is deflected by an angle proportional to the microwave drive frequency. The optical beam then impinges on one of a cluster of optical fibers, each of which has a binary delay value. Outputs of all the fibers are collected in an optical power combiner, which in turn feeds a photo detector (Jemison and Herczfeld, 1993; Jemison et al., 1994). Figure 6.47 sketches such a configuration.

6.4.1.3 Modulators and Photodetectors. Laser light modulators are either direct or external. In direct modulation the RF signal is impressed upon the DC current source for the laser. The resulting optical intensity is proportional to the bias current. Distributed feedback (DFB) semiconductor diode lasers are usually used for direct modulation as their insertion loss can be low, with good dynamic range, and low noise. Wavelengths are usually 1310 or 1515 nm. The laser module may contain, in addition to the DFB laser, a thermistor and thermoelectric cooler for monitoring and controlling the laser

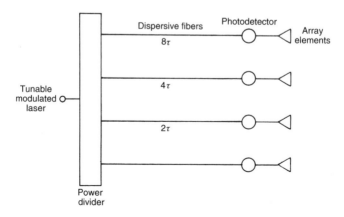

Figure 6.47 Beam scanning with dispersive delay.

temperature, a photodiode for monitoring power, an optical isolator to reduce reflected light from the fiber, and impedance matching components. DFB laser bandwidth appears to be limited to 20–30 GHz; however, quantum well lasers may offer much wider bandwidths.

External modulators need a higher power (for good linearity) and low-noise light source; most use a diode-pumped solid-state NdYAG device. The laser output is coupled to a fiber, with the external modulator placed on the fiber. Modulators can be electro-optic or acousto-optic. Most common is the Mach–Zehnder (M–Z) electro-optic device fabricated in $LiNbO_3$ or in GaAs, or possibly in an electro-optical polymer. The M–Z is an interferometer in which the fiber is split into two paths, with the electro-optic material phase modulating one path due to RF bias; the recombined paths contain the RF-modulated optical carrier.

Acousto-optic modulators (Bragg cell) utilize an RF acoustic travelling wave along the crystal, which may be indium phosphide. An optical beam receives a single-sideband Doppler shift due to phonon–phonon interactions. Thus the optical carrier is modulated by the RF signal.

PIN photodiodes are primarily used as broadside-coupled (instead of edge-coupled) detectors for fiber optics, and there is a trade between quantum efficiency and modulation frequency; higher frequencies reduce efficiency. The frequency response is limited by the transit time of carriers out of the depletion region, and by the time constant of the junction region capacitance plus stray capacitance. Thinner depletion layers for higher frequencies absorb less light and hence are less efficient. At 10 GHz efficiencies may be high, but at 60 GHz the efficiency may be as low as 30% (Wentworth et al., 1992). For such high microwave frequencies, waveguide structures with orthogonal carrier transport and absorption obviate the need for thin depletion layer. The photodiode collection aperture must match the fiber spot size to maximize collection efficiency. Antireflection coatings can be used to reduce reflected light. Finally, the photodiode circuit must be impedance matched to the appropriate system characteristic impedance. See Tamir, Griffel and Bertoni (1995) for papers on components.

6.4.2 Wavelength Division Fiber Delay

Delay can be selected by choosing the proper light wavelength, through use of dispersive fibers, or through use of frequency selective gratings.

6.4.2.1 Dispersive Fiber Delay. The RF-modulated optical beam is sent through a length of highly dispersive fiber; the incremental delay is proportional to the wavelength shift of the laser carrier (Soref, 1992). A representative chromatic dispersion is 50 ps per millimeter of wavelength shift per kilometer of fiber. This dispersion is approximately constant over the laser tuning range. A variable delay can be obtained simply by tuning the laser wavelength, with the delay produced by a length of dispersive fiber. A simple (but expensive) array feed would have a laser oscillator for each array element, with each laser output going through a dispersive fiber then to a photodetector to the element. Dispersive fiber lengths are of the order of 1–10 km. As mentioned elsewhere, the delay must be stable and accurate to a small part of the smallest phaser bit in the array.

A more elegant architecture utilizes one tunable laser (per plane of scan) with the modulated output fanned out to a cluster of dispersive fibers. These would be designed for delays of τ, 2τ, 4τ, $(2N-1)\tau$, and each would drive one array element. Tuning the laser would change τ and thus scan the beam; see Fig. 6.48. For large arrays the fan-out loss can be appreciable. For 2-D scan, two tunable lasers would be used, and the azimuth and elevation delays would be combined at the element photodetectors (Johns et al., 1993; Frankel et al., 1996).

6.4.2.2 Bragg Fiber Grating Delay.

A grating can be made in a length of fiber by focusing ultraviolet energy onto a periodic sequence of spots along the fiber, thereby producing a grating of spots of different index of refraction. Such a grating will reflect an optical wave at the resonance frequency of the grating. A series of Bragg gratings can be spaced along a fiber, with each reflecting at a different optical wavelength. An optical signal enters the fiber with gratings through an optical circulator; the reflected signal path length (delay) depends upon which grating is selected by the optical frequency. The circulator output then goes to an element, through a photodiode.

A more elaborate arrangement uses a fiber grating prism, where the prism consists of a set of Bragg grating lines, one fiber line per array element. Each line contains as many gratings as beam scan positions, so that each laser wavelength will give delays appropriate for that scan angle. To obtain the proper delays across the array, each fiber line has its gratings spaced farther apart than those of the previous line. Each line needs an optical circulator, and

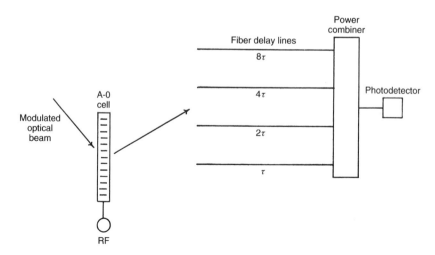

Figure 6.48 Acousto-optic cell delay switching.

the tunable laser feeds into an optical divider. Again, fan-out loss must be considered (Soref, 1996; Tong and Wu, 1996; Zmuda et al., 1997; Chang et al., 1997). See Fig. 6.49.

6.4.2.3 Travelling Wave Fiber Delay. Time delay can be achieved over a wide microwave frequency band through use of a travelling wave fiber that contains a bidirectional coupler for each array element (Lee et al., 1997). Into one end of the TW fiber is fed the optical carrier modulated by f_1, while one modulated by f_2 is fed into the other end. At each coupler the outputs are mixed, with $f_2 - f_1$ providing the desired microwave signal. The beam scan is related to $f_2 + f_1$ so that, by changing f_2 and f_1, the signal can have the proper frequency and delay.

6.4.3 Optical Delay

Time delay can of course be produced by paths in media other than fibers, such as optical resonators. However, the relatively short path lengths, even with many bounces, are more suitable for phase rather than time delay. A spatial light modulator (SLM), which might be a Bragg cell or a liquid crystal used in the birefringent mode, can be used to switch the optical beam out of the path toward a prism. The prism reflects the beam back to another series SLM, thus incurring a delay of twice the distance to the prism (Dolfi et al., 1992, 1994). Here again the path lengths are of necessity relatively short.

6.4.4 Optical Fourier Transform

Since the array amplitude and phase distribution is just the discrete Fourier transform (FT) of the desired pattern, the element excitations can be found directly by sampling the optical FT. For a fixed beam array, the modulated laser light beam illuminates a nonreflecting diaphragm containing a pinhole;

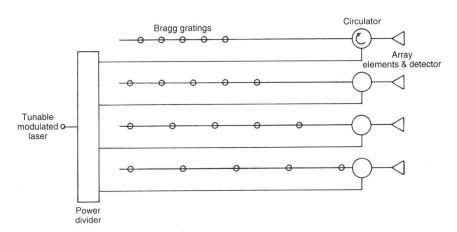

Figure 6.49 Bragg grating delay prism.

208 ARRAY FEEDS

the pinhole represents a narrow antenna beam. The diaphragm and pinhole can be replaced by a partially transparent screen with transmission that represents, for example, a Taylor main beam at the center and the Taylor sidelobe envelope around the main beam. The pinhole or screen illuminates an FT lens, with a 2-D microlens–fiber array in the output plane. Each fiber then drives an array element.

To obtain a scanning beam, a second laser source is introduced with a convenient IF difference frequency. The second source beam is deflected, either by an electro-optic spatial light modulator (Koepf, 1984) or by an acousto-optic cell SLM (Zmuda and Toughlian, 1994), and formed into a plane wave, which is then combined with the output of the FT lens; see Fig. 6.50. The beam deflection produces beam scan. However, the optical structure must be large to provide the time delay paths needed for a large array.

6.5 SYSTEMATIC ERRORS

A narrowband, fixed beam array feed may be designed to be well matched. However, when the array is scanned in one or both planes, the mutual impedance contributions change owing to the changing relative phase between elements. These impedance changes will produce reflections internal to the feed, and may produce second-order beams caused by these systematic errors.

6.5.1 Parallel Phasers

When parallel phasers are located in the branch or element feed lines, the scan impedance changes will produce a small reflection at each element. This reflected energy passes through the phasers, and at the feed line is re-reflected, owing to the light coupling inherent in an array of many elements.

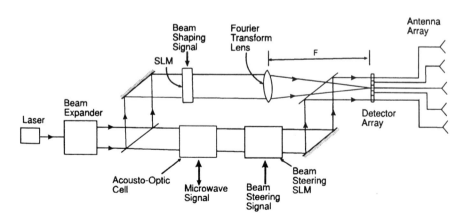

Figure 6.50 Fourier transform beam scanning. (Courtesy Zmuda, H. and Toughlian, E. N., Eds., *Photonic Aspects of Modern Radar*, Artech House, 1994.)

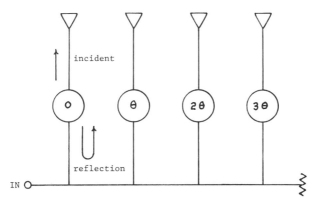

Figure 6.51 Second-order beam generation.

The re-reflected energy now traverses the phaser again, as sketched in Fig. 6.51, and is radiated as a second-order beam, with amplitude reduced by the scan impedance reflection coefficient Γ. Since there are two extra phaser trips, this beam will appear at an angle corresponding to an interelement phase three times that of the main beam. Not all this re-reflected energy is radiated; some is reflected back toward the phaser. Thus a set of second-order beams is produced, with phase delays of 3ϕ, 5ϕ, 7ϕ, etc., and amplitudes of $|\Gamma|$, $|\Gamma|^2$, $|\Gamma|^3$, etc. Beam positions are given by

$$\sin\theta_m = m\sin\theta_0 \pm \frac{p}{d/\lambda}. \qquad (6.35)$$

Here the main beam is at angle θ_0, and the last term is a modulo 2π contribution. A typical example, measured on a 16-element array, but with all impedance terminations alike, is shown in Fig. 6.52. The main beam is at 33.75 deg, with second-order beams at -60.80, -13.75, -4.55 deg. With a reflection coefficient of 0.17, these beams are down by $|\Gamma|^m$; the element spacing is 0.7λ. Table 6.1 gives the calculated values, which compare well with the measurement results of Spradley and Odlum (1956).

TABLE 6.1 Systematic Error Second-order Beams

Beam Angle (deg)	Beam level (dB)	m	p
33.75	0	0	0
−60.81	−7.7	1	−1
−13.78	−15.4	3	−1
−4.55	−23.1	5	−2

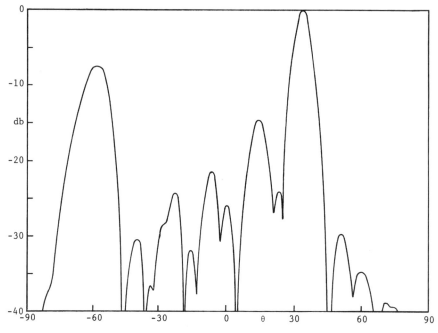

Figure 6.52 Second-order beams: parallel phasers.

6.5.2 Series Phasers

When series phasers are used in the feed line, systematic errors can exist, but they tend to be smaller than those just discussed. Again multiple reflections occur, but now those in the feed line are of interest. Owing to multiple reflections in the feed line, the input impedance and branch port voltages are given by an A–B–C–D matrix cascade. When all branch impedances are equal, and the main feed is matched at each junction, the impedance can be written as a double sum involving *scan impedance*, characteristic impedances, intercoupler phase shifts, etc. (Kummer, 1966). The result is that the systematic errors produce a second-order beam of amplitude Γ/N, where there are N elements. In general, these systematic errors are not critical.

6.5.3 Systematic Error Compensation

Systematic errors may be reduced by special array designs that minimize *scan impedance* changes; these are discussed in Chapter 7. Compensation in the feed network is also possible; this allows the element face to be unchanged. Coupling circuits can be designed into the feed network to compensate for the systematic phase errors. These would connect between adjacent branch line feeds (Hannan et al., 1965). A simpler technique uses a coupling hole between adjacent waveguide feed lines (Cook and Pecina, 1963). Both of these techniques are relatively narrowband, as it is difficult to compensate for *scan impedance* and frequency changes together.

ACKNOWLEDGMENT

Photographs courtesy of Dr. Sembian Rengarajan, Doug Comstock, Jeff Nemit, Dr. Winifred Kummer, Dr. E. L. Christensen, Dean Paschen, Geoffrey O. White, and Dr. Chris Hemmi.

REFERENCES

Ajioka, J. S., "Frequency-Scan Antennas," in *Antenna Engineering Handbook*, R. C. Johnson, Ed., McGraw-Hill, 1993, Chapter 19.

Altman, J. L., *Microwave Circuits*, Van Nostrand, 1964.

Bailin, L. L., "Fundamental Limitations of Long Arrays," Rep. TM330, Hughes Aircraft, Culver City, Calif., 1956.

Begovich, N. A., "Frequency Scanning," in *Microwave Scanning Antennas*, Vol. III, R. C. Hansen, Ed., Academic Press, 1966 [Peninsula Publishing, 1985], Chapter 2.

Brachat, P. and Baracco, J. M., "Dual-Polarization Slot-Coupled Printed Antennas Fed by Stripline," *Trans. IEEE*, Vol. AP-43, July 1995, pp. 738–742.

Chang, Y. et al., "Optically-Controlled Serially-Fed Phased Array Transmitter," *IEEE Microwave Guided Wave Lett.*, Vol. 7, Mar. 1997, pp. 69–71.

Cheston, T. C. and Frank, J., "Array Antennas," in *Radar Handbook*, M. I. Skolnik, Ed., McGraw-Hill, 1970, Chapter 11.

Cohen, E. D., "Trends in the Development of MMICs and Packages for Active Electronically Scanned Arrays (AESAs)," *Proceedings IEEE Symposium on Phased Array Systems and Technology*, Oct. 1996, Boston, Mass., pp. 1–4.

Cook, J. S. and Pecina, R. G., "Compensation Coupling between Elements in Array Antennas," *IRE APS Symposium Digest*, Boulder, Colo., 1963.

Croney, J. and Foster, D., "New Techniques in the Construction of Frequency-Scanning Arrays," *Microwave J.*, Vol. 7, May 1964, pp. 72–74.

Dolfi, D. et al., "Two-Dimensional Optical Architecture for Phase and Time-Delay Beam Forming in a Phased Array Antenna," *Proc. SPIE*, Vol. 1703, Apr. 1992, Orlando, Fla., pp. 481–489.

Dolfi, D. et al., "Two Dimensional Optical Beam-Forming Networks," *Proc. SPIE*, Vol. 2155, Jan. 1994, Los Angeles, Calif., pp. 205–217.

Elachi, C., "Space Borne Radar Remote Sensing," *IEEE Press*, 1988, p. 121.

Elliott, R. S., "On the Design of Traveling-Wave-Fed Longitudinal Shunt Slot Arrays," *Trans. IEEE*, Vol. AP-27, Sept. 1979, pp. 717–720.

Elliott, R. S., "An Improved Design Procedure for Small Arrays of Shunt Slots," *Trans. IEEE*, Vol. AP-32, Jan. 1983, pp. 48–53.

Elliott, R. S. and Kurtz, L. A., "The Design of Small Slot Arrays," *Trans. IEEE*, Vol. AP-26, Mar. 1978, pp. 214–219.

Fitzsimmons, G. W. et al., "A Connectorless Module for an EHF Phased-Array Antenna," *Microwave J.*, Jan. 1994, pp. 114–126.

Frankel, M. Y., Matthew, P. J., and Esman, R. D., "Two-Dimensional Fiber-Optic Control of a True Time-Steered Array Transmitter," *Trans. IEEE*, Vol. MTT-44, Dec. 1996, pp. 2969–2702.

Fry, D. W. and Goward, F. K., *Aerials for Centimetre Wavelengths*, Cambridge University Press, 1950.

Goutzoulis, A. P. and Davies, D. K., "Hardware-Compressive 2-D Fiber Optic Delay Line Architecture for Time Steering of Phased-Array Antennas," *Appl. Opt.*, Vol. 29, Dec. 20, 1990, pp. 5353–5359.

Granholm, J. et al., "Microstrip Antenna for Polarimetric C-Band SAR," *IEEE APS Symposium Digest*, Seattle, Wash., June 1994, pp. 1844–1847.

Gustafson, L. A., "S-Band Two-Dimensional Slot Array," Tech. Memo. 462, Hughes Aircraft Company, Culver City, Calif., Mar. 1957.

Hall, P. S. and Hall, C. M., "Coplanar Corporate Feed Effects in Microstrip Patch Array Design," *IEE Proc.*, Vol. 135H, June 1988, pp. 180–186.

Hall, P. S., Dahele, J. S., and James, J. R., "Design Principles of Sequentially Fed Wide Bandwidth Circularly Polarised Microstrip Antennas," *Proc. IEE*, Vol. 136H, 1989, pp. 381–388.

Hall, P. S. and Smith, M. S., "Sequentially Rotated Arrays with Reduced Sidelobe Levels," *Proc. IEE—Microwave Antennas Propag.*, Vol. 141, Aug. 1994, pp. 321–325.

Hannan, P. W., Lerner, D. S., and Knittel, G. H., "Impedance Matching a Phased-Array Antenna over Wide Scan Angles by Connecting Circuits," *Trans. IEEE.*, Vol. AP-13, Jan. 1965, pp. 28–34.

Hadge, E., "Compact Top Wall Hybrid Junction," *Trans. IRE*, Vol. MTT-1, Jan. 1953, pp. 29–30.

Hansen, R. C., "Communications Satellites Using Arrays," *Proc. IRE*, Vol. 49, June 1961, pp. 1067–1074.

Hansen, R. C. and Brunner, G., "Dipole Mutual Impedance for Design of Slot Arrays," *Microwave J.*, Vol. 22, Dec. 1979, pp. 54–56.

Hockham, G. A. and Wolfson, R. I., "Frequency Scanning Antenna Using Evanescent-Mode Waveguide," *Proceedings IEE Conference on Antennas*, Nov. 1978, London, pp. 21–24.

Holzman, E. L. and Agrawal, A. K., "A Comparison of Active Phased Array, Corporate Beamforming Architectures," *Proceedings IEEE Symposium on Phased Array Systems and Technology*, Oct. 1996, Boston, Mass., pp. 429–434.

Huang, J., "A Technique for an Array to Generate Circular Polarization with Linearly Polarized Elements," *Trans. IEEE*, Vol. AP-34, Sept. 1986, pp. 1113–1124.

Huang, J., "A Ka-Band Circularly Polarized High-Gain Microstrip Array Antenna," *Trans. IEEE*, Vol. AP-43, Jan. 1995, pp. 113–116.

Ince, W. J. and Temme, D. H., "Phasers and Time Delay Elements," in *Advances in Microwaves*, Vol. 4, L. Young, Ed., Academic Press, 1969.

Iwasaki, H., Nakajima, T., and Suzuki, Y., "Gain Improvement of Circularly Polarized Array Antenna Using Linearly Polarized Elements," *Trans. IEEE*, Vol. AP-43, June 1995, pp. 604–608.

Jemison, W. D. and Herczfeld, P. R., "Acoustooptically Controlled True Time Delays," *IEEE Microwave and Guided Wave Lett.*, Vol. 3, Mar. 1993, pp. 72–74.

Jemison, W. D., Yost, T., and Herczfeld, P. R., "Acoustooptically Controlled True Time Delays: Experimental Results," *IEEE Microwave and Guided Wave Lett.*, Vol. 6, Aug. 1996, pp. 283–285.

REFERENCES

Johns, S. T. et al., "Variable Time Delay of Microwave Signals Using High Dispersion Fiber," *Electron. Lett.*, Vol. 29, Mar. 18, 1993, pp. 555–556.

Khac, T. V., A Study of Some Slot Discontinuities in Rectangular Waveguides, PhD thesis, Monash University, 1974.

Koepf, G. A., "Processor for Optical Beam-Forming System Phased Array Antenna Beam Formation," *Proc. SPIE*, Vol. 477, 1984, pp. 75–81.

Koul, S. K. and Bhat, B., *Microwave and Millimeter Wave Phase Shifters*, Vols. I, II, Artech House, 1991.

Kumar, A., *Antenna Design with Fiber Optics*, Artech House, 1996.

Kummer, W. H., "Feeding and Phase Scanning," in *Microwave Scanning Antennas*, Vol. III, R. C. Hansen, Ed., Academic Press, 1966 [Peninsula Publishing, 1985], Chapter 1.

Lange, J., "Interdigitated Stripline Quadrature Hybrid," *Trans. IEEE*, Vol. MTT-17, Dec. 1969, pp. 1150–1151.

Lee, J. J., "G/T and Noise Figure of Active Array Antennas," *Trans. IEEE*, Vol. AP-41, Feb. 1993, pp. 241–244.

Lee, J. J. et al., "Photonic Wideband Array Antennas," *Trans. IEEE*, Vol. AP-43, Sept. 1995, pp. 966–982.

Lee, J. J. et al., "A Multibeam Array Using RF Mixing Feed," *IEEE APS Symposium Digest*, Montreal, July 1997.

Levy, R., "Directional Couplers," in *Advances in Microwaves*, Vol. 1, L. Young, Ed., Academic Press, 1966.

Ng, W. W. et al., "The First Demonstration of an Optically Steered Microwave Phased Array Antenna Using True-Time-Delay," *J. Lightwave Technol.*, Vol. 9, Sept. 1991, pp. 1124–1131.

Ostroff, E. D. et al., *Solid-State Radar Transmitters*, Artech House, 1985.

Park, P. K., "Planar Shunt Slot Array with L-Shaped Series/Series Coupling Slot," *IEEE APS Symposium Digest*, Newport Beach, Calif., June 1995, pp. 564–567.

Riblet, H. J., "Short-Slot Hybrid Junction," *Proc. IRE*, Vol. 40, Feb. 1952, pp. 180–184.

Schiffman, B. M., "A New Class of Broadband Microwave 90° Phase Shifters," *Trans. IRE*, Vol. MTT-6, Apr. 1958, pp. 232–237.

Shimizu, J. K. and Jones, E. M. T., "Coupled-Transmission Line Directional Couplers," *Trans. IRE*, Vol. MTT-6, Oct. 1958, pp. 403–410.

Smith, M. S. and Hall, P. S., "Analysis of Radiation Pattern Effects in Sequentially Rotated Arrays," *Proc. IEE—Microwave Antennas Propag.*, Vol. 141, Aug. 1994, pp. 313–320.

Soref, R. A., "Programmable Time-Delay Devices," *Appl. Opt.*, Vol. 23, Nov. 1, 1984, pp. 3736–3737.

Soref, R. A., "Optical Dispersion Technique for Time-Delay Beam Steering," *Appl. Opt.*, Vol. 31, Dec. 10, 1992, pp. 7395–7397.

Soref, R. A., "Fiber Grating Prism for True Time Delay Beamsteering," *Fiber Integrated Opt.*, Vol. 15, 1996, pp. 325–333.

Spradley, J. L. and Odlum, W. J. "Systematic Errors Caused by the Scanning of Antenna Arrays: Phase Shifters in the Main Feed Line," Rep. AFCRL-56-795, SR11/1317, Hughes Aircraft Co., Culver City, Calif., 1956.

Stark, P. A., *Introduction to Numerical Methods*, Macmillan, 1970.

Tai, C. T. and Long, S. A. "Dipoles and Monopoles," in *Antenna Engineering Handbook*, R. C. Johnson, Ed., McGraw-Hill, 1993.

Tamir, T., Griffel, G., and Bertoni, H. L., *Guided-Wave Optoelectronics—Device Characterization, Analysis, and Design*, Plenum Press, 1995.

Tong, D. T. K. and Wu, M. C., "Programmable Dispersion Matrix Using Bragg Fibre Grating for Optically Controlled Phased Array Antennas," *Electron. Lett.*, Vol. 32, Aug. 15, 1996, pp. 1532–1533.

Tseng, F. I. and Cheng, D. K., "Antenna Pattern Response to Arbitrary Time Signals," *Can. J. Phys.*, Vol. 42, July 1964, pp. 1358–1368.

Tyrell, W. A., "Hybrid Circuits for Microwaves," *Proc. IRE*, Vol. 35, Nov. 1947, pp. 1294–1306.

Watson, W. H., *The Physical Principles of Waveguide Transmission and Antenna Systems*, Oxford University Press, 1947.

Wentworth, R. H., Bodeep, G. E., and Darcie, T. E., *IEEE J. Lightwave Technol.*, Vol. 10, 1992, pp. 84–89.

Whicker, L. R., Ed., Special Issue on Microwave Control Devices for Array Antenna Systems, *Trans. IEEE*, Vol. MTT-22, June 1974, pp. 589–708.

Wilkinson, E. J., "An N-Way Hybrid Power Divider," *Trans. IEEE*, Vol. MTT-8, Jan. 1960, pp. 116–118.

Yee, H. Y., "Impedance of a Narrow Longitudinal Shunt Slot in a Slotted Waveguide Array," *Trans. IEEE*, Vol. AP-22, July 1974, pp. 589–592.

Young, L. B., "Impedance Bridges," in *Technique of Microwave Measurements*, C. G. Montgomery, Ed., MIT Radiation Lab. Series, Vol. 11, McGraw-Hill, 1947, Chapter 9.

Young, L., *Parallel Coupled Lines and Directional Couplers*, Artech House, 1972.

Zmuda, H. and Toughlian, E. N., Eds., *Photonic Aspects of Modern Radar*, Artech House, 1994.

Zmuda, H. et al., "Photonic Beamformer for Phased Array Antennas Using a Fiber Grating Prism," *IEEE Photon. Technol. Lett.*, Vol. 9, Feb. 1997, pp. 241–243.

CHAPTER SEVEN

Mutual Coupling

7.1 INTRODUCTION

It will be shown below that mutual coupling is responsible for all the unique characteristics of phased arrays. The task, then, is to understand it, and to make advantageous use of it.

This chapter is concerned with infinite arrays for two reasons. First, the essential characteristics of all scanning arrays exist in infinite arrays, and are most easily calculated there. Second, most array design starts with an infinite array, with finite array (edge) effects included near the end of the design process. These edge effects are the subject of Chapter 8.

Mutual coupling fundamentals are discussed first. There follow discussions of the basic analysis methods: spatial domain (element-by-element), spectral domain (periodic cell), scatting matrix; all may include moment method techniques. Finally, methods for compensation of *scan impedance* are covered.

7.2 FUNDAMENTALS OF SCANNING ARRAYS

7.2.1 Current Sheet Model

The simplest concept of a phased array is an infinite flat current sheet carrying a uniform current flow parallel to one of the coordinate axes, and is due to Wheeler (1948, 1965). This current is phased to radiate at an angle away from broadside. Such a current sheet can be used as a *gedanken* to derive key scanning properties of phased arrays. The sheet may be either receiving or transmitting, with the two situations giving related behavior. First consider an incident plane wave (receiving) case. Figure 7.1 shows the current sheet in side view, both for a wave incident in the E-plane and in the H-plane. If the current sheet is matched for normal incidence ($\theta = 0$) to η ohms per square, then at other scan angles there is a mismatch. For H-plane incidence the incoming wave "sees" a section of current sheet which is wider than the section

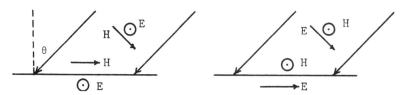

Figure 7.1 Electric current sheet: (a) *H*-plane; (b) *E*-plane.

of wavefront; thus the apparent resistance is lower by $\cos\theta$. The reflection coefficient is now

$$\Gamma = \frac{\cos\theta - 1}{\cos\theta + 1} = -\tan^2\frac{\theta}{2}. \tag{7.1}$$

E-plane incidence gives the opposite result; the incoming wave sees a section of current sheet which is longer than the section of wavefront, and thus the apparent resistance is higher by $1/\cos\theta$. The reflection coefficient is

$$\Gamma = \frac{\sec\theta - 1}{\sec\theta + 1} = \tan^2\frac{\theta}{2}. \tag{7.2}$$

When the current sheet transmits instead of receives, the reflection coefficient signs will be reversed. Figure 7.2 shows the variation of current-sheet resistance for *E*- and *H*-plane scans, for transmitting, with $R_0 = \eta$. Of course, this simple current sheet model gives no information about reactance. The electric current

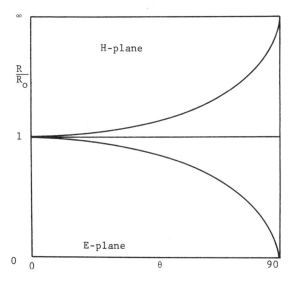

Figure 7.2 Current sheet scan resistance. (Courtesy Wheeler, H. A., "Simple Relations Derived from a Phased-Array Antenna Made of an Infinite Current Sheet," *Trans. IEEE*, Vol. AP-13, July 1965, pp. 506–514.)

sheet requires an open-circuit reflector behind the sheet, where open circuit means $\epsilon = 0$, $\eta = \infty$. Magnetic current sheets may also be considered, with E and H interchanged. Now the reflector behind the sheet is a short circuit $\sigma = \infty$. Thus the magnetic current sheet, unlike the electric current sheet, may be approximated physically. An array of short slots in a metal sheet, for example, provides such an approximation. This simple concept indicates the basic behavior of phased arrays: *scan resistance* increases in one scan plane, but decreases in the other. This trend will be observed in later sections where precise *scan impedances* or admittances are calculated.

Further insight can be obtained from another concept due to Wheeler (1965) namely, that of an ideal element pattern. Again, consider a transmitting current sheet, but let the current sheet be composed of short dipoles with patterns

$$\begin{aligned} H\text{-plane:} \quad & F(\theta) = 1; \\ E\text{-plane:} \quad & F(\theta) = \cos\theta. \end{aligned} \quad (7.3)$$

The patterns are affected by the effective (projected) aperture, which varies as $\cos\theta$, where θ is measured from the normal to the array. Since the effective aperture broadens the beam, the radiation resistance is proportional to $\sec\theta$, so that the net effect is

$$\frac{R}{R_0} = \frac{F^2}{\cos\theta}. \quad (7.4)$$

If the element power pattern were $\cos\theta$, the resistive part of the array impedance would be matched. Thus the "ideal" element pattern is $F(\theta) = \cos^{1/2}\theta$. This is a conical pattern, symmetric about the axis. The ideal element pattern has been approximated by a Huygens source, which is a crossed electric dipole and magnetic dipole, sometimes realized by a dipole and loop. However, the pattern of the Huygens source is

$$F(\theta) = \tfrac{1}{2}(1 + \cos\theta) = \cos^2\frac{\theta}{2} \quad (7.5)$$

and this is only a fair approximation to $\cos^{1/2}\theta$. In the section on scan compensation it will appear that both electric and magnetic modes are needed. The Huygens source is then a crude approximation, giving a single electric and a single magnetic mode.

7.2.2 Free and Forced Excitations

Arrays may be analyzed by either of two viewpoints: free or forced excitation. In the forced excitation model a constant driving voltage (current) is applied to each element, with the element phases adjusted to provide the desired scan angle. Each element has a *scan impedance* (admittance), which is the impedance of an element in an infinite array at the scan angle, and an associated *scan reflection coefficient*. The array element currents (voltages) are the solution of an impedance (admittance) matrix equation

$$[V] = [Z][I]. \tag{7.6}$$

This impedance matrix contains all the interelement mutual impedances Z_{ij}; the matrix is symmetric so $Z_{ij} = Z_{ji}$. The mutual impedances, in principle, are calculated between two elements, with all other elements open circuited. In practice, however, these impedances are almost always calculated with only the two elements present. Results are usually very good; take half-wave dipole elements which when open circuited become quarter-wave wires. These have very small scattering (coupling) cross sections. The overall array pattern is given by the sum over the array elements with the currents as coefficients, all multiplied by the isolated element pattern. From this the *scan element pattern* can be obtained by factoring out the array factor. This type of analysis is relatively easy to carry out. However, implementation of such an array is difficult as each element must be fed by a constant-voltage (current) source. Simple feed networks, in contrast, are of the constant-available-power type where an element impedance mismatch reduces the applied voltage.

Free excitation assumes that each area of the feed network is equivalent to a voltage (current) source in series with R_0. These are thus constant-available-power sources. Such sources, with constant incident power, are suited to a scattering analysis:

$$[V_r] = [S][V_i]. \tag{7.7}$$

Here V_i and V_r are the incident and reflected voltage (current) vectors and S is the scattering matrix of coupling coefficients. The *scan reflection coefficient* is given by

$$\Gamma_s = \sum_p \sum_q S_{00,pq} \frac{A_{pq}}{A_{00}}, \tag{7.8}$$

where S is the coupling coefficient between the "00" element and the "pq" element. The excitation coefficients differ by the progressive scan phase between the two elements, and may also differ in amplitude if a finite array with a tapered excitation is used. Although the scattering approach accurately represents most arrays, and is conceptually simple, there is no direct way of calculating the coupling coefficients. They may be obtained from the impedance (admittance) matrix:

$$[S] = \frac{[Z] - Z_0}{[Z] + Z_0}. \tag{7.9}$$

where the Z_0 applies only to diagonal terms. Measurement of the scattering matrix is performed with all elements present and terminated in matched loads. For arrays of moderate gain elements (e.g., horns in a reflector feed array) this measurement is generally satisfactory. Resonant size elements (dipoles) have coupling in the H-plane that decreases as $1/r^2$, while coupling in the E-plane decreases as $1/r$ (Wheeler, 1959). In an array, however, the coupling decay becomes asymptotic to $1/r^2$ as shown by Hannan (1966), Galindo and Wu (1968), and Steyskal (1974). This is borne out by measurements on large arrays (Amitay et al., 1964; Debski and Hannan, 1965). Phase measurements show that the coupled energy has the phase velocity of free space provided there is no external loading.

In some arrays there is a beam position (other than $\theta = 90$ deg) where the *scan element pattern* has a zero, or in different terms where the *scan reflection coefficient* has magnitude of unity. Such an angle is called a "blind spot." It will be shown later that the appearance of these can be precluded by proper choice of array parameters.

7.2.3 Scan Impedance and Scan Element Pattern

A most important and useful parameter is *scan impedance*[1]; it is the impedance of an element as a function of scan angles, with all elements excited by the proper amplitude and phase. From this the *scan reflection coefficient* is immediately obtained. Array performance is then obtained by multiplying the isolated element power pattern (normalized to 0 dB max) times the isotropic array factor (power) times the impedance mismatch factor $(1 - |\Gamma|^2)$. The isolated element pattern is measured with all other elements open-circuited. This is not quite the same as with all other elements absent, except for canonical minimum scattering antennas (see next section). Here it is assumed that the array is sufficiently large that edge effects are negligible and that *scan impedance* is that for an infinite array. This simple performance expression allows the contributions of array lattice and element spacing, element type, and mutual coupling to be discerned.

An equivalent array performance expression combines the isolated element pattern and the impedance mismatch factor into a new parameter, the *scan element pattern* (SEP) (formerly active element pattern). Now the overall array performance is the product of the isotropic array factor and the *scan element pattern*. The former shows the array beamwidth and sidelobe structure for the scan angle of interest; the latter, like the isolated element pattern, is slowly varying, and shows array gain versus scan angle. Unlike *scan impedance*, which is difficult to measure as all elements must be properly excited, *scan element pattern* is measured with one element excited and all other elements terminated in Z_0. It is important to note that *scan element pattern* provides the radar or communications system designer array gain, at the peak of the scanned beam, versus scan angles. Figure 7.3 pictorially indicates the difference between the two approaches. The *scan element pattern* can be calculated

---

[1] The obsolete term "active impedance" is confusing, and is deprecated.

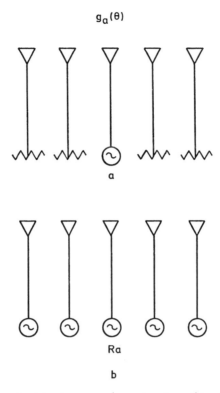

Figure 7.3 Array gain: (a) using scan element pattern; (b) using scan impedance.

from the *scan impedance* and the isolated element pattern as follows. At the peak of the scanned beam the array gain is

$$G(\theta) = Ng_s(\theta) = \frac{4\pi|E(\theta)|^2}{\eta P_{\text{avail}}}, \qquad (7.10)$$

where g_s is the scan element (power) pattern, and P_{avail} is the available power. The electric field can be written in terms of the isolated element pattern, normalized by the scan element current $I_s(\theta)$:

$$E(\theta) = \frac{NI_s(\theta)E_{\text{iso}}(\theta)}{I_{\text{iso}}}. \qquad (7.11)$$

The isotropic element gain is expressed as the ratio of field squared to power:

$$g_{\text{iso}}(\theta) = \frac{4\pi|E_{\text{iso}}(\theta)|^2}{\eta|I_{\text{iso}}|^2 R_{\text{iso}}}. \qquad (7.12)$$

Power available to the array is written in terms of scan current and impedances:

$$P_{\text{avail}} = \frac{NV^2}{4R_g} = \frac{N|I_s(\theta)|^2|Z_s(\theta) + Z_g|^2}{4R_g}. \tag{7.13}$$

Combining these results yields the *scan element pattern* expressed in terms of isolated element gain and impedances (Allen et al., 1962):

$$g_s(\theta) = \frac{4R_g R_{\text{iso}} g_{\text{iso}}(\theta)}{|Z_s(\theta) + Z_g|^2}. \tag{7.14}$$

When the *scan element pattern* is normalized to unity at $\theta = 0$, the SEP can be expressed in terms of only the isolated element power pattern and the *scan impedance*:

$$g_s(\theta) = \frac{4R_s(0) g_{\text{iso}}(\theta)}{|Z_s(\theta) + Z_g|^2 g_{\text{iso}}(0)}. \tag{7.15}$$

For the general case where the generator reactance is not zero, it is appropriate to use the conjugate reflection coefficient, defined as

$$\Gamma_* = \frac{Z_s^*(\theta) - Z_g}{Z_s(\theta) + Z_g}. \tag{7.16}$$

This gives

$$1 - |\Gamma_*|^2 = \frac{4R_g R_s(\theta)}{|Z_s(\theta) + Z_g|^2}, \tag{7.17}$$

which allows the *scan element pattern* to be expressed in terms of the *scan reflection coefficient*, the isolated element pattern, and the *scan resistance*:

$$g_s(\theta) = \frac{R_{\text{iso}} g_{\text{iso}}(\theta)}{R_s(\theta)} (1 - |\Gamma_*|^2). \tag{7.18}$$

Now a generator (source) conjugate matched to the *scan resistance* at $\theta = 0$ allows the final form of SEP to be realized:

$$g_s(\theta) = \frac{R_s(0) g_{\text{iso}}(\theta)}{R_s(\theta) g_{\text{iso}}(0)} [1 - |\Gamma_*(\theta)|^2]. \tag{7.19}$$

Note that only for the special case where the generator impedance is real the SEP can be written as

$$g_s(\theta) = \frac{R_{iso}\, g_{iso}(\theta)}{R_g} |1 - \Gamma(\theta)|^2. \tag{7.20}$$

The previous form is preferred as it explicates the roles of *scan resistance*, isolated element pattern, and *scan reflection coefficient*.

An often quoted approximate result for SEP needs to be examined. This purports to show a $\cos\theta$ variation (Hannan, 1964):

$$g_s(\theta) \simeq \frac{4\pi A_{elem}}{\lambda^2} \cos\theta \bigl(1 - |\Gamma(\theta)|^2\bigr), \tag{7.21}$$

where A_{elem} is the unit cell area of the element. Use of the spectral domain results of Section 7.4.1 for slots and dipoles, and similar results for short slots and dipoles (Oliner and Malech, 1966) show that this expression is exact, provided that no grating lobes exist, the elements are thin and straight, and no higher modes are engendered by the feed. However, the scan behavior of the mismatch factor must not be overlooked. In Section 7.4 it is shown that the *scan element pattern* behavior is closer to $\cos^{1.5}\theta$, the extra power contributed by the impedance mismatch. So it is better to use the generally applicable formula for SEP (Eqn. 7.19) to get accurate and useful results. Extensive *scan impedance* data for dipole arrays are given in Section 7.4.1.

7.2.4 Minimum Scattering Antennas

In understanding the relationship between actual and ideal array element impedances, the concepts of minimum scattering antennas (MSA) and canonical minimum scattering antenna (CMS) are useful (Montgomery, 1947; Kahn and Kurss, 1965; Gately et al., 1968; Wasylkiwskyj and Kahn, 1970). Consider an antenna with N ports; for example, these ports might be the dominant mode and higher modes (usually evanescent) in an open-ended waveguide element. When each port of an MSA is terminated in the proper reactance, the scattered power S_{ii} is zero. The CMS antenna is a special type of MSA in that an open circuit at each port produces $S_{ii} = 0$. The CMS antenna is lossless, which implies that the scattering matrix is unitary. The scattering matrix can be written as

$$[S] = \left[\begin{array}{c|c} 0 & S_{ij}^+ \\ \hline S_{ij} & 1 - S_{ij} S_{ij}^+ \end{array} \right]. \tag{7.22}$$

(Note that $+$ denotes the conjugate transpose.) This N-port antenna possess N orthogonal radiation patterns, which are the S_{ij}. In the absence of non-reciprocal components, e.g., ferrites, the antenna is reciprocal, which makes all patterns real and symmetric about the origin. Scattered and radiated patterns are equal. No impedance or admittance matrix exists, and the N eigenvalues of the scattering matrix are all -1. Unlike most scatterers (or antennas) the scattered field pattern of a CMS antenna is independent of the incident-field direction, although the amplitude of the pattern will depend upon the incident wave. If, then, the pattern of an element is taken in the array environment with all other elements open circuited, this pattern is exactly the isolated array pattern only if the elements are CMS antennas. For example, when dipoles in an array are open-circuited, there is a small effect due to the half-length conducting rods that remain when each dipole is open-circuited. As expected, half-wave dipoles are not CMS antennas. Similarly, when slots in a ground plane are shorted across the feed terminals, the remaining half-length slots affect the current distributions on the ground plane.

Wasylkiwskyj and Kahn (1970) showed that the mutual impedance (admittance) between two identical MSAs can be written as an integral of the power pattern of an isolated element over certain real and complex angles (see also Bamford et al., 1993):

$$Z_{12} = 2 \int_{\text{path}} \exp(-jkd \cos w) P(w) \, dw, \qquad (7.23)$$

where $w = u + jv$, and the power pattern is defined as $P(w) = f(w)f(w + \pi)$ and the integration path in the complex w plane is from $(-\pi/2, -\infty)$ to $(-\pi/2, 0)$ to $(3/2, 0)$ to $(3/2, \infty)$. The power pattern is normalized so that its integration over all real angles is unity. This integral can also be expressed in wavenumber coordinates. For example, with two antennas in the xy plane, the mutual impedance is given by

$$Z_{12} = \frac{2}{k} \int_{-\infty}^{\infty} \int_{-\infty}^{\infty} \frac{\exp[-j(k_x d_x + k_y d_y)]}{\sqrt{k^2 - k_x^2 - k_y^2}} P(k_x, k_y) \, dk_x \, dk_y. \qquad (7.24)$$

Mutual coupling, then, is specified by the element pattern and the lattice, and is completely independent of the means utilized to produce that pattern. This development for MSA is a generalization of that of Borgiotti which is discussed in the next section under grating lobe series.

Most multimode elements do not have orthogonal patterns, and hence are not MSAs. However, many single-mode antennas approximate to an MSA. Short dipoles, where the current distribution is essentially linear, are closely MSA. A resonant (near half-wave) dipole is approximately MSA if the radius is very thin; this thinness forces the current to be nearly sinusoidal. Mutual impedance between thin collinear half-wave dipoles is shown to agree with the Carter results (Wasylkiwskyj and Kahn, 1970). Andersen et al. (1974) show that crossed dipoles are approximately MSA in one pattern plane, and that the mutual impedance calculated from MSA theory is good if the center of one cross lies on a line bisecting the arms of the other cross. Small helices are also approximately MSA.

7.3 SPATIAL DOMAIN APPROACHES TO MUTUAL COUPLING

Except for arrays with only a few elements per side, array design is based on large or infinite array theory. This predicts the behavior of all except the edge elements; these are treated separately and subsequently; see Chapter 8. In the spatial domain, the array is simulated by an impedance (admittance) matrix that relates the voltages (currents) applied to the dipoles (slots) and the resulting currents (voltages). For most scan angle ranges a modest array size will provide *scan impedance* and *scan element pattern* that are nearly independent of that size. Before proceeding to this element-by-element approach, the mutual impedance between two elements will be examined; as seen in the preceding sections, mutual impedance is both the cause and explanation of all scanned array effects. These mutual impedances become the elements of the array impedance matrix.

7.3.1 Canonical Couplings

7.3.1.1 Dipole and Slot Mutual Impedance. There are several intuitive rules that concern mutual coupling between two antenna elements. First, the magnitude of mutual impedance decreases with distance between the elements. Second, if one antenna is in the pattern maximum of the other, coupling will be strong compared to that which the pattern null produces. If the antenna is in the radiating near-field of the other, then the near-field pattern is used. Third, if the electric fields are parallel, coupling will be stronger than if they are collinear. For wire antennas this can be restated in terms of shadowing: large shadowing will correlate with large coupling. Fourth, larger antennas have smaller coupling. For example, large horns have lower mutual impedance than small horns. These rules, although useful, are no substitute for obtaining actual values of mutual impedance or admittance. The mutual impedance between dipoles is covered here in depth, as dipoles and slots are common array elements. These impedances are typical of resonant element mutual impedances, such as microstrip patches[2], so the dipole mutual impedance behavior is of general usefulness. Thin cylindrical dipoles are equivalent to flat strip dipoles, where the strip width is twice the wire diameter, and slot mutual admittance data can be derived from strip dipole results via Babinet's principle.

For mutual impedance the zero-order (sinusoidal current distribution) theory is usually adequate. For this case, the formulas of King (1957) based on the original work of Carter (1932) can be used. They give mutual impedance between parallel coplanar dipoles of unequal length in echelon. However, these formulas contain 24 Sine and Cosine Integral terms, of various complicated and diverse arguments. Although the results of King could be rearranged and grouped into a computer algorithm, it is simpler to derive the formula

---

[2] For large substrate thickness, patch mutual impedance has a surface wave component so that the dipole results are no longer similar.

directly, using the exact near-electric-field formulations of Schelkunoff and Friis (1952). The exact field from a sinusoidal current is written as three spherical waves, the field components in cylindrical coordinates. Mutual impedance is then the integral of this electric field (from each half-dipole) times the current distribution at the other dipole. With the geometry of Fig. 7.4 the mutual impedance expression is

$$Z = \frac{-j30}{S_1 S_2} \int_0^{d_1} [\Psi_1 - 2\Psi_2 \cos(kd_2) + \Psi_3 + \Psi_4 - 2\Psi_5 \cos(kd_2) + \Psi_6]$$
$$\times \sin[k(d_1 - x)] \, dx, \qquad (7.25)$$

where

$$S_1 = \sin(kd_1), \qquad S_2 = \sin(kd_2), \qquad \Psi_i = \exp(-jkR_i)/R_i.$$

The R values are

$$\begin{aligned}
R_1^2 &= y_0^2 + (x_0 + d_2 - x)^2, \\
R_2^2 &= y_0^2 + (x_0 - x)^2, \\
R_3^2 &= y_0^2 + (x_0 - d_2 - x)^2, \\
R_4^2 &= y_0^2 + (x_0 - d_2 + x)^2, \\
R_5^2 &= y_0^2 + (x_0 + x)^2, \\
R_6^2 &= y_0^2 + (x_0 + d_2 + x)^2.
\end{aligned} \qquad (7.26)$$

The integral could be evaluated numerically, and for large dipole separations this method is preferable. However, the adjacent and nearby dipoles are most important, and for these an exact evaluation is necessary.

The exact solution can be written as two sums (Hansen and Brunner, 1979), where n steps by increments of 2:

$$Z = \frac{-15}{S_1 S_2} \sum_{m=-1}^{1} \sum_{n=-1}^{1,2} C_m \{\exp(jkU)[E(kA + kU) - E(kB + kV)]$$
$$+ \exp(-jkU)[E(kA - kU) - E(kB - kV)]\}. \qquad (7.27)$$

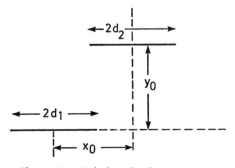

Figure 7.4 Echelon dipole geometry.

The coefficients are

$$C_{-1} = C_1 = 1, \qquad C_0 = 2\cos kd_2,$$

and $E(x)$ is the Exponential Integral (Abramowitz and Stegun, 1970) with arguments:

$$\begin{aligned}
U &= d_1 + n(x_0 + md_2), \\
V &= n(x_0 + md_2), \\
A^2 &= y_0^2 + (x_0 + nd_2 + md_2)^2, \\
B^2 &= y_0^2 + (x_0 + md_2)^2.
\end{aligned} \qquad (7.28)$$

When the two dipoles are the same length the formula simplifies (Hansen, 1972) to

$$Z = \frac{15}{S^2} \sum_{m=-2}^{2} \sum_{n=-1}^{1,2} A_m \exp[-jkn(x_0 + md)] E(k\beta), \qquad (7.29)$$

where

$$\beta = \sqrt{(x_0 + md)^2 + y_0^2} - n(x_0 + md)$$

and the coefficients are

$$\begin{aligned}
A_{-2} &= A_2 = 1, & A_{-1} &= A_1 = -4\cos kd, \\
A_0 &= 2(1 + 2\cos^2 kd), & S &= \sin kd.
\end{aligned} \qquad (7.30)$$

$E(x)$ can be expressed as the Cosine and Sine Integrals, Ci and Si. When $y_0 = 0$, i.e., the dipoles are collinear, y_0 is replaced by the dipole radius a. When the dipoles are half-wave, the $m = 2$ and 4 terms disappear and the exponential simplifies. This formulation is similar to that of Richmond (1970) for two equal-length thin dipoles with axes at an angle. Computer subroutines for Ci, Si are readily available in most computer libraries, with computation time comparable to that of trig functions. These subroutines use the economized series developed by Wimp and Luke (see Wimp, 1961).

Figure 7.5 shows Z_{ij} between two parallel half-wave dipoles on an impedance plot. The curve, with spacing/wavelength as a parameter, is similar to a Cornu spiral. It can be seen that the magnitude of impedance decreases as spacing increases. Data are normalized by the self-resistance, which, for a zero-order (zero-thickness) dipole, is 73.13 ohms. Figure 7.6 is a similar plot for collinear dipoles. The spacing here is between dipole tips; for center-to-center spacing add 0.5. As expected, the collinear coupling is less owing both to lower shadowing and a pattern null. Recall, however, that the dipole near-field

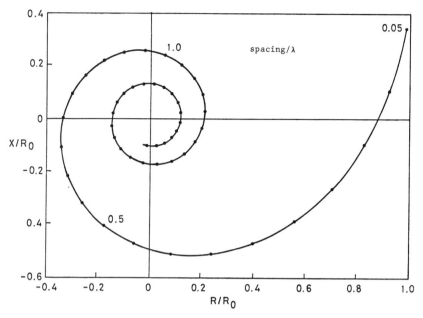

Figure 7.5 Parallel dipole mutual impedance.

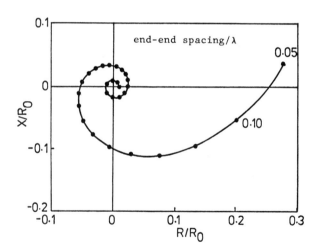

Figure 7.6 Collinear dipole mutual impedance.

has both axial and radial electric fields and that the latter does not have a pattern null along the axis.

When slots are located in a stripline surface or in contiguous waveguides the mutual admittance between two slots behaves as though the slots are located in an infinite ground plane. The slots can then be replaced by strip dipoles by Babinet's principle developed by Booker (1946) for both impedance and mutual impedance (Begovich, 1950). The slot mutual impedance Z_{12} and strip dipole mutual admittance, Y_{12} are related:

$$2Z_{12} = \eta^2 Y_{12}, \tag{7.31}$$

where $\eta = 120\pi$. The factor of 2 has been reduced from 4 as the slots radiate one side only.

When the slots are narrow, which is usually the case, the equivalent strip dipoles are also narrow; they can be accurately replaced by cylindrical dipoles of radius equal to one-fourth the slot or strip dipole width (Lo, 1953). Thus the extensive development of dipole mutual impedance can be used for slots.

7.3.1.2 Microstrip Patch Mutual Impedance.
Calculation of mutual impedance between two microstrip patches must include the effects of the grounded (dielectric) substrate, typically via a Green's function. This requires one, or often several, integrations of the Sommerfeld type, plagued by oscillating and unbounded behavior on the real axis, and branch cuts, surface wave, and other poles for the complex integration. The simplest procedure is based on the transmission line model; however, the patch edge gap formulation is critical. In general this model has been superseded by more accurate models. These are of two types: cavity model and moment method model. Both utilize the rigorous Green's function for the slab.

The moment method model establishes current expansion functions on the patch, such as rooftop functions; these and the test functions allow the integral equation to be discretized (Pozar, 1982; Newman et al., 1983). Several Green's function integrations are usually involved.

In the cavity model, the antennas are replaced by a grounded dielectric slab with magnetic current distributions at each patch cavity open wall. Calculations by Haneishi and Suzuki (1989) show the real and imaginary parts of mutual admittance oscillating and decaying with increasing separation, with the nulls of one occurring at the peaks of the other. The curves are close to those for dipole mutual impedance. An advantage of this model is that a single Green's function integration is needed, and this can be expedited via FFT (Mohammadian et al., 1989). This approach appears to be as accurate as the moment method approach, although comparative data for higher ϵ have not been available. Both sets of results compare well with L-band measurements of Jedlicka et al. (1981). Only a small change was seen as the thickness was doubled.

These measurements were repeated by Mohammadian et al. (1989) with good agreement; measurements were also made at 5 GHz on patches $0.282\lambda_0 \times 0.267\lambda_0$ with substrate of $\epsilon = 2.55$ and $t = 0.0252\lambda_0$, and were compared with calculations. Figure 7.7 again shows good agreement. It may therefore be concluded that a relatively simple and accurate method exists for calculation of patch mutual admittance.

There have been many papers, over many years, on the subject of evaluation of Green's function integrals. Here a few references are given: Pearson (1983);

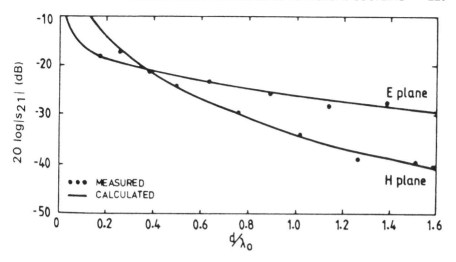

Figure 7.7 Patch coupling. (Courtesy Mohammadian, A. H., Martin, N. M., and Griffin, D. W., "A Theoretical and Experimental Study of Mutual Coupling in Microstrip Antenna Arrays," *Trans. IEEE*, Vol. AP-37, Oct. 1989, pp. 1217–1223.)

Johnson and Dudley (1983); Marin et al. (1989); Chew (1989); Barkeshli et al. (1990); Barkeshli and Pathak (1990); and Marin and Pathak (1992).

7.3.1.3 Horn Mutual Impedance. The limiting case of a waveguide horn is the open-end waveguide radiator. A single-mode excitation in one element tends to excite evanescent modes in addition to propagating modes in the second guide. Most analyses utilize aperture integration; see Luzwick and Harrington (1982) and Bird (1987) for rectangular guide elements; Bird (1990) for rectangular elements of different sizes; Bird and Bateman (1994) for rotated rectangular elements; Bailey and Bostian (1974), Bailey (1974), and Bird (1979) for circular guide elements.

Coupling between E-plane sectoral horns has been measured by Lyon et al. (1964). Horns of 8 dB gain with a separation of 0.73λ showed low coupling as seen in Fig. 7.8. The E- and H-plane couplings alternate peaks and valleys as the orientation angle is changed. For large spacing the parallel polarization decays as $1/r^2$, or 12 dB for each doubling of distance r, while the collinear coupling decays as $1/r$. Coupling data for pyramidal horns of a separation of 5.93λ are shown in Fig. 7.9.

In a different approach, Hamid (1967) used Geometric Theory of Diffraction (GTD) to calculate mutual coupling between sectoral horns with 3λ apertures; the coupling showed a typical peak and null behavior with center-to-center separations, with the envelope varying roughly as coupling $\simeq (-49 - s/\lambda)$ dB. Extensive waveguide horn data will be provided in a forthcoming book *Mutual Coupling Between Antennas* (Peter Peregrinus) by Dr. Trevor Bird of CSIRO.

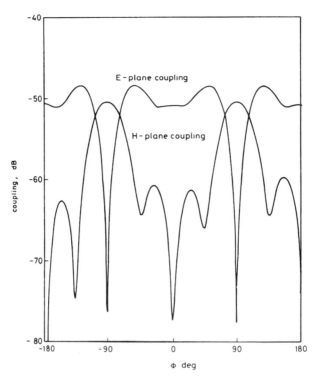

Figure 7.8 *E*-plane sectoral horn coupling. (Courtesy Lyon, J. A. M. et al., "Interference Coupling Factors for Pairs of Antennas," *Proceedings 1964 USAF Antenna Symposium,* Allerton, Ill., AD-609 104.)

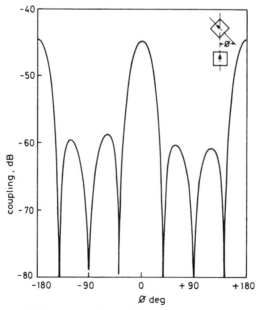

Figure 7.9 Pyramidal horn coupling. (Courtesy Lyon, J. A. M. et al., "Interference Coupling Factors for Pairs of Antennas," *Proceedings 1964 USAF Antenna Symposium,* Allerton, Ill., AD-A609 104.)

7.3.2 Impedance Matrix Solution

Finite arrays of dipoles, slots, or patches can be analyzed simply by setting up a mutual impedance matrix. Carter-type mutual impedances (from the previous sections) are usually adequate. If an array uses dipoles over a ground plane, it is replaced by an image array, which makes the impedance matrix $2N \times 2N$ instead of $N \times N$. With a symmetric array and no scan the matrix can be folded to reduce its size to roughly half (exactly half if N is even). The scan angle appears in the progressive phase of the drive vector, of which the component from the nth row and mth column is

$$V_i = A_{nm} \exp\left[-j2\pi(nu + mv)\right]. \tag{7.32}$$

As before, the steering vector components are

$$u = \frac{d_x}{\lambda} \sin\theta \cos\phi, \qquad v = \frac{d_y}{\lambda} \sin\theta \sin\phi. \tag{7.33}$$

The mutual impedance Z_{ij} is between the ith element and the jth element. When the complex simultaneous equations are solved[3] the result is the current vector

$$[V_i] = [I_j][Z_{ij}] \tag{7.34}$$

and from the current vector the *scan impedance* of each element in the array is determined. When the solution is desired for several scan angles, so that only the drive vector changes but not the Z_{ij} matrix, a simultaneous equation solver can be used that stores a diagonalized matrix, allowing a rapid solution for new vectors after the first. From the current vector, of course, array patterns can be calculated. The impedance matrix solution has an intrinsic advantage: array excitation tapers can be incorporated directly into the drive vector, unlike the infinite array methods of Section 7.4 where the excitation must be uniform.

For large arrays the *scan impedance* results are essentially the same as those for infinite arrays (see Section 7.4) except for scan angles near 90 deg. Diamond (1968) calculated an array of 65×149 half-wave dipoles with screen; no difference was noticed until the scan angle exceeded 70 deg, probably because there are more "edge elements" for larger scan angles.

Multimode elements in a finite array can, in principle, be handled by increasing the size of the matrix from $N \times N$ to $NM \times NM$, where each element has M modes, for example. Since the computer time for solution of simultaneous equations varies as the cube of the matrix size, large arrays may be difficult. An alternative solution has been offered by Goldberg (1972). The mode voltage at each element is written as a sum (over the elements) of drive voltages times

---

[3]Note that solving simultaneous equations is roughly three times faster than matrix inversion (see Westlake, 1968).

232 MUTUAL COUPLING

coupling coefficients times scan phase factor. The coupling coefficients are approximately independent of array size (strictly true only for CMS elements) and may be determined from an infinite array. An infinite array with the proper spacing, scan range, and bandwidth is then designed as described elsewhere in this chapter; the results are mode voltages and *scan reflection coefficient* versus u. A Fourier inversion is used to obtain the coupling coefficients, which are then summed to give a set of mode voltages and reflection coefficient for each element in the array. Only the center elements in a large array experience essentially all the coupling; elements at or near the edge employ fewer strong coupling coefficients as they decay rapidly with element separation except near grating lobe incidence. Effects of amplitude taper can be determined simply by including the amplitude coefficients in each sum. From the reflection coefficient of each element, the *scan impedance* can be written, and the pattern of the array is found by adding the modal array patterns which use element modal patterns and mode voltages.

For large arrays the *scan impedance* and *scan element pattern* are those of infinite arrays (see Section 7.4), with small oscillations; larger arrays have more oscillations and they are of smaller amplitude. Data for small arrays are given in Chapter 8.

7.3.3 The Grating Lobe Series

An intuitive understanding of the role of mutual coupling in array scanning is facilitated by the grating lobe series, which links the current sheet concepts of this section with the Floquet series approach of the next section. It could be discussed along with the periodic cell, Floquet material, but the kernel of the grating lobe series relates to an array element; thus it is located here.

It was observed by Wheeler (1966) that each term in the double series for dipole *scan impedance* is associated with a point in the u,v grating lobe lattice, e.g., Fig. 7.10. This is true for all types of elements in a regular lattice, but is most easily visualized by considering short dipoles. For short dipoles, the formulas for dipoles of general length simplify because the sinc and cosc factors become unity. If now Eqns. (7.41) and (7.42) (derived later in Section 7.4) are divided by the broadside *scan resistance*, the normalized impedance becomes

$$\frac{R(u,v)}{R(0,0)} = H_{00} = \frac{k^2 - k_{x0}^2}{k\sqrt{k^2 - k_{x0}^2 - k_{y0}^2}},$$

$$\frac{X(u,v)}{R(0,0)} = -\sum\sum H_{nm} = \sum\sum \frac{k^2 - k_{xn}^2}{k\sqrt{k_{xn}^2 + k_{ym}^2 - k^2}}.$$

(7.35)

Here the expressions for H_{00} and H_{nm} are identical with Eqn. (7.40). Wheeler correlated these terms with the u,v grating lobe points, and identified each term as an impedance crater, a concept developed by Rhodes (1974). The normalized impedance becomes

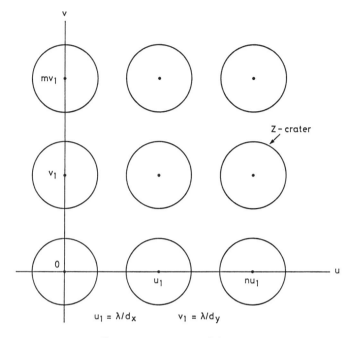

Figure 7.10 Grating lobe craters.

$$\frac{Z_s(u,v)}{R_s(0,0)} = \sum\sum \frac{Z_c(u - n/d_x, v - m/d_y)}{R_s(0,0)}. \tag{7.36}$$

Here Z_s/R_s is the impedance crater (Frazita, 1967), which is a plot of the normalized impedance of an element in u,v space. Figure 7.11 shows the crater for a short dipole and Fig. 7.12 is the crater for a half-wave slot. Wheeler discovered that mutual-coupling effects with scan, and the effects associated with grating lobe onset, could be visualized by placing an impedance crater at each (inverse) lattice point in the u,v plane. The term "crater" arises because the resistance presents the topological form of a crater. Thus the grating lobe series is a link between the current sheet concepts discussed earlier and the periodic structure approach.

The grating lobe series can also be expressed in terms of transforms of aperture fields (Borgiotti, 1968). *Scan admittance* of an infinite array on a rectangular lattice is written as an integral in wavenumber space of the Fourier transform of the aperture field times the FT of the conjugate field for each element. The doubly infinite sum representing element phases can be expressed as a doubly periodic delta function, which reduces the integrals to

$$Y_s = \frac{4\pi\lambda^2}{k\eta d_x d_y} \sum_{-\infty}^{\infty}\sum_{-\infty}^{\infty} \left[\frac{k^2}{w} E_\rho E_\rho^* + w E_\psi E_\psi^*\right], \tag{7.37}$$

234 MUTUAL COUPLING

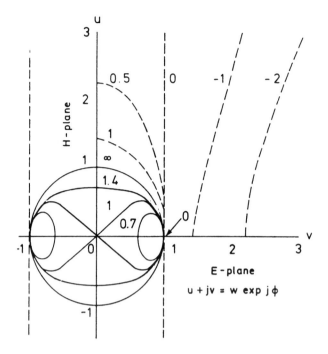

Figure 7.11 Impedance crater of short dipole. (Courtesy Wheeler, H. A., "The Grating-Lobe Series for the Impedance Variation in a Planar Phased-Array Antenna," *Trans. IEEE*, Vol. AP-14, Nov. 1966, pp. 707–714.)

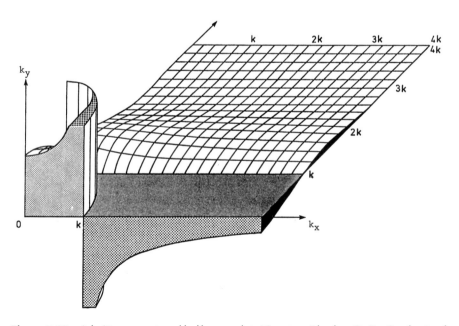

Figure 7.12 Admittance crater of half-wave slot. (Courtesy Rhodes, D. R., *Synthesis of Planar Antenna Sources*, Clarendon Press, 1974.)

where $w = -j\sqrt{u^2 + v^2 - k^2}$ and the fields are evaluated at the u,v grating lobe points. Mutual impedance between two elements can similarly be written as an integral in wavenumber space of aperture of distribution FTs, which can be reduced to the aperture complex power equations developed by Borgiotti (1963) and Rhodes (1964).

7.4 SPECTRAL DOMAIN APPROACHES

7.4.1 Dipoles and Slots

Floquet's theorem for arrays states that an infinite regular periodic structure will have the same fields in each cell except for a progressive exponential multiplier (Amitay et al., 1972; Catedra et al., 1995). Further, the fields may be described as a set of orthogonal modes. In essence the boundary conditions are matched in the Fourier transform domain, resulting in some cases in an integral equation reducing to an algebraic equation. Wheeler (1948) was probably the first to develop the unit cell or periodic cell for fixed beam arrays, where each element is contained in a cell, with all cells alike and contiguous. Edelberg and Oliner (1960) and Oliner and Malech (1966) extended the periodic cell approach to scanned arrays. It has proved to be the most powerful and perceptive technique for understanding and for designing sophisticated arrays. An equivalent development is by Diamond (1968). Because of the Floquet symmetry, the single unit cell contains the complete admittance behavior of each element in the array. The unit cells are normal to the array face, centered about the elements, and contiguous; see Fig. 7.13. For broadside radiation with no grating lobes the unit cell has two opposite electric walls (zero electric field) and two magnetic walls (zero magnetic field) with a TEM mode. However, when the beam is scanned, such a unit cell must be tilted along with the beam, and another unit cell must be added for each grating lobe.

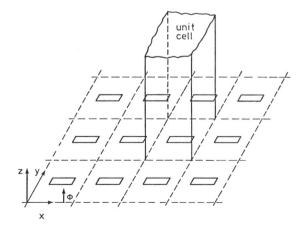

Figure 7.13 Slot array unit cell.

Because of this complexity, Oliner utilized a fixed normal unit cell that would accommodate any scan angle and any number of grating lobes. The unit cell walls are not in general electric or magnetic, and the modes inside are LSE and/or LSM, depending upon the scan plane. Opposite walls support fields that differ by the Floquet phase shift. Each main lobe or grating lobe is represented by a propagating mode, with the evanescent modes contributing energy storage (susceptance) at the array face.

Although short slots or dipoles are simpler to analyze by the periodic-cell method, the resonant dipole is of most importance; so this case will be carried through in some detail. Scanning in a principal plane is also simpler, but arbitrary scan will be considered here. For this case all four unit cell walls are scan dependent, with both E_z and H_z field components. A superposition of the dominant LSE and dominant LSM modes is necessary. The wavenumber in the z-direction is

$$k_z = \sqrt{k^2 - k_{x0}^2 - k_{y0}^2} = k \cos \theta_0, \qquad (7.38)$$

where

$$k_{x0} = k \sin \theta_0 \cos \phi_0, \qquad k_{y0} = k \sin \theta_0 \sin \phi_0.$$

The unit cell dimensions are $D_x = d_x/\lambda$ and $D_y = d_y/\lambda$, where the element lattice spacings are d_x and d_y. The generalized wave numbers are

$$k_{xn} = k\left(\sin \theta_0 \cos \phi_0 + \frac{n}{d_x}\right),$$
$$k_{ym} = k\left(\sin \theta_0 \sin \phi_0 + \frac{m}{d_y}\right), \qquad (7.39)$$

and the generalized scan factor is

$$H_{nm} = \frac{\left(\sin \theta_0 \cos \phi_0 + \dfrac{n}{d_x}\right)^2 - 1}{\sqrt{\left(\sin \theta_0 \cos \phi_0 + \dfrac{n}{d_x}\right)^2 + \left(\sin^2 \theta_0 \sin^2 \phi_0 + \dfrac{m}{d_y}\right)^2 - 1}}, \qquad (7.40)$$

$$H_{00} = \frac{1 - \sin^2 \theta_0 \cos^2 \phi_0}{\cos \theta_0}.$$

Assuming dipole feed terminals at the center of the dipole, the reactance is in parallel with the unit-cell waveguide, but in series with the feed terminals. Consideration of the ground plane displays one of the advantages of the unit-cell approach. Whereas the dipole array in free space is represented by a unit cell one each side of the dipole, with both unit cells semiinfinite in length, the ground plane is simply represented by a short across one of the unit cells at height h from the dipole. Since the ground plane to dipole spacing is usually of the order of $\lambda/4$, the higher-order unit-cell evanescent modes are sufficiently damped out in travelling from the dipole to the ground plane and back that they can be neglected. However, the ground plane affects both resistance and reactance through the dominant mode. Figure 7.14 shows the dimensions. The dipole current is assumed to flow only in the x-direction and to be constant in the y-direction. Along x, a cosine distribution is assumed. A match is then made between the electric field in the dipole plane produced by the dipole and that produced by a set of unit-cell waveguide modes. From this, the series for *scan reactance* is obtained. For the *scan resistance* the ground plane is replaced by an image dipole at distance $2h$. The *scan resistance* is

$$R_s = \frac{480 a_s^2}{\pi d_x d_y \lambda^2} \cosc^2 \frac{k_{x0} a_s}{2} \sinc^2 \frac{k_{y0} b_s}{2} \sin^2(k_{z0} h) H_{00}. \qquad (7.41)$$

When a ground plane is not present, the $\sin^2 k_{z0} h$ factor is replaced by $1/2$. The expression for reactance is

$$X_s = \frac{240 a_s^2}{\pi d_x d_y \lambda^2} \left[\cosc^2 \frac{k_{x0} a_s}{2} \sinc^2 \frac{k_{y0} b_s}{2} \sin^2(k_{z0} h) H_{00} \right.$$
$$\left. - \sum \sum \cosc^2 \frac{k_{xn} a_s}{2} \sinc^2 \frac{k_{ym} b_s}{2} H_{nm} \right]. \qquad (7.42)$$

The sinc and cosc factors can be recognized as the Fourier transforms of the uniform and cosine current distributions.

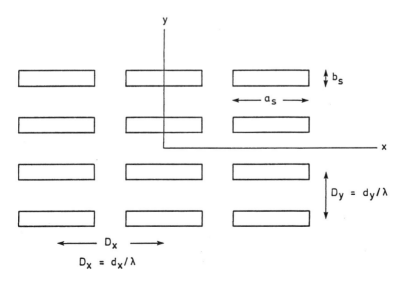

Figure 7.14 Strip dipole array geometry.

238 MUTUAL COUPLING

Graphs of *scan impedance* and *scan element pattern* give insight into how arrays work as scan angle and lattice spacing are changed. Figures 7.15 through 7.17 show SEP and *scan impedance* for thin half-wave dipoles on a half-wave square lattice. As expected the H-plane *scan resistance* and *scan reactance* increase with scan angle, while they decrease in the E-plane. Diagonal plane behavior is mixed. The SEP shows more decrease with scan angle in the E-plane, owing to the increasingly poor mismatch. A comparison of the SEP with powers of $\cos\theta$ shows that the best fit is $\cos^{1.5}\theta$; this power pattern lies roughly between the H-plane and E-plane curves. Note that slot parameters are obtained from those of the dipole by multiplying by a constant (Oliner and Malech, 1966). Adding a ground plane raises the broadside *scan resistance*, but most important removes the H-plane trend to infinity; see Figs. 7.18 through 7.20. This is offset by the screen pattern factor, with the result that the SEP for dipoles/screen (Fig. 7.20) is worse for all planes than that of the dipole array (Fig. 7.17). Again, the $\cos^{1.5}$ power pattern is a good fit out to about 50 deg scan; beyond this the SEP falls rapidly owing to the screen factor.

Impedance behavior at grating lobe incidence is illuminating; the cases just discussed are repeated for a square lattice of 0.7λ, which allows a grating lobe to occur at $\theta = 25.38$ deg. Figures 7.21 through 7.23 show *scan impedance* and SEP for a dipole array; both parts of *scan impedance* have an infinite singularity at the grating lobe angle. The SEP shows a blind angle there for H-plane scan. The E-plane and diagonal plane scans are only slightly affected. Adding a screen replaces the infinite singularities by a steep but finite jump in *scan resistance*, and a cusp in *scan reactance*; see Figs. 7.24 through 7.26. The

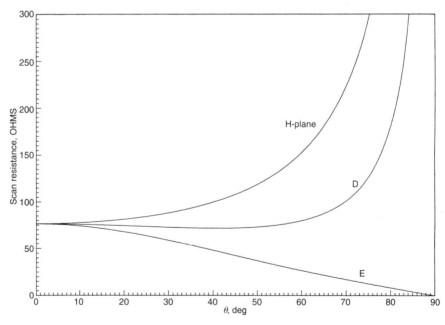

Figure 7.15 Dipole array, $D_x = D_y = 0.5$, $L = 0.5\lambda$.

SPECTRAL DOMAIN APPROACHES   239

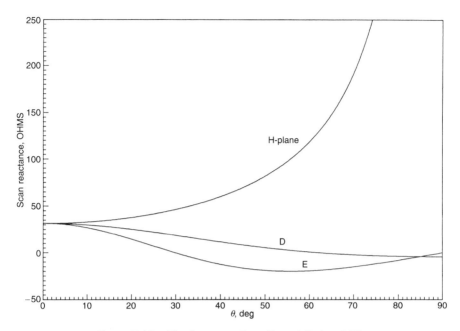

Figure 7.16 Dipole array, $D_x = D_y = 0.5$, $L = 0.5\lambda$.

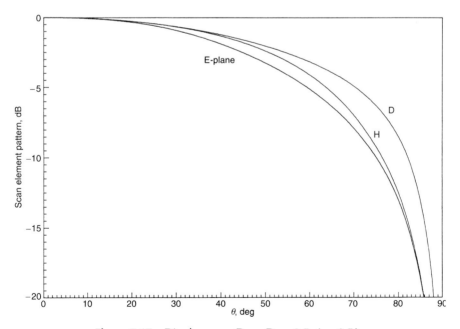

Figure 7.17 Dipole array, $D_x = D_y = 0.5$, $L = 0.5\lambda$.

240 MUTUAL COUPLING

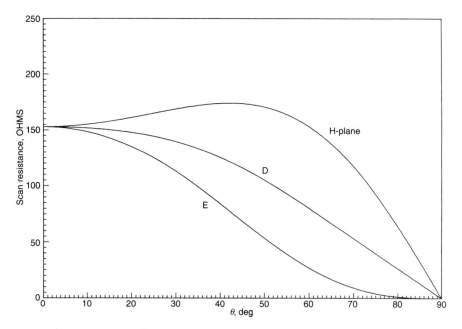

Figure 7.18 Dipole/screen array, $D_x = D_y = 0.5$, $L = 0.5\lambda$, $H = 0.25\lambda$.

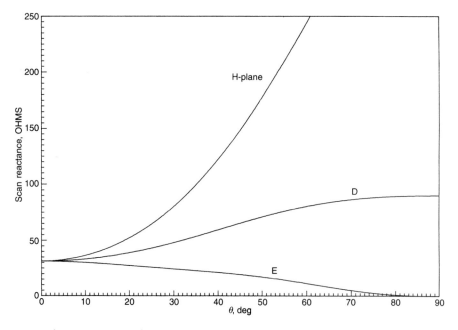

Figure 7.19 Dipole/screen array, $D_x = D_y = 0.5$, $L = 0.5\lambda$, $H = 0.25\lambda$.

SPECTRAL DOMAIN APPROACHES 241

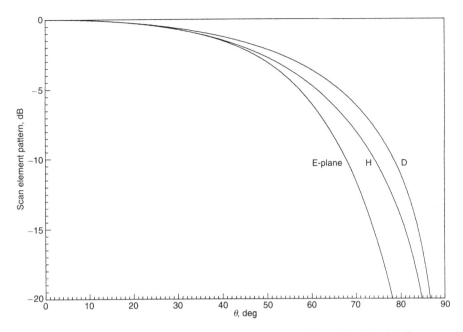

Figure 7.20 Dipole/screen array, $D_x = D_y = 0.5$, $L = 0.5\lambda$, $H = 0.25\lambda$.

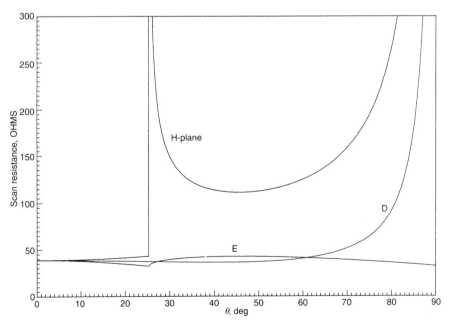

Figure 7.21 Dipole array, $D_x = D_y = 0.7$, $L = 0.5\lambda$.

242 MUTUAL COUPLING

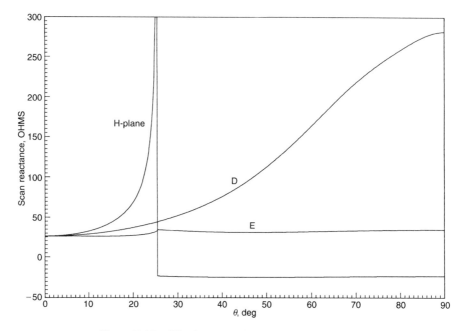

Figure 7.22 Dipole array, $D_x = D_y = 0.7$, $L = 0.5\lambda$.

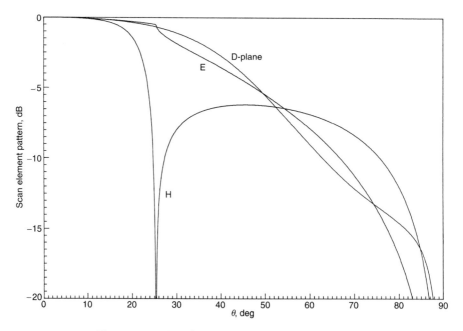

Figure 7.23 Dipole array, $D_x = D_y = 0.7$, $L = 0.5\lambda$.

SPECTRAL DOMAIN APPROACHES   **243**

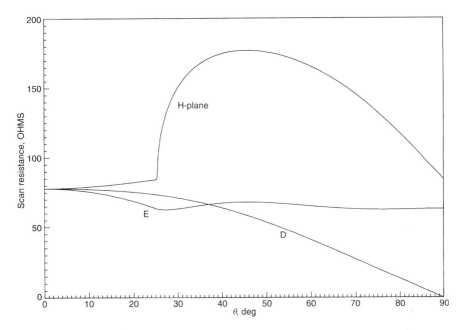

Figure 7.24   Dipole/screen array, $D_x = D_y = 0.7$, $L = 0.5\lambda$, $H = 0.25\lambda$.

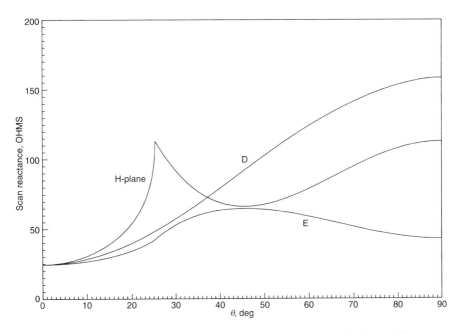

Figure 7.25   Dipole/screen array, $D_x = D_y = 0.7$, $L = 0.5\lambda$, $H = 0.25\lambda$.

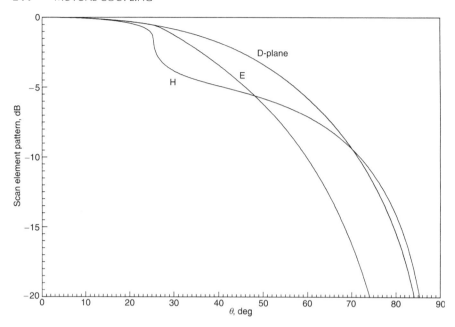

Figure 7.26 Dipole/screen array, $D_x = D_y = 0.7$, $L = 0.5\lambda$, $H = 0.25\lambda$.

SEP has a roughly 3 dB dip at grating lobe incidence but the other planes are not much affected. Clearly for wideband operation, a screen is essential. Behavior for other lattice spacings is similar, about the grating lobe angle(s). The blind angle occurs because the reflection coefficient there is of unity magnitude. These data have been compared with spatial domain calculations on a large array using Carter mutual impedance data, with perfect agreement (Oliner and Malech, 1966). Typically 900 (30 × 90) terms have been used in the reactance series. All of the curves presented are for a dipole radius of 0.001λ; fatter dipoles reduce the broadside values of *scan impedance*, but the changes decrease with scan angle.

These spectral domain results can be derived directly from the Carter dipole mutual impedances. Consider an array of collinear coplanar thin dipoles along the y-axis and spaced by d_x and d_y. *scan impedance* is obtained as the sum of mutual impedances between the array dipoles and another dipole spaced x_0 and y_0 from one of the array dipoles. The spatial *scan impedance* is

$$Z = \sum\sum \exp[jk(nu + mv)] Z_{nm} \qquad (7.43)$$

where Z_{nm} contains three spherical wave terms as derived by Carter, circa 1932 (see Balanis, 1982). With dipole radius of a, the three distances, from dipole ends and center, are

SPECTRAL DOMAIN APPROACHES 245

$$r_1^2 = (x_0 + a + nd_x)^2 + \left(y + y_0 + md_y - \frac{\ell_s}{2}\right)^2,$$
$$r_2^2 = (x_0 + a + nd_x)^2 + (y + y_0 + md_y)^2, \quad (7.44)$$
$$r_3^2 = (x_0 + a + nd_x)^2 + \left(y + y_0 + md_y + \frac{\ell_s}{2}\right)^2,$$

where the single and array dipole lengths are ℓ_s and ℓ_t. The mutual dipole–dipole impedance becomes

$$Z_{nm} = \frac{j30}{S_s S_t} \int_{-l_t/2}^{l_t/2} \left[\frac{\exp(-jkr_1)}{r_1} - 2C_s \frac{\exp(-jkr_0)}{r_0} + \frac{\exp(-jkr_2)}{r_2}\right]$$
$$\times \sin k(\ell_t/2 - |y|) \, dy, \quad (7.45)$$

with $S_s = \sin k\ell_s/2$, $S_t = \sin k\ell_t/2$, $C_s = \cos k\ell_s/2$. One of the three mutual impedance terms is shown here:

$$Z_1 = \frac{j30}{S_s S_t} \sum_n \exp(jnkd_x u) \int \sin[k(\ell_t/2 - |y|)]$$
$$\times \sum_n \exp(jmkd_y v) \frac{\exp(-jkr_1)}{r_1}. \quad (7.46)$$

The Poisson sum formula combined with the frequency shift theorem provides

$$\sum_m \exp(jmkd_y v) \frac{\exp(-jkr_1)}{r_1} = \frac{j\pi}{d_y} \sum_m \exp[jk\beta(y + y_0 - \ell_s/2)]$$
$$\times H_0^{(2)}\left(k\sqrt{1-\beta^2}\,(x_0 + a + nd_x)\right), \quad (7.47)$$

where $\beta = v + m\lambda/d_y$. This allows the integral in the Z_1 expression to be collected, and integrated:

$$Z_1 = \frac{-30\pi}{S_s S_t d_y} \sum_n \exp(jnkd_x u) \sum_m \exp[-jk\beta(y_0 - \ell_s/2)]$$
$$\times H_0^{(2)}\left(k\sqrt{1-\beta^2}\,(x_0 + a + nd_x)\right) \int_{-\ell_t/2}^{\ell_t/2} \exp(jk\beta y) \sin[k(\ell_t/2 - |y|)] \, dy.$$
$$(7.48)$$

The integral becomes a generalized pattern function:

$$-\frac{2[\cos(k\beta\ell_t/2) - \cos(k\ell_t/2)]}{k(\beta^2 - 1)}. \quad (7.49)$$

MUTUAL COUPLING

The process for the other two Carter terms is the same. The coefficients and exponentials become

$$\exp(jk\beta\ell_s/2) - 2C_s + \exp(-jk\beta\ell_s/2) = 2[(\cos(k\beta\ell_s/2) - \cos(k\ell_s/2)]. \quad (7.50)$$

Thus two generalized pattern functions are multiplied, giving

$$P_s P_t = \frac{[\cos(k\beta\ell_s/2) - \cos(k\ell_s/2)][\cos(k\beta\ell_t/2) - \cos(k\beta\ell_t/2)]}{1 - \beta^2}. \quad (7.51)$$

The *scan impedance* becomes

$$Z = \frac{-120\pi}{S_s S_t k d_y} \sum_n \exp(jnkd_x u) \sum_m P_s P_t \exp(jk\beta y_0)$$
$$\times H_0^{(2)}\left[k\sqrt{1-\beta^2}(x_0 + a + nd_x)\right]. \quad (7.52)$$

The previous Poisson transformation can be used to change the Hankel sum into a spectral sum:

$$Z = \frac{2\eta_0}{S_s S_t k^2 d_x d_y} \sum_m P_s P_t \exp(jk\beta y_0)$$
$$\times \sum_n \exp(jk\alpha x_0) \frac{\exp(-jka\sqrt{1-\alpha^2-\beta^2})}{\sqrt{1-\alpha^2-\beta^2}}, \quad (7.53)$$

where $\alpha = u + n\lambda/d_x$. This symmetric formula is the spectral domain result for *scan impedance*, developed by Munk and colleagues (see Luebbers and Munk, 1975, 1978; Munk and Burrell 1979; Munk et al., 1979). Similar techniques can be used for parallel dipoles (array of monopoles), or for dipoles at any 3-D skew angle (Munk et al., 1979; Munk and Burrell, 1979). Slots are also readily handled (Luebbers and Munk, 1978). Moment method techniques can use this methodology as it applies directly to the highly efficient piecewise sinusoidal expansion and test functions (Larson and Munk, 1983). The resulting codes, such as the *periodic moment method*, have found wide use. This spectral domain method has also been applied to finite-by-infinite arrays; see Chapter 8.

7.4.2 Microstrip Patches

Large (and infinite) arrays of patch elements are also amenable to periodic cell, Floquet series-based analyses. This was pioneered by Liu, Shmoys, and Hessel (1982) for an infinite array of patch strips with *E*-plane scan; a moment method approach was used in the spectral domain with sinusoidal current expansion functions. For a discussion of spectral moment methods for layered media, see Mosig (1989). It is not surprising that the *scan impedance* variation with angle follows closely that for dipole arrays (see Section 7.4.1). These computer simulations showed a leaky wave resonance that produces a peak in *scan resistance* versus angle; the peak becomes sharper and higher, and occurs at a smaller scan angle as the substrate thickness increases. This work was extended to printed dipole arrays and then to arrays of rectangular patches by Pozar and Schaubert (1984). Entire domain trigonometric expansion functions were used in the spectral domain moment method. Blind angles due to both grating lobe incidence and surface wave excitation were found; several cases are of interest. A thin ($0.02\lambda_0$) substrate with $\epsilon_r = 2.55$ and $\lambda_0/2$ spacings showed well-behaved *scan resistance* except for an *E*-plane blind angle at $\theta = 82.9$ deg, which corresponds to the surface wave $\beta/k = 1.0076$, where $\sin\theta = k/\beta$. A thicker substrate ($0.06\lambda_0$) gives a surface wave $\beta/k = 1.028$, with blind angle of 68.8 in the *E*-plane, as seen in Fig. 7.27. This calculation uses an *E*-plane element spacing of $0.51\lambda_0$, so a grating lobe appears at 73.9 deg. When a higher dielectric constant is used ($\epsilon = 12.8$) for the thick substrate, with half-wave lattice, the blind angle is now 45.6 deg, as seen in Fig. 7.28. This analysis by Pozar and Schaubert did not use probe current attachment modes, direct probe radiation, or edge condition expansion functions. The latter affects convergence, but the probe factors may here produce artifactual *H*-plane blind angles. Thus these have been eliminated in Figs. 7.27 and 7.28.

Liu, Hessel, and Shmoys (1988) used a moment method approach with probe attachment modes and with edge singularity matching expansion functions. Figure 7.29 shows, for square patches on a square lattice, that *scan resistance* becomes flatter and lower for thicker substrates. These resonant peaks are due to a phase match between the grating lobe and the leaky wave:

$$\sin\theta_{\ell w} = \left| \frac{\beta}{k} - \frac{\lambda}{d_E} \right|, \qquad (7.54)$$

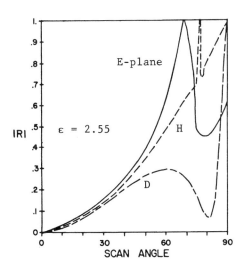

Figure 7.27 Patch array scan reflection coefficient. (Courtesy Pozar, D. M. and Schaubert, D. H., "Scan Blindness in Infinite Phased Arrays of Printed Dipoles," *Trans. IEEE*, Vol. AP-32, June 1984, pp. 602–610.)

248 MUTUAL COUPLING

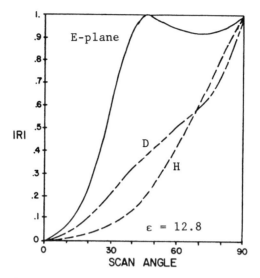

Figure 7.28 Patch array scan reflection coefficient. (Courtesy Pozar, D. M. and Schaubert, D. H., "Scan Blindness in Infinite Phased Arrays of Printed Dipoles," *Trans. IEEE*, Vol. AP-32, June 1984, pp. 602–610.)

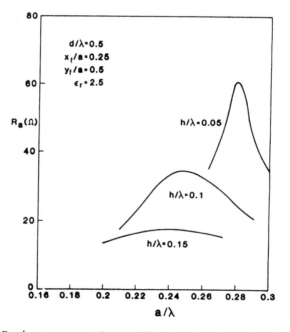

Figure 7.29 Patch array scan resistance. (Courtesy Liu, C. C., Hessel, A., and Shmoys, J., "Performance of Probe-Fed Microstrip-Patch Element Phased Arrays," *Trans. IEEE*, Vol. AP-36, Nov. 1988, pp. 1501–1509.)

where $\alpha + j\beta$ is the leaky wave propagation constant. The Q of the resonance as seen in the figure is given by

$$Q = \frac{\pi \sin \theta_{\ell w}}{\alpha \lambda}. \tag{7.55}$$

Design guidelines from Liu et al. (1988) are as follows:

(a) Too low an ϵ requires a large patch with light loading and rapid scan variation. Too high an ϵ gives a narrow bandwidth. A good range is $2 \leq \epsilon \leq 5$.
(b) A small thickness, such as $0.05\lambda_0$, reduces scan variation.
(c) The lattice should preclude E-plane blind angles, and allow a reasonable slope there. H-plane grating lobes should be precluded.
(d) A transverse width roughly equal to half the transverse spacing will avoid H-plane TE resonances.
(e) Select patch length to resonate at optimum angle (may be broadside).
(f) Examine *scan reflection coefficient* versus angle over frequency band.
(g) Iterate design as needed.

Guidelines for improving the bandwidth performances of scanned patch arrays are given by Nehra et al. (1995).

In Chapter 5 it was shown how a parasitic (stacked) patch could roughly double the bandwidth. Arrays of patches with parasitic patches are analyzed by Lubin and Hessel (1991), again via moment method in the spectral domain. The behavior of a stacked patch may be understood as follows. The driven and parasitic patches produce two resonances which should be separated (for good bandwidth). The driven patch with both dielectric layers present but no parasite approximately controls the lower resonant frequency. The parasitic patch on the upper dielectric layer, but parallel to the metallic sheet, approximately controls the upper resonant frequency. The patch size ratio affects mostly the resonant peak relative amplitudes. Element spacing should be selected to place resonance peaks beyond the upper frequency, at the maximum scan angle. Bandwidth can be adjusted by changing lower and upper ϵ, with average held roughly constant. Adjusting the layer thicknesses allows the two resonance Qs to be equalized, to achieve a smoother performance. Spectral convergence has been studied by Aberle et al. (1994); they also utilize waveguide simulators (see Chapter 12).

Scan performance may be improved through use of a shorting pin on each patch; this obviates surface waves and thus eliminates blind angles due to them (Waterhouse, 1996). The cost of this is a larger patch, but still less than $\lambda/2$, so these elements may be arrayed.

It may be said that the basic *scan impedance* behavior is controlled primarily by element lattice (grating lobe blindness) and by substrate ϵ and t (surface wave blindness); the type of element is less important. In all of these moment method array simulations, the convergence of the series is affected both by the number of spectral terms in x and in y and by the number of expansion functions in x and in y. As shown by Munk and others, these must be carefully chosen to assure accurate results. Acceleration techniques can often significantly reduce computation time. Two useful ones are by Shanks (1955) and Levin (1973)—the latter does not accumulate roundoff errors; see also Singh and Singh (1993). This author has had excellent success with Levin acceleration in spectral domain simulations.

7.4.3 Printed Dipoles

For convenience in manufacture, strip dipoles and feed lines are sometimes printed on dielectric sheets, with the sheets normal to the ground plane. Figure 7.30 sketches a typical configuration. The dielectric sheets, finite in height above the ground plane, must be taken into account. This has been done in the spectral domain using LSE and LSM hybrid waveguide modes by Chu and Lee (1987), where a current distribution on the dipoles was assumed. A combined spectral domain, moment method approach solves for the dipole currents, as well as for *scan impedance* (Bayard et al., 1991). Excellent agreement with the spectral results of Section 7.4.1 is shown in Figs. 7.31 and 7.32, with $\epsilon = 1$; these figures are for half-wave dipoles on a $0.6\lambda \times 0.6\lambda$ lattice. As expected, scan blind angles are observed for E-plane scan. A blind angle can be produced by a surface wave excited along each dielectric sheet, with the location controlled by the sheet thickness and dielectric constant. More serious is the blind angle caused by excitation of an asymmetric mode in the dipole and feed lines; this occurs even for free-standing dipoles. Figure 7.33 sketches the two current modes. A remedy was found by Reale (1974), wherein the dipole

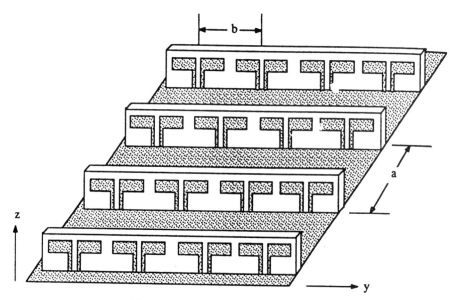

Figure 7.30 Array of printed dipoles.

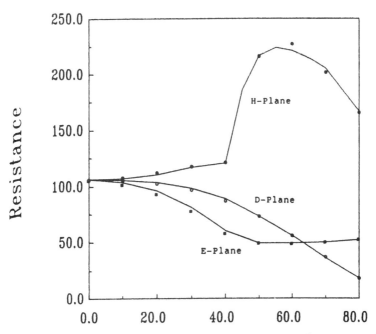

Figure 7.31 Printed dipole array scan resistance. (Courtesy Bayard, J.-P. R., Cooley, M. E., and Schaubert, D. H., "Analysis of Infinite Arrays of Printed Dipoles on Dielectric Sheets Perpendicular to a Ground Plane," *Trans. IEEE*, Vol. AP-39, Dec. 1991, pp. 1722–1732.)

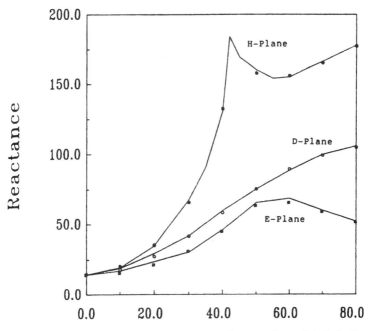

Figure 7.32 Printed dipole array scan reactance. (Courtesy Bayard, J.-P. R., Cooley, M. E., and Schaubert, D. H., Analysis of Infinite Arrays of Printed Dipoles on Dielectric Sheets Perpendicular to a Ground Plane," *Trans. IEEE*, Vol. AP-39, Dec. 1991, pp. 1722–1732.)

252 MUTUAL COUPLING

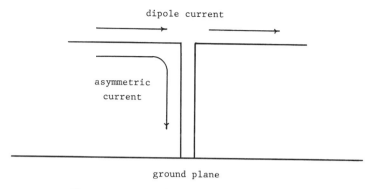

Figure 7.33 Dipole current modes due to scan.

arms are bent into a vee, with the dipole ends closer to the ground plane. A number of large air defense radars have been built using fat (not printed) vee dipoles. A moment method analysis of strip vee dipole arrays has been given by Bayard and Cooley (1995). Dielectric radome covers over printed dipole arrays have been analyzed also, using the spectral domain technique (Bayard, 1994).

7.4.4 Printed TEM Horns

The TEM horns of Chapter 5 can be arrayed in both planes. Although earlier work was on printed circuit horns with a linear taper (Cooley et al., 1991), as one might expect from waveguide horn experience, tapers that are roughly exponential give better results. Figure 7.34 depicts a TEM horn array; each horn includes a slotline cavity at the narrow end of the horn, and a stripline feed line and stub. A patch moment method has been used to model the horn and feed, as sketched in Fig. 7.35. Extensive parameter studies have been made by Schaubert and Shin (1995) and Shin and Schaubert (1995). Some typical results are very useful. For a broadside array, a VSWR ≤ 2 occurs over a two-octave bandwidth. At the lowest frequency, the E-plane lattice spacing is 0.139λ while the H-plane lattice spacing is 0.150λ. The horn length is roughly twice the horn mouth, while the latter is 0.095λ at the lowest frequency. Thus in an array, the horns can be smaller than a free-standing horn. A reduced array lattice allows scanning over a ± 45 deg range, as shown in Fig. 7.36. Here at the lowest frequency the lattice spacing and the mouth are 0.065λ; a remarkable result. A blind angle occurs in the H-plane, at roughly a lattice spacing of 0.312λ. This may be caused by a surface wave supported by the corrugated surface comprised of the metallic horn sheets; the blind angle is a strong function of E-plane spacing, and a less strong function of H-plane spacing (Shin and Schaubert, 1995). Another type of blindness occurs when the H-plane spacing exceeds half-wavelength. This E-plane scan blind angle is due to the corrugated nature along one horn sheet; the surface wave excitation depends on E-plane spacing and on the depth from horn mouth to ground screen (Schaubert, 1996).

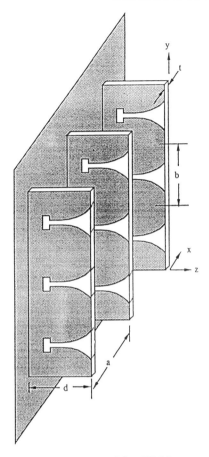

Figure 7.34 Array of flat TEM horns.

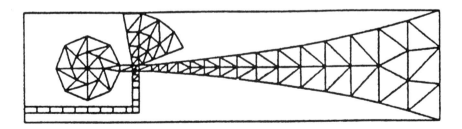

Figure 7.35 TEM horn modelling. (Courtesy Schaubert, D. H. and Shin, J., "Parameter Study of Tapered Slot Antenna Arrays," *IEEE AP 1995 Symposium Digest*, Newport Beach, Calif., June 1995, pp. 1376–1379.)

The horn throat cavity and the feedline stub have significant effects upon the low and high ends of the useful frequency band, while the middle of the band is most affected by the horn taper (Schaubert et al., 1996). Care must be taken

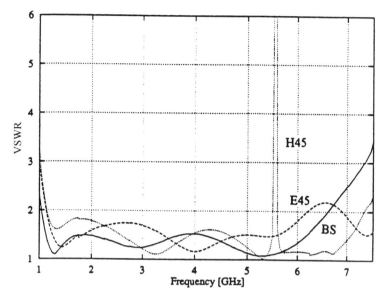

Figure 7.36 Flat TEM horn array. (Courtesy Schaubert, D. H. and Shin, J., "Parameter Study of Tapered Slot Antenna Arrays," *IEEE AP-1985 Symposium Digest*, Newport Beach, Calif., June 1995, pp. 1376–1379.)

with any gaps in the printed conductors as they can induce resonances, with high VSWR (Schaubert, 1994). Gaps are sometimes introduced for ease of fabrication of 2-D arrays, or to reduce mutual coupling.

7.4.5 Unit Cell Simulators

At certain scan angles and scan planes, the unit cell walls become perfectly conducting. This allows the unit cell, and thus the array performance to be simulated in a waveguide with an iris containing an excerpt of the unit cells. These powerful measurement tools are discussed in Chapter 12.

7.5 SCAN COMPENSATION AND BLIND ANGLES

7.5.1 Blind Angles

Some arrays radiate no power at a certain angle; this is called a blind angle. This phenomenon can, like array performance in general, be viewed from the standpoint of either *scan reflection coefficient* or *scan element pattern*. The *scan reflection coefficient* may develop a magnitude near to or equal to unity, or the *scan element pattern* may be near to or equal to zero. Taking first the element *scan reflection coefficient* case, the blind angle occurs when a higher mode cancels the dominant mode in the element. Since the relative phase changes rapidly with scan angle, the two modes produce a resonance, with a resonant phase angle that can be defined on the decrease of $|\Gamma|$ with scan angle. The higher mode can be produced by external or internal structure. Examples of external structure causing a blind angle are a dielectric radome covering each element or row of elements (Lechtreck, 1968), a dielectric sheet placed on the array face (Bates, 1965; Byron and Frank, 1968; Wu and Galindo, 1968; Gregorwich et al., 1968), and a dielectric plug protruding from the mouth of each waveguide element (Hannan, 1967). In all cases the active wave impedance at the array face is modified by the external structure so as to produce a dominant-mode and higher-mode resonance at a particular scan angle; the higher mode exists primarily outside the face. Examples of internal structure that can allow a higher mode that can cause a blind angle are dielectric plugs flush with the array face (Amitay and Galindo, 1968), dielectric-loaded waveguide elements (as opposed to a short plug) (Diamond, 1967), and a brick array of rectangular guide elements (Diamond, 1967). The latter is sketched in Fig. 7.37. For these various structures the waveguide element supports the next higher mode, which causes the resonance at the blind angle. For the brick array the higher mode is just below cutoff owing to the large guide size. An equivalent circuit for the two modes has been used by Oliner (1969) and Knittel et al. (1968), to predict the reflection coefficient behavior. Complete reflection corresponds to an active conductance that is zero. Simulators (see Section 7.4) have been used both to verify the presence of blind angles and to preclude their occurrence through design.

From the *scan element pattern* viewpoint, the *scan impedance* at the array face, including the effect of external or internal structure, allows a leaky wave to be excited. This leaky wave is usually backward, and its amplitude depends on the *scan impedance* and on the length of the impedance surface (the array face). Thus small arrays usually have no blind angle or only a small dip instead of a null, while very large arrays are more susceptible as it is easier to excite the leaky wave. The leaky wave has a complex wavenumber, analogous to Wood anomalies on an optical reflection grating (Hessel and Oliner, 1965). When the angle and phase of the leaky waves are suitable, the direct radiation at the angle will be canceled, providing a *scan element pattern* null, or blind angle; see Fig. 7.38. The leaky-wave pole gives an approximate behavior of cancelation

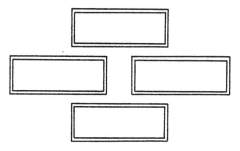

Figure 7.37. Brick waveguide array.

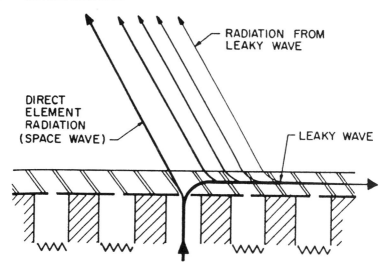

Figure 7.38 Blind angle due to wave cancellation. (Courtesy Knittel, G. H., Hessel, A., and Oliner, A. A., "Element Pattern Nulls in Phased Arrays and Their Relation to Guided Waves", *Proc. IEEE*, Vol. 56, Nov. 1968, pp. 1822–1836.)

around the blind angle, where Γ is expanded into a Taylor series about the pole. Figure 7.39 shows the *scan reflection coefficient* for an array of narrow slots fed by parallel-plate waveguides and covered by a dielectric slab, for E-plane scan. This curve was calculated using 21 modes. Slot spacing was $\lambda/2$, waveguide width was 0.3λ, while the slab had $\epsilon = 2.56$ and thickness 0.2λ. Using the leaky-wave pole Knittel et al. (1968) calculated the dots shown in the figure, where the leaky-wave Q is given by

$$Q \simeq \frac{u^2 \operatorname{Re}(k_x/k)}{2(1-u^2)\operatorname{I_m}(k_x/k)}. \qquad (7.56)$$

k_x is the wavenumber along the array. For this array the grating lobe occurs at 90 deg, and the *scan reflection coefficient* is also unity there. Another example used a brick array of rectangular guides $0.905 \times 0.4\lambda$, on a square brick lattice of 1.008λ spacing. An array of only 7×7 elements exhibits a blind angle over 26 dB deep in the H-plane at 27 deg (Farrell and Kuhn, 1966). A grating lobe occurs on the 60 deg circle in u,v space; the blind angle is usually just inside the grating lobe, but in this example it is well inside (Oliner, 1972).

The understanding of how blind angles arise also gives indications of corrective measures to take (Knittel, 1972; Oliner, 1972). These include reducing dimensions to preclude/eliminate higher propagating modes, alternating external structure or external impedance, and changing various parameters to obtain a suitably low VSWR over a wide scan angle range.

7.5.2 Scan Compensation

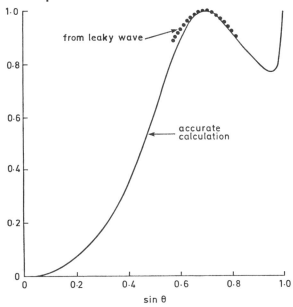

Figure 7.39 Blind angle of array of slits. (Courtesy Knittel, G. H., Hessel, A., and Oliner, A. A., "Element Pattern Nulls in Phased Arrays and Their Relation to Guided Waves," *Proc. IEEE*, Vol. 56, Nov. 1968, pp. 1822–1836.)

scan impedance changes can be compensated by any of four different methods. These are reduction of coupling, feed networks, multimode elements, and external wave transformers.

7.5.2.1 Coupling Reduction. The simplest way to reduce relative coupling is to pack the elements closely together in wavelengths. The results in Section 7.4 showed that the admittance variation with scan increases as grating lobe onset is approached. Reducing the element spacing below half-wave reduces the admittance variation owing to the grating lobe being farther away (in invisible space). This is, of course, difficult without going to narrowband elements. A more practical scheme uses H-plane baffles between slots or dipoles (Edelberg and Oliner, 1960). These baffles are simply metal strips perpendicular to the ground plane and running between the dipoles or slots rows in the H-plane as shown in Fig. 7.40. Figure 7.41 shows the improvement that can be realized. In this example the dipoles are spaced 0.5λ apart and 0.25λ above a ground plane. Use of baffles about 0.5λ high makes the E- and H-plane patterns closely alike. Over a 60 deg conical scan volume a VSWR ≤ 1.6 can be achieved. An analysis of the baffled array in terms of surface waves is given by Mailloux (1972). A moment method ana-

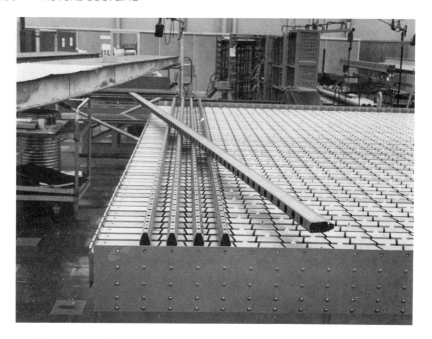

Figure 7.40 Edge slot array with baffles. (Courtesy Hughes Aircraft Co.)

lysis of a dipole array with fences has been given by Lee (1990). L-Shaped fences, with a flange parallel to the ground plane, are useful for a linear array (Forooraghi and Kildal, 1995).

Use of baffles is effective and inexpensive for arrays that scan only in one plane, provided the space, weight, and fabrication can be managed. However, some elements behave less well in the H-plane than the dipole/screen array of Fig. 7.41. Rengarajan (1996), using a spectral domain moment method, shows the reduction in coupling provided by baffles.

Another scheme changes the impedance of the ground plane between elements by introducing corrugations (Dufort, 1968). There is a tradeoff here in that it may be necessary to increase element spacing to allow room for the corrugations.

7.5.2.2 Compensating Feed Networks. A complex matrix-type feed network that adapts its outputs according to the driving phase shifts could, in principle, compensate for most of the mutual impedance effects. Such a network connected between the antenna elements and the generators can be used either to make reflectionless elements, or to make scatterless elements. The former is easier and will be discussed first. By use of successively more complex networks, the situation where $\Gamma = 0$ may be closely approached (Hannan, 1967). However, an exact match at all angles can only be approxi-

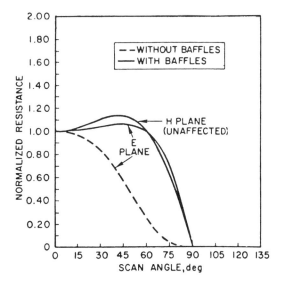

Figure 7.41 Scan resistance with baffles. (Courtesy Edelberg, S. and Oliner, A. A., "Mutual Coupling Effects in Large Antenna Arrays. Part I: Slot Arrays," and "Part II: Compensation Effects," *Trans. IRE*, Vol. AP-8, May and July 1960, pp. 286–297, 360–367.)

mated (Varon and Zysman, 1968). Both methods improve the match by making the scan element power pattern closer to $\cos\theta$ (for a planar array with no grating lobes).

As a means of understanding the matching network process, consider the network to consist of three networks in cascade. The first network is a $2N$-port that transforms the array excitation vectors into eigenvectors of the *scan impedance* matrix. The second network consists of N separate 2-ports that equalize the eigenvalues. This process is analogous to the convergence speedup in adaptive arrays where the vectors are first orthogonalized, then their amplitudes are made equal. The third network is a $2N$-port which is the inverse of the first; it restores the array excitation vectors, which have now been matched. An excitation in an infinite array producing a beam in invisible space is represented by a singular (zero or infinite) eigen excitation. For finite arrays such excitations produce very small or very large eigen excitations. Attempts to match such eigenvectors produce tolerance problems similar to those encountered in superdirectivity. Kahn (1977) has calculated *scan element pattern*s for a linear array of uniform line sources. Figure 7.42 is for an element spacing of $\lambda/2$, for the center element of 25-element array. The solid line is the ideal $\cos\theta$ and the dashed line is the actual element pattern. The fit is excellent, and for larger arrays the oscillations will be more numerous and smaller. Care must be exercised for smaller spacings, where inclusion of small eigenvalues may cause instability.

Descattering the array via a network requires that the element transmit and receive patterns be equal, i.e., that the elements are minimum scattering (see

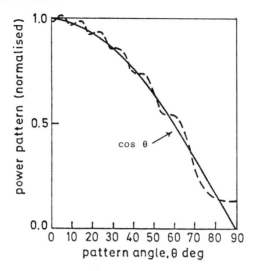

Figure 7.42 Scan element pattern of 25-element array. (Courtesy Kahn, W. K., "Impedance-Match and Element-Pattern Constraints for Finite Arrays," *Trans. IEEE*, Vol. AP-25, Nov. 1977, pp. 747–755.)

Section 7.2.4). In practice, descattering may be useful for single-mode elements such as slots and dipoles. A further requirement is that all mutual impedances be reactive (Andersen and Rasmussen, 1976). A corollary is that the mutual resistance, since it is proportional to the beam orthogonality integral, must equal zero (Wasylkiwskyj and Kahn, 1970). Now the mutual impedance is related to the element power pattern; so the zero mutual resistance becomes a constraint on the element pattern. For example, rotationally symmetric power patterns of $\cos^m \theta$ yield $R_{12} = 0$ for spacings that satisfy

$$j_{m/2}(kd) = 0, \tag{7.56}$$

where $j_{m/2}$ is the spherical Bessel function of order $m/2$. Pursuing the example further for isotropic elements ($m = 0$), the root gives $d = \lambda/2$, so that a half-wave spaced array may be completely decoupled and descattered. When a grating lobe just enters visible space, the *scan element pattern* has a null (at that angle), or in *scan impedance* terms the reflection coefficient magnitude is unity. Such a blind angle cannot be removed by a matching network (Wasylkiwskyj and Kahn, 1977). However, for those blind angles that occur at scan ranges inside the grating lobe scan contour (in u,v space), a suitable network can improve or even remove the null, provided that the open-circuited element pattern does not have a null at that angle.

A variation on the compensating network may be used with a receiving array with digital beam forming (Steyskal and Herd, 1990). Here the vector of array outputs is multiplied by a matrix designed to produce a scan element power pattern of $\cos \theta$. The matrix elements are based on a numerical simulation of the array *scan impedance* pattern.

Although the single-scan-plane compensation techniques can be effective, they are not often used. Instead, the array feeds can be matched to the *scan impedance* at an intermediate angle between broadside and maximum scan angle, so that the *scan element pattern* loss is equalized. This is frequently sufficiently good that no complicated techniques are warranted.

7.5.2.3 Multimode Elements. The ideal element pattern discussions presented earlier imply that both TE and TM modes are needed in the array unit cell to allow a $\cos\theta$ scan element power pattern. The additional modes can be produced by modifying the radiating element in any of several ways. Probably the most effective way of producing the optimum mix of modes uses open-end waveguide radiators, rectangular or circular, with the waveguide internal environment or the external environment suitably modified. Tang and Wong (1968) and Wong et al. (1972) use a dielectric plug at the rectangular guide mouth and a matching post. Tsandoulas (1972) uses rectangular waveguide width and height steps (in different planes). Tsandoulas and Knittel (1973) use both waveguide size steps and dielectric plugs at the mouth. Wu (1972) uses plugs in circular guide elements. The mix of modes that is needed depends upon the waveguide and lattice dimensions. A typical mode set might be TE_{10}, TE_{20}, TE_{30}, TE_{11}, TE_{01}, TM_{11}. With a dielectric plug in the waveguide the TE_{20} mode will likely be propagating; all other higher modes will be evanescent. The plug should be half-wave long (in loaded guide wavelengths) at center frequency to properly locate the *scan admittance* on the Smith chart. Then the guide dimensions are adjusted to give a high susceptance for the TE_{20} mode looking into the matching network. Steps in guide height can then be used for center angle impedance matching; additional dielectric plugs can also be used for matching. Figure 7.43 sketches a typical guide configuration, including two sheets over the array face. All of these are designed by a mode matching technique, where propagating and evanescent waveguide modes are matched at each boundary. This is conveniently performed in the spectral domain. Since the unit cell at scan angles other than broadside has impedance walls, the modes are LSE and LSM rather than TE and TM. It is necessary to assume values of all physical parameters to start the process: slab dielectric constants and thicknesses, plug dielectric constants and lengths, waveguide heights and widths and, of course, the element lattice dimensions. A mode set for the waveguide element is then assembled. Choosing a reasonably good set of all these parameters requires considerable experience, gained mostly by running various cases and then examining the results. With all parameters set, the unit-cell admittance is calculated for each waveguide mode, assuming unit amplitude of the latter. Many modes will be required in the unit cell to match the waveguide mode field at the aperture, and for each unit cell mode the admittance must be calculated. Starting with the value for large distance (from the slabs), this admittance is transformed through the dielectric slabs, and then combined with the other unit cell modes. Numbers of modes typically are 100 to 300. Longitudinal-section modes must be converted to TE and TM modes so that mode admittances can be matched. Next, starting

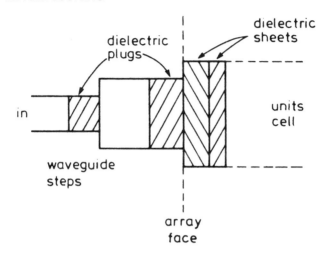

Figure 7.43 Typical open guide matching. (Courtesy Tsandoulas, G. N. and Knittel, G. H., "The Analysis and Design of Dual-Polarization Square-Waveguide Phased Arrays," *Trans. IEEE*, Vol. AP-21, Nov. 1973, pp. 796–808.)

with an incident TE_{10} mode in the waveguide, the aperture admittance of each waveguide mode is calculated. Although the mode admittances must be equal across the aperture, the mode amplitudes depend upon the total admittance. Thus, a set of simultaneous equations one for each waveguide mode, can be solved. Typical results are shown in Fig. 7.44, which is for $0.403\lambda_0$ square waveguides in an equilateral triangular lattice (Tsandoulas and Knittel, 1973). Bandwidth for VSWR ≤ 3 is 25%, and the scan range is roughly a quarter hemisphere. The solid line is E-plane; the dashed line is H-plane. An interesting design principle resulting from many calculations by these authors is that polarization coupling (for a dual polarized array) is reduced as the element size is increased. Thus when multimode elements are used they should be as large as the lattice allows. For a single polarization, bandwidth of greater than 50% with VSWR ≤ 4 and scan over nearly a quarter hemisphere has been obtained (Tsandoulas, 1972, 1980).

An iris can be used in each waveguide radiator, below the aperture, in addition to dielectric plugs, as an aid to matching. Excellent results have been obtained by Lee and Jones (1971) and by Lee (1971). Excellent results have been demonstrated over wide scan angles. One or more dielectric sheets may be placed on the array face as part of the mode matching process (Wu, 1972).

Another scheme combines the TE and TM radiators. Figure 7.45 shows a longitudinal shunt slot array with parasitic monopoles with L-shaped extensions developed by Clavin et al. (1974). The monopoles reduce the E-plane beamwidth and add an electric-type mode to complement the slot magnetic-type mode. Patterns in E- and H-planes can be made equal (Hansen, 1985). Mutual coupling between elements is reduced 14 dB, and the array wide-angle

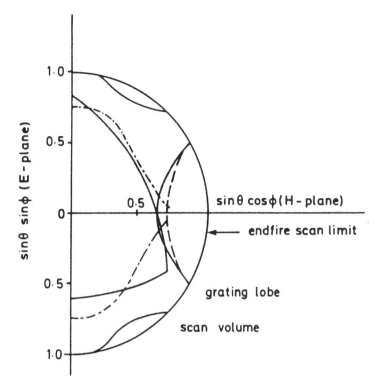

Figure 7.44 VSWR = 3 contour for waveguide array. (Courtesy Tsandoulas, G. N. and Knittel, G. H., "The Analysis and Design of Dual-Polarization Square-Waveguide Phased Arrays," *Trans. IEEE*, Vol. AP-21, Nov. 1973, pp. 796–808.)

sidelobes and backlobe are significantly reduced. These arrays have been analyzed by the periodic cell spectral domain technique, using a piecewise expansion on the monopole wires (Ng and Munk, 1988). The parasitic array can support surface waves and these can cause blind angles in some cases (Ng and Munk, 1989). With the spectral domain analysis, these problems can be obviated.

A different use of a dielectric sheet involves placing a high dielectric constant thin sheet near the array face. This technique, called wide-angle impedance matching (WAIM) by Wheeler (Magill and Wheeler, 1966), uses the susceptances of the sheet to equalize part of the aperture susceptance variation with scan. This is possible because the sheet susceptance varies differently in the E-plane and H-plane. Using the well-known formulas (Collin, 1991) for reflection from lossless dielectric sheet, the susceptance is obtained for a thin sheet, where this means $kt\sqrt{\epsilon_r} \ll 1$. Thus higher dielectric constant sheets must be thinner. The normalized broadside conductance is simply $G_0 = kt(\epsilon_r - 1)$. For H-plane and E-plane scans, the sheet susceptances are

Figure 7.45 Slots with parasitic elements. (Courtesy Clavin, A., Huebner, D. A., and Kilburg, F. J., "An Improved Element for Use in Array Antennas," *Trans. IEEE*, Vol. AP-22, July 1974, pp. 521–526.)

$$\text{H-plane:} \quad \frac{B}{G_0} = \frac{1}{\cos\theta},$$
$$\text{E-plane:} \quad \frac{B}{G_0} = \cos\theta - \frac{\sin^2\theta}{\epsilon_r \cos\theta}. \tag{7.57}$$

An optimum match is obtained by adjusting the sheet thickness t (for a given ϵ_r) and the sheet spacing from the array face to minimize *scan element pattern* variations over the scan volume. The Smith chart procedure described by Magill and Wheeler (1966; Kelly, 1966) may be suitable for some well-behaved elements; in general numerical optimization based on a periodic cell spectral model or on extensive measurements is necessary. A review of scan compensation techniques is given by Knittel (1972).

Another multimode technique employs dielectric coatings on slots that can support a partial surface wave; the effect is to make the *E*- and *H*-plane patterns more alike, and to extend them toward endfire (Villeneuve et al., 1982).

7.5.2.4 External Wave Filter. The array *scan element pattern* may be altered through use of a spatial filter or transformer placed in front of the array face, but sufficiently far away that the filter does not directly affect element impedances. The transfer function of the spatial filter is designed to compensate the SEP variations; polarization effects may also be improved.

With dielectric sheets the transmission coefficient varies differently with angle in E- and H-planes, so that in principle one or more sheets could improve the SEP. With several sheets in cascade, matrices for each sheet may be cascaded. Note that impedance, admittance, and scattering matrices do not cascade. Two that do are the A–B–C–D matrix (Altman, 1964) and the transmission matrix (Collin, 1991; Section 3.4). Conversion of either with Z, Y, or S matrices is, of course, readily performed. Mailloux (1976) applied the T matrix to a stack of dielectric slabs to reduce far-out sidelobes. Munk, Kornbau, and Fulton (1979) cascaded in the spectral domain; this approach is more powerful as many arrays are best analyzed therein. As an example an array of short slots over a screen was matched by a single dielectric slab of $0.367\lambda_0$ thickness and $\epsilon = 1.33$; the resulting VSWR was less than 1.5 for both planes of scan to 80 deg (Munk, Kornbau, and Fulton, 1979). Other useful references on stratified wave filters that can contain thin metallic inclusions as well as dielectrics are Ng and Munk (1988) and Munk and Kornbau (1988).

ACKNOWLEDGMENT

Photograph courtesy of Dr. J. J. Lee.

REFERENCES

Aberle, J. T., Pozar, D. M., and Manges, J., "Phased Arrays of Probe-Fed Stacked Microstrip Patches," *Trans. IEEE*, Vol. AP-42, July 1994, pp. 920–927.

Abramowitz, M. and Stegun, I. A., *Handbook of Mathematical Functions*, NBS, AMS-55, 1970.

Allen, J. L., "Gain and Impedance Variation in Scanned Dipole Arrays," *Trans. IRE*, Vol. AP-10, Sept. 1962, pp. 566–572.

Altman, J. L., *Microwave Circuits*, Van Nostrand, 1964.

Amitay, N. et al., "On Mutual Coupling and Matching Conditions in Large Planar Phased Arrays," *IEEE AP International Symposium Digest*, Sept. 1964, Long Island, N.Y., pp. 150–156.

Amitay, N. and Galindo, V., "Analysis of Circular Waveguide Phased Arrays," *BSTJ*, Vol. 47, Nov. 1968, pp. 1903–1931.

Amitay, N., Galindo, V., and Wu, C. P., *Theory and Analysis of Phased Array Antennas*, Wiley-Interscience, 1972.

Andersen, J. B., Lessow, H. A., and Schjaer-Jacobsen, H., Coupling Between Minimum Scattering Antennas, *Trans. IEEE*, Vol. AP-22, Nov. 1974, pp. 832–835.

Andersen, J. B. and Rasmussen, H. H., "Decoupling and Descattering Networks for Antennas," *Trans. IEEE*, Vol. AP-24, Nov. 1976, pp. 841–846.

Bailey, M. C., "Mutual Coupling between Circular Waveguide-Fed Apertures in a Rectangular Ground Plane," *Trans. IEEE*, Vol. AP-22, July 1974, pp. 597–599.

Bailey, M. C. and Bostian, C. W., "Mutual Coupling in a Finite Planar Array of Circular Apertures," *Trans. IEEE*, Vol. AP-22, Mar. 1974, pp. 178–184.

Balanis, C. A., "Near-Fields of Dipoles," in *Antenna Theory*, Harper & Row, 1982, Section 7.2.

Bamford, L. D., Hall, P. S., and Fray, A., "Calculation of Antenna Mutual Coupling from Far Radiated Fields," *Electron. Lett.*, Vol. 29, July 8, 1993, pp. 1299–1301.

Barkeshli, S. and Pathak, P. H., "Radial Propagation and Steepest Descent Path Integral Representations of the Planar Microstrip Dyadic Green's Function." *Radio Sci.*, Vol. 25, Mar.-Apr. 1990, pp. 161–174.

Barkeshli, S., Pathak, P. H., and Marin, M., "An Asymptotic Closed-Form Microstrip Surface Green's Function for the Efficient Moment Method Analysis of Mutual Coupling in Microstrip Antennas," *Trans. IEEE*, Vol. AP-38, Sept. 1990, pp. 1374–1383.

Bates, R. H. T., "Mode Theory Approach to Arrays," *Trans. IEEE*, Vol. AP-13, Mar. 1965, pp. 321–322.

Bayard, J.-P. R., "Analysis of Infinite Arrays of Microstrip-Fed Dipoles Printed on Protruding Dielectric Substrates and Covered with a Dielectric Radome," *Trans. IEEE*, Vol. AP-42, Jan. 1994, pp. 82–89.

Bayard, J.-P. R. and Cooley, M. E., "Scan Performance of Infinite Arrays of Microstrip-FED Dipoles with Bent Arms Printed on Protruding Substrates," *Trans. IEEE*, Vol. AP-43, Aug. 1995, pp. 884–888.

Bayard, J.-P. R., Cooley, M. E., and Schaubert, D. H., "Analysis of Infinite Arrays of Printed Dipoles on Dielectric Sheets Perpendicular to a Ground Plane," *Trans. IEEE*, Vol. AP-39, Dec. 1991, pp. 1722–1732.

Begovich, N. A., "Slot Radiators," *Proc. IRE*, Vol. 38, July 1950, pp. 803–806.

Bird, T. S., "Mode Coupling in a Planar Circular Waveguide Array," *IEE J. Microwaves, Opt. Acoust.*, Vol. 3, 1979, pp. 172–180.

Bird, T. S., "Mutual Coupling in Finite Coplanar Rectangular Waveguide Arrays," *Electron. Lett.*, Vol. 23, Oct. 22, 1987, pp. 1199–1201.

Bird, T. S., "Analysis of Mutual Coupling in Finite Arrays of Different-Sized Rectangular Waveguides," *Trans. IEEE*, Vol. AP-38, Feb. 1990, pp. 166–172.

Bird, T. S. and Bateman, D. G., "Mutual Coupling between Rotated Horns in a Ground Plane," *Trans. IEEE*, Vol. AP-42, July 1994, pp. 1000–1006.

Booker, H. G., "Slot Aerials and Their Relation to Complementary Wire Aerials (Babinet's Principle)," *J. IEE*, Vol. 93, Pt. IIIA, 1946, pp. 620–626.

Borgiotti, G. V., "Radiation and Reactive Energy of Aperture Antennas," *Trans. IEEE*, Vol. AP-11, Jan. 1963, pp. 94–95.

Borgiotti, G. V., "A Novel Expression for the Mutual Admittance of Planar Radiating Elements," *Trans. IEEE*, Vol. AP-16, May 1968, pp. 329–333.

Byron, E. V. and Frank, J., "'Lost Beams' from a Dielectric Covered Phased-Array Aperture," *Trans. IEEE*, Vol. AP-16, July 1968, pp. 496–499.

Carter, P. S., "Circuit Relations in Radiating Systems and Applications to Antenna Problems," *Proc. IRE*, Vol. 20, June 1932, pp. 1004–1041.

Catedra, M. F. et al. *The CGT-FFT Method—Application of Signal Processing Techniques to Electromagnetics*, Artech House, 1995.

Chew, W. C., "Some Observations on the Spatial and Eigenfunction Representations of Dyadic Green's Functions," *Trans. IEEE*, Vol. AP-37, Oct. 1989, pp. 1322–1327.

Chu, R.-S. and Lee, K.-M., "Radiation Impedance of a Dipole Printed on Periodic Dielectric Slabs Protruding over a Ground Plane in an Infinite Phased Array," *Trans. IEEE*, Vol. AP-35, Jan. 1987, pp. 13–25.

Clavin, A., Huebner, D. A., and Kilburg, F. J., "An Improved Element for Use in Array Antennas," *Trans. IEEE*, Vol. AP-22, July 1974, pp. 521–526.

Collin, R. E., *Field Theory of Guided Waves*, IEEE Press, 1991.

Cooley, M. E. et al., "Radiation and Scattering Analysis of Infinite Arrays of Endfire Slot Antennas with a Ground Plane," *Trans. IEEE*, Vol. AP-39, Nov. 1991, pp. 1615–1625.

Debski, T. R. and Hannan, P. W., "Complex Mutual Coupling Measurements in a Large Phased Array Antenna," *Microwave J.*, Vol. 8, June 1965, pp. 93–96.

Diamond, B. L., "Resonance Phenomena in Waveguide Arrays," *IEEE AP Symposium Digest*, Oct. 1967, Ann Arbor, Mich., pp. 110–115.

Diamond, B. L., "A Generalized Approach to the Analysis of Infinite Planar Array Antennas," *Proc. IEEE*, Vol. 56, Nov. 1968, pp. 1837–1850.

DuFort, E. C., "A Design Procedure for Matching Volumetrically Scanned Waveguide Arrays," *Proc. IEEE*, Vol. 56, Nov. 1968, pp. 1851–1860.

Edelberg, S. and Oliner, A. A., "Mutual Coupling Effects in Large Antenna Arrays. Part I: Slot Arrays", and "Part II: Compensation Effects," *Trans. IRE*, Vol. AP-8, May and July 1960, pp. 286–297, 360–367.

Farrell, G. F., Jr. and Kuhn, D. H., "Mutual Coupling Effects in Infinite Planar Arrays of Rectangular Waveguide Horns," *IEEE AP Symposium Digest*, Dec. 1966, Palo Alto, Calif., pp. 392–397.

Forooraghi, K. and Kildal, P. S., "Transverse Radiation Pattern of a Slotted Waveguide Array Radiating between Finite Height Baffles in Terms of a Spectrum of Two-Dimensional Solutions," *Proc. IEE*, Part H, Vol. 140, Feb. 1993, pp. 53–58.

Frazita, R. F., "Surface-Wave Behavior of a Phased Array Analyzed by the Grating-Lobe Series," *Trans. IEEE*, Vol. AP-15, Nov. 1967, pp. 823–824.

Galindo, V. and Wu, C. P., "Surface Wave Effects on Phased Arrays of Rectangular Waveguides Loaded with Dielectric Plugs," *Trans. IEEE*, Vol. AP-16, May 1968, pp. 358–360.

Gately, A. C., Jr. et al., "A Network Description for Antenna Problems," Proc. IEEE, Vol. 56, July 1968, pp. 1181-1193.

Goldberg, J. J., Study of a Phased Array Finite in Its Plane of Scan, MS Thesis, Polytech. Inst. of Brooklyn, 1972.

Gregorwich, W. S. et al., "A Waveguide Simulator for the Determination of a Phased-Array Resonance," *IEEE AP Symposium Digest*, Sept. 1968, Boston, Mass., pp. 134–141.

Hamid, M. A. K., "Mutual Coupling between Sectoral Horns Side by Side," *Trans. IEEE*, Vol. AP-15, May 1967, pp. 475–477.

Haneishi, M. and Suzuki, Y., "Circular Polarisation and Bandwidth," in *Handbook of Microstrip Antennas*, Vol. 1, J. R. James and P. S. Hall, Eds., IEE/Peregrinus, 1989, Chapter 4.

Hannan, P. W., "The Element-Gain Paradox for a Phased Array Antenna," *Trans. IEEE*, Vol. AP-12, July 1964, pp. 423–433.

Hannan, P. W., "The Ultimate Decay of Mutual Coupling in a Planar Array Antenna," *Trans. IEEE*, Vol. AP-14, 1966, pp. 246–248.

Hannan, P. W., "Proof That a Phased Array Antenna Can Be Impedance Matched for All Scan Angles," *Radio Sci.*, Vol. 2, Mar. 1967, pp. 361–369.

Hannan, P. W., "Discovery of an Array Surface Wave in a Simulator," *Trans. IEEE*, Vol. AP-15, July 1967, pp. 574–576.

Hansen, R. C., "Formulation of Echelon Dipole Mutual Impedance for Computer," *Trans. IEEE*, Vol. AP-20, Nov. 1972, pp. 780–781.

Hansen, R. C., "Scan Compensated Active Element Patterns," *Proceedings Phased Arrays Symposium*, RADC-TR-85-171, Aug. 1985, AD-A169 316.

Hansen, R. C. and Brunner, G., "Dipole Mutual Impedance for Design of Slot Arrays," *Microwave J.*, Vol. 22, Dec. 1979, pp. 54–56.

Hessel, A. and Oliner, A. A., "A New Theory of Wood's Anomalies on Optical Gratings," *Appl. Opt.*, Vol. 4, 1965, pp. 1275–1297.

Jedlicka, R. P., Poe, M. T., and Carver, K. R., "Measured Mutual Coupling between Microstrip Antennas," *Trans. IEEE*, Vol. AP-29, Jan. 1981, pp. 147–149.

Johnson, W. A. and Dudley, D. G., "Real Axis Integration of Sommerfeld Integrals: Source and Observation Points in Air," *Radio Sci.*, Vol. 18, Mar.-Apr. 1983, pp. 175–186.

Kahn, W. K., "Impedance-Match and Element-Pattern Constraints for Finite Arrays," *Trans. IEEE*, Vol. AP-25, Nov. 1977, pp. 747–755.

Kahn, W. K. and Kurss, H., "Minimum-Scattering Antennas," *Trans. IEEE*, Vol. AP-13, Sept. 1965, pp. 671–675.

Kelly, A. J., "Comments on 'Wide-Angle Impedance Matching of a Planar Array Antenna by a Dielectric Sheet,'" *Trans. IEEE*, Vol. AP-14, Sept. 1966, pp. 636–637.

King, H. E., "Mutual Impedance of Unequal Length Antennas in Echelon," *Trans. IRE*, Vol. AP-5, July 1957, pp. 306–313.

Knittel, G. H., "Wide-Angle Impedance Matching of Phased-Array Antennas: A Survey of Theory and Practice," in *Phased Array Antennas*, A. A. Oliner and G. H. Knittel, Eds., Artech House, 1972.

Knittel, G. H., Hessel, A., and Oliner, A. A., "Element Pattern Nulls in Phased Arrays and Their Relation to Guided Waves," *Proc. IEEE*, Vol. 56, Nov. 1968, pp. 1822–1836.

Larson, C. J. and Munk, B. A., "The Broad-Band Scattering Response of Periodic Arrays," *Trans. IEEE*, Vol. AP-31, Mar. 1983, pp. 261–267.

Lechtreck, I. W., "Effect of Coupling Accumulation in Antenna Arrays," *Trans. IEEE*, Vol. AP-16, Jan. 1968, pp. 31–37.

Lee, J. J., "Effects of Metal Fences on the Scan Performance of an Infinite Dipole Array," *Trans. IEEE*, Vol. AP-38, May 1990, pp. 683–692.

Lee, S. W., "Aperture Matching for an Infinite Circular Polarized Array of Rectangular Waveguides," *Trans. IEEE*, Vol. AP-19, May 1971, pp. 332–342.

Lee, S. W. and Jones, W. R., "On the Suppression of Radiation Nulls and Broadband Impedance Matching of Rectangular Waveguide Phased Arrays," *Trans. IEEE*, Vol. AP-19, Jan. 1971, pp. 41–51.

Levin, D., "Development of Non-Linear Transformations for Improving Convergence of Sequences," *Int. J. Computer Math.*, Sec. B, Vol. 3, 1973, pp. 371–388.

Liu, C. C., Hessel, A., and Shmoys, J., "Performance of Probe-Fed Microstrip-Patch Element Phased Arrays," *Trans. IEEE*, Vol. AP-36, Nov. 1988, pp. 1501–1509.

Liu, C. C., Shmoys, J., and Hessel, A., "*E*-Plane Performance Trade-Offs in Two-Dimensional Microstrip-Patch Element Phased Arrays," *Trans. IEEE*, Vol. AP-30, Nov. 1982, pp. 1201–1206.

Lo, Y. T., "A Note on the Cylindrical Antenna of Noncircular Cross Section," *J. Appl. Phys.*, Vol. 24, 1953, pp. 1338–1339.

Lubin, Y. and Hessel, A., "Wide-Band, Wide-Angle Microstrip Stacked-Patch-Element Phased Arrays," *Trans. IEEE*, Vol. AP-39, Aug. 1991, pp. 1062–1070.

Luebbers, R. J. and Munk, B. A., "Cross Polarization Losses in Periodic Arrays of Loaded Slots," *Trans. IEEE*, Vol. AP-23, Mar. 1975, pp. 159–164.

Luebbers, R. J. and Munk, B. A., "Some Effects of Dielectric Loading on Periodic Slot Arrays," *Trans. IEEE*, Vol. AP-26, July 1978, pp. 536–542.

Luzwick, J. and Harrington, R. F., "Mutual Coupling Analysis in a Finite Planar Rectangular Waveguide Antenna Array," *Electromagnetics*, Vol. 2, 1982, pp. 25–42.

Lyon, J. A. M. et al., "Interference Coupling Factors for Pairs of Antennas," *Proceedings 1964 USAF Antenna Symposium*, Allerton, Ill., AD-A609 104.

Magill, E. G. and Wheeler, H. A., "Wide-Angle Impedance Matching of a Planar Array Antenna by a Dielectric Sheet," *Trans. IEEE*, Vol. AP-14, Jan. 1966, pp. 49–53.

Mailloux, R. J., "Surface Waves and Anomalous Wave Radiation Nulls on Phased Arrays of TEM Waveguides with Fences," *Trans. IEEE*, Vol. AP-20, Mar. 1972, pp. 160–166.

Mailloux, R. J., "Synthesis of Spatial Filters with Chebyshev Characteristics," *Trans. IEEE*, Vol. AP-24, Mar. 1976, pp. 174–181.

Marin, M. A. and Pathak, P. H., "An Asymptotic Closed-Form Representation for the Grounded Double-Layer Surface Green's Function," *Trans. IEEE*, Vol. AP-40, Nov. 1992, pp. 1357–1366.

Marin, M., Barkeshli, S., and Pathak, P. H., "Efficient Analysis of Planar Microstrip Geometries Using a Closed-Form Asymptotic Representation of the Grounded Dielectric Slab Green's Function," *Trans. IEEE*, Vol. MTT-37, Apr. 1989, pp. 669–679.

Mohammadian, A. H., Martin, N. M., and Griffin, D. W., "A Theoretical and Experimental Study of Mutual Coupling in Microstrip Antenna Arrays," *Trans. IEEE*, Vol. AP-37, Oct. 1989, pp. 1217–1223.

Montgomery, C. G., Ed., *Technique of Microwave Measurements*, McGraw-Hill, 1947.

Mosig, J. R., "Integral Equation Technique," in "Numerical Techniques for Microwave and Millimeter-Wave Passive Structures," T. Itoh, Ed., Wiley-Interscience, 1989, Chapter 3.

Munk, B. A. and Burrell, G. A., "Plane-Wave Expansions for Arrays of Arbitrarily Oriented Piecewise Linear Elements and Its Application in Determining the Impedance of a Single Linear Antenna in a Lossy Half-Space," *Trans. IEEE*, Vol. AP-27, May 1979, pp. 331–343.

Munk, B. A. and Kornbau, T. W., "Comments on 'On the Use of Metallized Cavities in Printed Slot Arrays with Dielectric Substrates'," *Trans. IEEE*, Vol. AP-36, July 1988, pp. 1036–1041.

Munk, B. A., Kornbau, T. W., and Fulton, R. D., "Scan Independent Phased Arrays," *Radio Sci.*, Vol. 14, Nov.-Dec. 1979, pp. 979–990.

Nehra, C. P. et al., "Probe-Fed Strip-Element Microstrip Phased Arrays: *E*- and *H*-Plane Scan Resonances and Broadbanding Guidelines," *Trans. IEEE*, Vol. AP-43, Nov. 1995, pp. 1270–1280.

Newman, E. H., Richmond, J. H., and Kwan, B. W., "Mutual Impedance Computation between Microstrip Antennas," *Trans. IEEE*, Vol. MTT-31, Nov. 1983, pp. 941–945.

Ng, K. T. and Munk, B. A., "Scan-Independent Slot Arrays with Parasitic Wire Arrays in a Stratified Medium," *Trans. IEEE*, Vol. AP-36, Apr. 1988, pp. 483–495.

Ng, K. T. and Munk, B. A., "Surface-Wave Phenomena in Phased Slot Arrays with Parasitic Wire Arrays," *Trans. IEEE*, Vol. AP-37, Nov. 1989, pp. 1398–1406.

Oliner, A. A., "On Blindness in Large Phased Arrays," *Alta Freq.*, Vol. 38, 1969, pp. 221–228.

Oliner, A. A., "Surface-Wave Effects and Blindness in Phased-Array Antennas," in *Phased Array Antennas*, A. A. Oliner and G. H. Knittel, Artech House, 1972.

Oliner, A. A. and Malech, R. G., "Mutual Coupling in Infinite Scanning Arrays," in *Microwave Scanning Antennas*, Vol. II, R. C. Hansen, Ed., Academic Press, 1966 [Peninsula Publishing, 1985], Chapter 3.

Pearson, L. W., "On the Spectral Expansion of the Electric and Magnetic Dyadic Green's Functions in Cylindrical Harmonics," *Radio Sci.*, Vol. 18, Mar.–Apr. 1983, pp. 166–174.

Pozar, D. M., "Input Impedance and Mutual Coupling of Rectangular Microstrip Antennas," *Trans. IEEE*, Vol. AP-30, Nov. 1982, pp. 1191–1196.

Pozar, D. M. and Schaubert, D. H., "Analysis of an Infinite Array of Rectangular Microstrip Patches with Idealized Probe Feeds," *Trans. IEEE*, Vol. AP-32, Oct. 1984a, pp. 1101–1107.

Pozar, D. M. and Schaubert, D. H., "Scan Blindness in Infinite Phased Arrays of Printed Dipoles," *Trans. IEEE*, Vol. AP-32, June 1984b, pp. 602–610.

Reale, J. D., "PAR Hardened Crossed-Dipole Array (U),' *20th TriService Radar Symp.*, July 1974, pp. 351–364.

Rengarajan, S. R., "Mutual Coupling between Waveguide-Fed Longitudinal Broad Wall Slots Radiating between Baffles," *Electromagnetics*, Vol. 16, Nov.-Dec. 1996, pp. 671–683.

Rhodes, D. R., "On a Fundamental Principle in the Theory of Planar Antennas," *Proc. IEEE*, Vol. 52, Sept. 1964, pp. 1013–1021.

Rhodes, D. R., *Synthesis of Planar Antenna Sources*, Clarendon Press, 1974.

Richmond, J. H., "Mutual Impedance between Coplanar-Skew Dipoles," *Trans. IEEE*, Vol. AP-18, May 1970, pp. 414–416.

Schaubert, D. H., "A Gap-Induced Element Resonance in Single-Polarized Arrays of Notch Antennas," *IEEE 1994 AP Symposium Digest*, Seattle, Wash., 1994, pp. 1264–1267.

Schaubert, D. H., "A Class of E-Plane Scan Blindnesses in Single-Polarized Arrays of Tapered-Slot Antennas with a Ground Plane," *Trans. IEEE*, Vol. AP-44, July 1996, pp. 954–959.

Schaubert, D. H. and Shin, J., "Parameter Study of Tapered Slot Antenna Arrays," *IEEE AP 1995 Symposium Digest*, Newport Beach, Calif., June 1995, pp. 1376–1379.

Schaubert, D. H., Shin, J., and Wunsch, G., "Characteristics of Single-Polarized Phased Array of Tapered Slot Antennas," *IEEE Proceedings of the Symposium on Phased Array Systems and Technology*, Boston, Mass., Oct. 1996, pp. 102–106.

Schelkunoff, S. A. and Friis, H. T., *Antenna Theory and Practice*, Wiley, 1952, pp. 368, 401.

Shanks, D., "Non-Linear Transformations of Divergent and Slowly Convergent Sequences," *J. Math. Phys.*, Vol. 34, 1955, pp. 1–42.

Shin, J. and Schaubert, D. H., "Toward a Better Understanding of Wideband Vivaldi Notch Antenna Arrays," *Proceedings 1995 Antenna Applications Symposium*, Vol. 2, Sept. 1995, Allerton Park, Ill., AD-A309 723, pp. 556–585.

Singh, S. and Singh, R., "On the Use of Levin's T-Transform in Accelerating the Summation of Series Representing the Free-Space Periodic Green's Functions," *Trans. IEEE*, Vol. MTT-41, May 1993, pp. 884–886.

Steyskal, H., "Mutual Coupling Analysis of a Finite Planar Waveguide Array," *Trans. IEEE*, Vol. AP-22, July 1974, pp. 594–597.

Steyskal, H. and Herd, J. S., "Mutual Coupling Compensation in Small Array Antennas," *Trans. IEEE*, Vol. AP-38, Dec. 1990, pp. 1971–1975.

Tang, R. and Wong, N. S., "Multimode Phased Array Element for Wide Scan Angle Impedance Matching," *Proc. IEEE*, Vol. 56, Nov. 1968, pp. 1951–1959.

Tsandoulas, G. N., "Wideband Limitations of Waveguide Arrays," *Microwave J.*, Vol. 15, Sept. 1972, pp. 49–56.

Tsandoulas, G. N., "Unidimensionally Scanned Phased Arrays," *Trans. IEEE*, Vol. AP-28, Jan. 1980, pp. 86–99.

Tsandoulas, G. N. and Knittel, G. H., "The Analysis and Design of Dual-Polarization Square-Waveguide Phased Arrays," *Trans. IEEE*, Vol. AP-21, Nov. 1973, pp. 796–808.

Varon, D. and Zysman, G. I., "On the Mismatch of Electronically Steerable Phased-Array Antennas," *Radio Sci.*, Vol. 3, May 1968, pp. 487–489.

Villeneuve, A. T., Behnke, M. C., and Kummer, W. H., "Radiating Elements for Hemispherically Scanned Arrays," *Trans. IEEE*, Vol. AP-30, May 1982, pp. 457–462.

Wasylkiwskyj, W. and Kahn, W. K., "Theory of Mutual Coupling among Minimum-Scattering Antennas," *Trans. IEEE*, Vol. AP-18, Mar. 1970, pp. 204–216.

Wasylkiwskyj, W. and Kahn, W. K., "Element Pattern Bounds in Uniform Phased Array," *Trans. IEEE*, Vol. AP-25, Sept. 1977, pp. 597–604.

Waterhouse, R. B., "The Use of Shorting Posts to Improve the Scanning Range of Probe-Fed Microstrip Patch Phased Arrays," *Trans. IEEE*, Vol. AP-44, Mar. 1996, pp. 302–309.

Westlake, J. R., *A Handbook of Numerical Matrix Inversion and Solution of Linear Equations*, John Wiley, 1968, Chapters 2 and 7.

Wheeler, H. A., "The Radiation Resistance of an Antenna in an Infinite Array or Waveguide," *Proc. IRE*, Vol. 36, Apr. 1948, pp. 478–488.

Wheeler, H. A., "The Radiansphere around a Small Antenna," *Proc. IRE*, Vol. 47, Aug. 1959, pp. 1325–1331.

Wheeler, H. A., "Simple Relations Derived from a Phased-Array Antenna Made of an Infinite Current Sheet," *Trans. IEEE*, Vol. AP-13, July 1965, pp. 506–514.

Wheeler, H. A., "The Grating-Lobe Series for the Impedance Variation in a Planar Phased-Array Antenna," *Trans. IEEE*, Vol. AP-14, Nov. 1966, pp. 707–714.

Wimp, J., "Polynomial Approximations to Integral Transforms," *Math. Tables Other Aids Comput.*, Vol. 15, 1961, pp. 174–178.

Wong, N. S. et al., "Multimode Phased Array Element for Wide Scan Angle Impedance Matching," in *Phased Array Antennas*, A. A. Oliner and G. H. Knittel, Eds., Artech House, 1972, pp. 178–186.

Wu, C. P., "The Effects of Dielectrics", in *Theory and Analysis of Phased Array Antennas*, N. Amitay, V. Galindo, C. P. Wu, Wiley-Interscience, 1972, Chapter 6.

Wu, C. P. and Galindo, V., "Surface Wave Effects on Dielectric Sheathed Phased Arrays of Rectangular Waveguides," *BSTJ*, Vol. 47, Jan. 1968, pp. 117–142.

CHAPTER EIGHT

Finite Arrays

8.1 METHODS OF ANALYSIS

8.1.1 Overview

Although the infinite array techniques of Chapter 7 are excellent for system trades and preliminary design, final design requires a finite array simulation. Direct impedance (admittance) matrix methods were described in Section 7.3.2. These were developed by Oliner and Malech, 1966a; Galindo, 1972; Bailey, 1974; Bailey and Bostian, 1974; Cha and Hsiao, 1974; Steyskal, 1974; Bird, 1979; Luzwick and Harrington, 1982; Clarricoats et al., 1984; Pozar, 1985, 1986; Fukao et al., 1986; Deshpande and Bailey, 1989; Silvestro, 1989; Usoff and Munk, 1994; and others. When moment methods are not necessary (thin half-wave dipoles, for example), sizeable planar arrays may be solved. When the element current distribution is complicated, the number of moment method expansion functions needed for good convergence will restrict this method to small arrays.

An elegant solution to the semi-infinite array, that is, an array extending to infinity on three sides, is given by Wasylkiwskyj (1973). The Weiner–Hopf factorization procedure is extended to finite Fourier transforms, resulting in an expression for *scan reflection coefficient* of the semi-infinite array in terms of *scan reflection coefficient* for the corresponding infinite array, and an integral of a phased sum of the infinite array *scan impedance*s.

Still another approach embeds the finite array in a matrix of identical arrays with blank space between (Roederer, 1971; Ishimaru et al., 1985; Skriverik and Mosig, 1993; Roscoe and Perrott, 1994; Cátedra et al., 1995). For single-mode elements (thin dipoles, thin slots, thin patches) the procedure is simple. First, the *scan element pattern* (SEP) is computed for the corresponding infinite array. This SEP is then Fourier transformed [usually by discrete Fourier transform (DFT)] back to the aperture. Third, a periodic structure consisting of equally spaced finite arrays, with each array having the desired amplitude distribution and scan phase, is transformed. Fourth, these two transforms are convolved. Next the result is inverse transformed to get the SEPs over the periodic structure. Last, the SEP for one finite array is extracted, and multiplied by the isolated element pattern, yielding the final finite array *scan element pattern*.

When the type of element allows multiple modes or nonsimple currents, the changing mutual coupling over the finite array may change the mode mix or change the current distribution. For these cases, each element needs a moment method (or modal) expansion, with all of the equations representing the expansion functions and number of elements solved simultaneously. A term of the impedance matrix for one expansion function consists of a double sum (over the elements) containing an integral of the appropriate Green's function times the expansion and test functions, times the exponential scan factor. Through the Poisson sum formula this is transformed to the spectral domain. An example of this formula is given in Section 7.4.1. The term is now written as a convolution product of this spectral domain result and the transform of the array amplitude distribution. The simultaneous equations are solved in the spectral domain, separately for each array element, resulting in the *scan impedance* for each element. To reduce errors introduced by aliasing between nearby arrays, it is necessary for the blank spaces to be greater than the width of the finite array. But as the ratio of blank to array becomes larger, the convergence deteriorates. This method, like all the others, has both good and bad aspects.

A simple approximate analysis method is called the "large array method," wherein the element currents are assumed to be the Floquet infinite array currents, and the mutual impedances multiplied by the scan phase factors are summed over the finite array to get *scan impedance*. Amplitude tapers are readily included. Impedance of the center element is given by

$$Z = \sum\sum A_{nm} Z_{nm} \exp[jk(nd_x u + md_y v)]. \qquad (8.1)$$

The *scan element pattern* is obtained from this *scan impedance*, as shown in Section 7.2.3:

$$g(u, v) = \frac{R(0, 0)g_{iso}(u, v)}{R(u, v)g_{iso}(0, 0)} \left(1 - |\Gamma(u, v)|^2\right). \qquad (8.2)$$

Here $R(u, v)$ and $\Gamma(u, v)$ are the *scan resistance* and *scan reflection coefficient*; $g_{iso}(u, v)$ is the isolated element (and backscreen) power pattern. Both resistance and pattern are normalized to broadside values.

Results from the large-array method show finite array oscillations superimposed onto infinite array results (Hansen, 1990). Figures 8.1 and 8.2 show *scan element pattern* for the 0.5λ lattice without screen, for array sizes 41×41 and 101×101. As expected, the number of oscillations in the pattern is about half the number of elements along x or y, but only half the symmetric pattern is shown. Somewhat surprisingly, the wide-angle (largest) oscillations appear to have an amplitude independent of the array size. Because of the oscillatory effect of adding a row or column of elements to the array, it might be expected

METHODS OF ANALYSIS 275

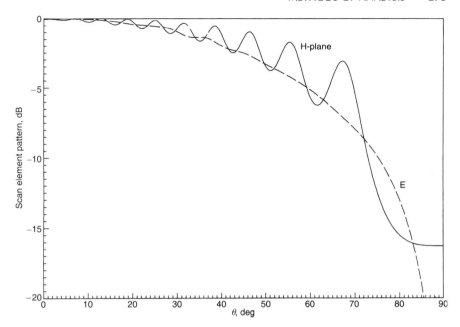

Figure 8.1 Dipole array, large array approximation, 41×41, $d_x = d_y = 0.5\,\lambda$, $L = 0.5\,\lambda$.

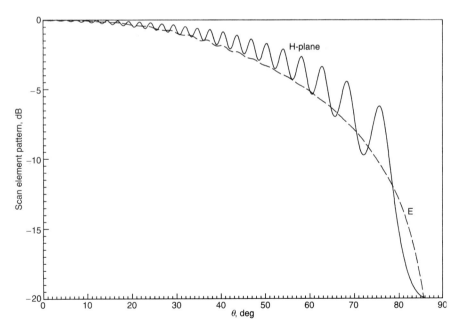

Figure 8.2 Dipole array, large array approximation, 101×101, $d_x = d_y = 0.5\,\lambda$, $L = 0.5\,\lambda$.

that adding half the mutual impedances for the perimeter elements would reduce the oscillations. This proves to be the case, a 1 dB reduction is obtained, so all results shown utilize half-perimeter mutual impedances. Figures 8.3 and 8.4 show SEP for the 0.7λ lattice case. Again oscillations are superimposed on the infinite array results and, as expected, the number of oscillations is governed by the number of elements in the scan direction. The 41 × 41 array of Fig. 8.3 provides a coarse approximation to the infinite array results; the 101 × 101 array of Fig. 8.4 is better. The array size in the transverse direction is less critical; representation of deep grating lobe blind angles requires 41 or more elements.

Addition of a back screen appears to fill in the cyclic behavior of the dipole–dipole impedances, with the result that the large-array approximation gives very small oscillations. Figure 8.5 shows that very good correlation with infinite array results occurs for 41 × 41 array elements. For 21 × 21 array, the oscillations are less than 1 dB. The large-array method appears useful for arrays where the element pattern is of dipole/screen type.

8.1.2 Finite-by-Infinite Arrays

Computational cost is reduced when a finite-by-infinite array is considered, with scan across the finite width of the array (Denison and Scharstein, 1995). Only principal plane scans are allowed here. Even when moment methods are applied, relatively large arrays can be simulated. From this type of

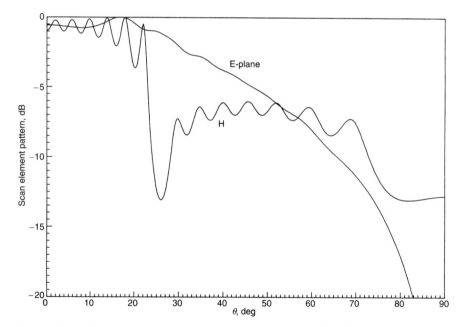

Figure 8.3 Dipole array, large array approximation, 41 × 41, $d_x = d_y = 0.7\,\lambda$, $L = 0.5\,\lambda$.

METHODS OF ANALYSIS 277

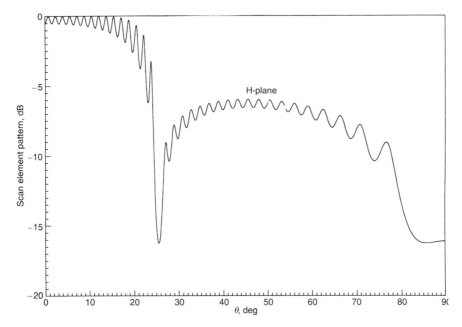

Figure 8.4 Dipole array, large array approximation, 101×101, $d_x = d_y = 0.7\,\lambda$, $L = 0.5\,\lambda$.

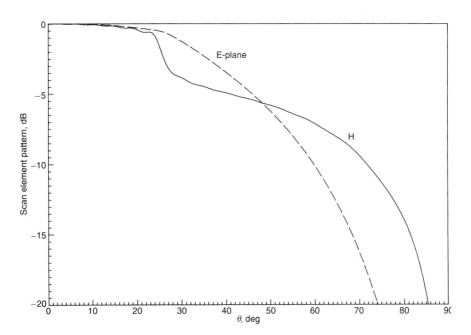

Figure 8.5 Dipole/screen array, large array approximation, 41×41, $d_x = d_y = 0.7\,\lambda$, $L = 0.5\,\lambda$, $H = 0.25\,\lambda$.

278 FINITE ARRAYS

simulator much can be gleaned about the behavior of edge elements in an array. As has been shown many times, half-wave thin dipole arrays exhibit all the pertinent features of mutual coupling. Thus complex array elements and moment method solutions are not needed at this stage. By considering collinear or parallel dipoles, both *H*-plane and *E*-plane scans can be accommodated. The computational model then consists of N infinite linear arrays as sketched in Figures 8.6 and 8.7. Each infinite array ("stick") is composed of half-wave thin dipoles. Thus mutual impedance between a dipole and the dipoles in one infinite "stick" is then computed. The set of these for all the sticks is used to form the usual impedance matrix, with the solution providing the *scan impedance* of a dipole in each stick.

The reference *scan impedance* for the infinite array is provided by the spectral domain formulation pioneered by Oliner and Malech (1996b). This give the *scan resistance* as a single term for each main lobe or grating lobe and a spectral summation of trigonometric type functions for the *scan reactance*. For array spacings close to $\lambda/2$ the formulation converges extremely rapidly.

For both the *H*-plane and *E*-plane scan cases, the dipole stick impedances have been computed both in the spatial domain and in the spectral domain. In the spatial domain a subroutine based on Carter's mutual impedance between two thin wire dipoles (Hansen, 1972) is used to obtain dipole–dipole impedance Z_a. This involves Sine and Cosine Integrals which are calculated in double precision, to allow many terms to be summed. For the collinear dipole sticks (*H*-plane scan), the dipole to linear array impedance is just

$$Z = \sum Z_a\left(x_0 + nd, \sqrt{a^2 + y^2}\right). \tag{8.3}$$

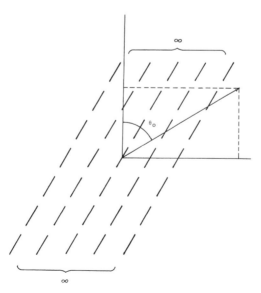

Figure 8.6 Finite-by-infinite dipole array with *H*-plane scan.

METHODS OF ANALYSIS   279

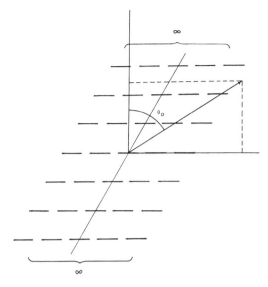

Figure 8.7 Finite-by-infinite dipole array with E-plane scan.

As shown in Figure 8.8 the isolated dipole offset is a, the parallel shift is x_0, the perpendicular spacing is y_0, the array spacing is d, and dipole half-lengths are h. For parallel dipole sticks (E-plane scan), the dipole to linear array impedance is

$$Z = \sum Z_a\left(\sqrt{(x_0 + nd)^2 + a^2}, y_0\right). \tag{8.4}$$

Dimensions are shown in Fig. 8.9. Difficulty occurs in that the sums oscillate, with this worsening as the dipole separation decreases. As the period of the

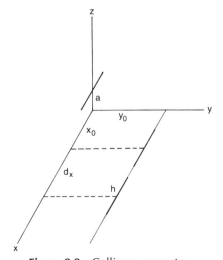

Figure 8.8 Collinear geometry.

sum oscillations changes with dimensions, no simple compensation method works. With these summations, no convergence was found after more than 1000 terms. Convergence acceleration techniques include those of Shanks (1955) and Levin (1973). Both have been used, but the Levin is superior as it does not concatenate roundoff errors. The Levin transformation is a rearrangement of partial sums (Singh et al., 1990; Singh and Singh, 1993):

$$S^n = \frac{\sum_{i=0}^{m}(-1)^i\binom{m}{i}\left(\frac{n+i}{n+m}\right)^{m-1}\frac{S_{n+1}}{S_{n+i+1}-S_{n+i}}}{\sum_{i=0}^{m}(-1)^i\binom{m}{i}\left(\frac{n+1}{n+m}\right)^{m-1}\frac{1}{S_{n+i+1}-S_{n+i}}}. \tag{8.5}$$

This gives the transform of m order. $\binom{m}{i}$ is the binomial coefficient.

Typically, 21 Levin cycles gave excellent accuracy for all cases; because parallel dipole mutual impedance decays asymptotically as $1/r$, instead of the collinear value of $1/r^2$, more Levin cycles are needed to achieve convergence for that case. For both scan planes, larger element spacing gives faster convergence.

The spectral domain summation technique was also used; it is a simplification of the *periodic moment method*, developed by Ben Munk and colleagues at Ohio State University (Schubert and Munk, 1983). The mutual impedance is written as a spectral (Floquet) summation over the wavenumber, where the wavenumber includes the element spacing. For the H-plane scan array, the summation contains Hankel functions and generalized pattern factors:

$$Z_i = -\frac{30}{d\sin^2 kh}\sum P^2 \exp(jkx_0 u)H_0^{(2)}\left(k\sqrt{(a^2+y_0^2)(1-u^2)}\right) \tag{8.6}$$

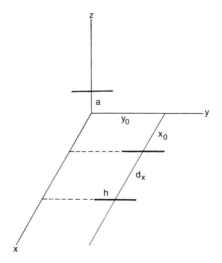

Figure 8.9 Parallel geometry.

with

$$P = \frac{\cos khu - \cos kh}{k\sqrt{1-u^2}}, \qquad u = \sin\theta_0. \tag{8.7}$$

For this H-plane case, 401 terms gave a mutual impedance accuracy of 0.01 ohm. Larger element spacings require more terms.

For the parallel (E-plane) case, a significant difference exists from the collinear case: the Hankel function is now inside the generalized pattern function integral, as shown by Skinner and Munk (1992) and Skinner et al. (1995), and the integral cannot be integrated in closed form. The formulation, with details of the derivation omitted, is

$$Z_i = -\frac{30}{d_x \sin^2 kh} \sum \exp(jkx_0 u) \int_{-h}^{h} \sin[k(h-|y|)]$$
$$\times H_0^{(2)}\left(k\sqrt{a^2 + (y+y_0-h)^2}\sqrt{1-u^2}\right) dy. \tag{8.8}$$

If d_x does not allow a grating lobe, the argument of $H_0^{(2)}$ is imaginary except for $n = 0$; $H_0^{(2)}$ is replaced by $(2/\pi)K_0(\text{arg})$. Numerical integration was used for each term in the series above. For $y_0 > 1$, a complex 32-point Gaussian was used, with a 128-point Gaussian for the closer spacings (Stroud and Secrest, 1966). Excellent convergence was achieved with 501 terms, but the result was a slow program.

With proper choice of limits of both spectral and spatial summations, the plots of *scan impedance* from these two techniques lay on top of one another. In general the spectral summation was slower and less satisfactory. Results are given in Section 8.3.1.

8.2 SCAN PERFORMANCE OF SMALL ARRAYS

Using the impedance matrix approach, Diamond (1972) computed *scan impedance* and *scan element pattern* of several small arrays of dipoles. Figure 8.10 shows *scan resistance* for the center element of an array of 7×9 half-wave dipoles on a $\lambda/2$ square lattice, with the infinite array results, shown dashed, for comparison. In Figure 8.11 the same array is over a ground screen at $\lambda/4$ spacing. The screen removes the high values of the H- and diagonal planes. Small arrays exhibit behavior that oscillates around that of a large array. Similar oscillations occur in the *scan reactance* of small arrays. In general, as might be expected, larger arrays have more oscillations, but of smaller amplitude.

Scan element patterns are of interest for small arrays although it is no longer valid to multiply the array factor by a single *scan element pattern* in order to get the overall array pattern. Figure 8.12 shows calculated and

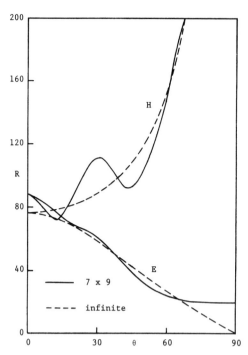

Figure 8.10 Scan resistance of center element of half-wave dipole array. (Courtesy Diamond, B. L., "Theoretical Investigation of Mutual Coupling Effects," Chapter 3 of TR381, Lincoln Laboratory, March 1965.)

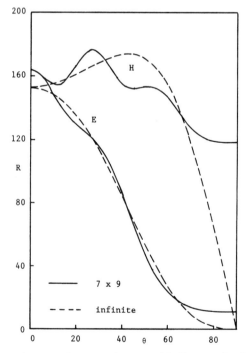

Figure 8.11 Scan resistance of center element of half-wave dipole array with screen. (Courtesy Diamond, B. L., "Theoretical Investigation of Mutual Coupling Effects," Chapter 3 of TR381, Lincoln Laboratory, March 1965.

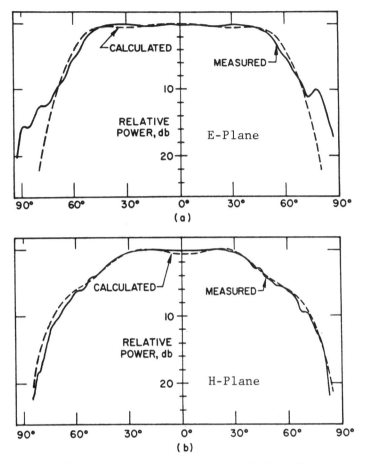

Figure 8.12 Scan element patterns of center element of a 5 × 5 dipole/screen array. (Courtesy W. E. Rupp.)

measured *scan element pattern*s of the center dipole in a 5 × 5 array of half-wave dipoles in a half-wave lattice, and quarter-wave over a ground plane. The broadening and flattening effect on the E-plane pattern is quite noticeable. Figures 8.13 and 8.14 are *scan element patterns* for the center element of a 7 × 9 array of dipoles for various lattice spacings. From these curves it is apparent that the *scan element pattern* limits the scan angle range. As more elements are added to the array, the number of oscillations in the pattern increases and the curves become smoother. Also the falloff becomes steeper. Figure 8.15 and 8.16 compare center, edge center, and corner *scan element patterns* for a similar 9 × 11 array of dipoles. The corner elements do not show the symmetric mutual coupling effects that tend to flatten the *scan element pattern*. Dummy elements can be added around the edges of an array to be scanned if space is available.

284 FINITE ARRAYS

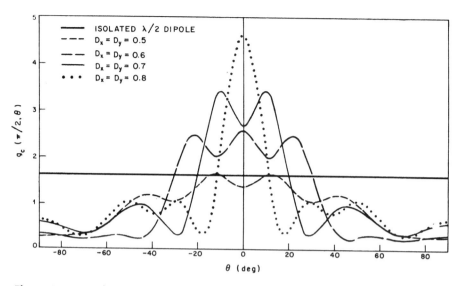

Figure 8.13 *H*-plane scan element patterns of center element of 7 × 9 dipole array. (Courtesy Diamond, B. L., "Theoretical Investigation of Mutual Coupling Effects," Chapter 3 of TR381, Lincoln Laboratory, March 1965.)

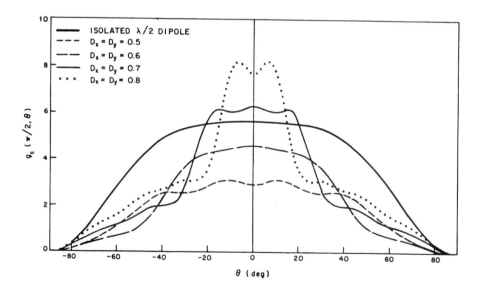

Figure 8.14 *H*-plane scan element patterns of center element of 7 × 9 dipole/screen array. (Courtesy Diamond, B. L., "Theoretical Investigation of Mutual Coupling Effects," Chapter 3 of TR381, Lincoln Laboratory, March 1965.)

SCAN PERFORMANCE OF SMALL ARRAYS   285

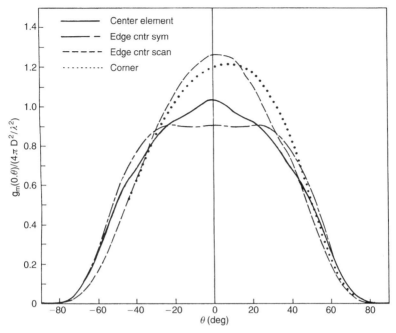

Figure 8.15 *E*-plane scan element patterns for 9 × 11 dipole/screen array. (Courtesy Diamond, B. L., "Theoretical Investigation of Mutual Coupling Effects," Chapter 3 of TR381, Lincoln Laboratory, March 1965.)

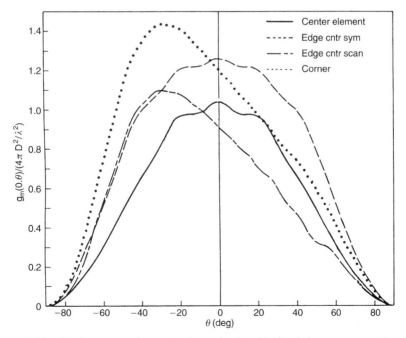

Figure 8.16 *H*-plane scan element patterns for 9 × 11 dipole/screen array. (Courtesy Diamond, B. L., "Theoretical Investigation of Mutual Coupling Effects," Chapter 3 of TR381, Lincoln Laboratory, March 1965.)

An interesting example of edge effects in a linear array is given by Gallegro (1969), for half-wave dipoles with and without a ground plane. Spacing between elements is 0.586λ, which gives a just-visible grating lobe at $\theta = 90$ deg for 45 deg scan; dipole orientation is both parallel and collinear. The center element is matched at 0 deg scan angle, with all elements using the same matching network. All arrays have 51 elements, and the ground plane spacing is 0.25λ. The collinear arrays are well behaved even at grating lobe incidence; the parallel arrays are somewhat less well behaved up to grating lobe (GL) incidence, but at incidence there was a marked edge effect, especially for the array without ground plane. Figures 8.17 and 8.18 show element impedance by number with and without ground plane for scan angles of 0 deg, 40 deg, and 45 deg for parallel dipoles. It appears that the edge elements in the scan direction are more affected than the edge elements in the reverse direction, but the VSWR of the reverse edge elements is higher owing to the grating lobe pointing in that direction. A rough tabulation of the number of edge elements where the edge element impedance is outside the center cluster is shown in Table 8.1. It must not be supposed that these results apply to planar arrays, but the trends should be useful.

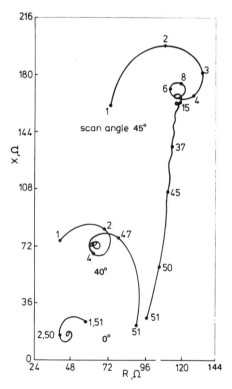

Figure 8.17 Scan impedance of linear array of parallel dipoles. (Courtesy Gallegro, A. D., Mutual Coupling and Edge Effects in Linear Phased Arrays, MS thesis, Polytechnic Institute of Brooklyn, 1969.

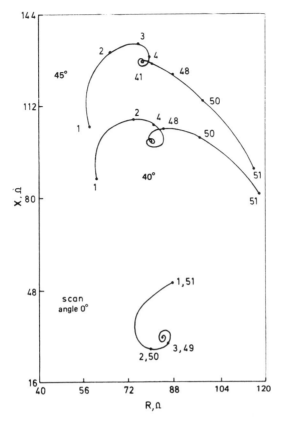

Figure 8.18 Scan impedance of a linear array of parallel dipoles/screen. (Courtesy Gallegro, A. D., Mutual Coupling and Edge Effects in Linear Phased Arrays, MS thesis, Polytechnic Institute of Brooklyn, 1969.)

TABLE 8.1 Number of "Edge Elements": Grating Lobe for 45 deg Scan

Dipoles	Ground Plane	Scan Direction	Reverse Direction
Parallel	No	1–5–30	1–2–3
Parallel	Yes	1–3–6	1–2–3
Collinear	No	1–2–2	1–1–1
Collinear	Yes	1–1–2	1–1–2

8.3 FINITE-BY-INFINITE ARRAY GIBBSIAN MODEL

8.3.1 Salient Scan Impedance Characteristics

The finite-by-infinite array simulators described in Section 8.1.2 were used to compute *scan impedance* across the array, for several arrays from large to small. Results for *H*-plane for arrays of 201 resonant dipoles are shown as

288 FINITE ARRAYS

the solid lines in Figures 8.19 through 8.21. Scan angles are 0, 45, and 60 deg. The dotted line curves are the Gibbsian model that is discussed later. The dipole lengths were adjusted to give a resonant impedance in an infinite broadside array, and these dipole lengths were used. The array lattice was $\lambda/2 \times \lambda/2$, and dipole radii were 0.005λ (Hansen and Gammon, 1995). *Scan impedances* are normalized by the infinite array value at each scan angle. The dipoles are numbered 1 through 201 starting at the edge, with beam scan toward element 1. These figures show apparently continuous curves, but note that they have been plotted from 201 points. For Figure 8.19 only half of the array has been plotted; the broadside curves are symmetric.

Figures 8.22 and 8.23 show *scan impedance* across a smaller array of 51 sticks, for scan angles of 0 and 60 deg. In Fig. 8.24 is shown *scan impedance* for a 21-stick array at broadside. Note that the smaller arrays fit the same pattern as the large array. The *scan impedance* oscillations are very small in the center of the array at broadside, and become larger as the scan angle increases. Edge values are larger for all scan angles, with 20% typical.

These dipole array results are directly applicable to arrays of slots in a metallic sheet, through Babinet's principle. With slots in a waveguide wall, the internal susceptance must be added to the slot admittance.

Figures 8.25 through 8.27 are for the same 201-element array but with a back screen at $\lambda/4$ spacing. Again dipole lengths were adjusted for resonance in an infinite broadside array. In these graphs, the oscillations decay more rapidly toward the array center, although the edge values are still large. Broadside results for smaller arrays of 51 and 21 elements are given in Figs. 8.28 and 8.29. Again these follow the pattern for the 201-element array.

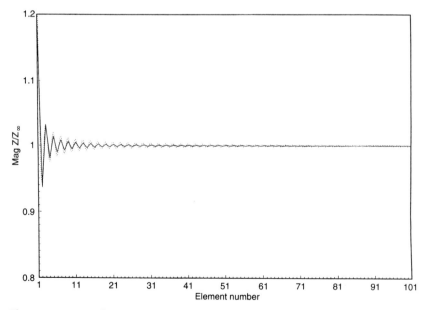

Figure 8.19 201 linear infinite arrays of resonant dipoles, *H*-plane scan at 0 deg.

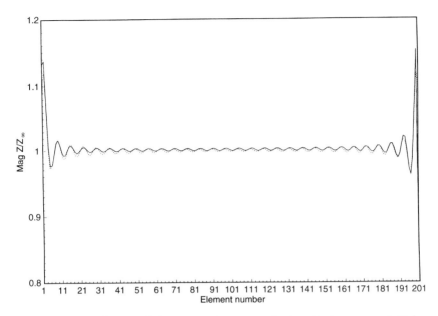

Figure 8.20 201 linear infinite arrays of resonant dipoles, *H*-plane scan at 45 deg.

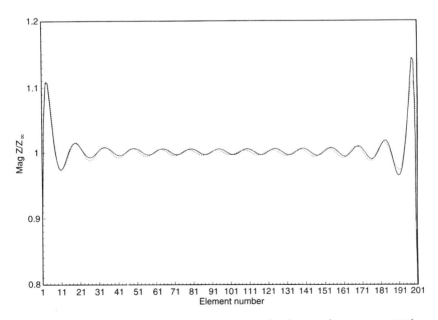

Figure 8.21 201 linear infinite arrays of resonant dipoles, *H*-plane scan at 60 deg.

290 FINITE ARRAYS

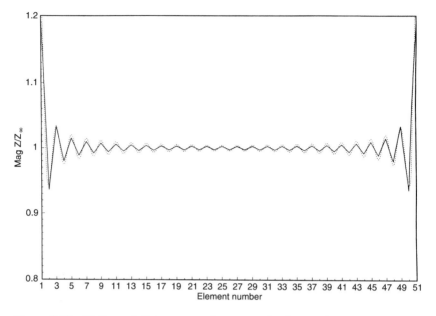

Figure 8.22 51 linear infinite arrays of resonant dipoles, *H*-plane scan at 0 deg.

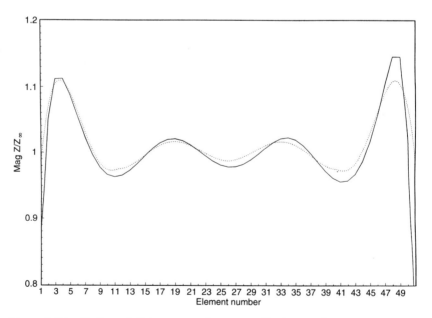

Figure 8.23 51 linear infinite arrays of resonant dipoles, *H*-plane scan at 60 deg.

FINITE-BY-INFINITE ARRAY GIBBSIAN MODEL

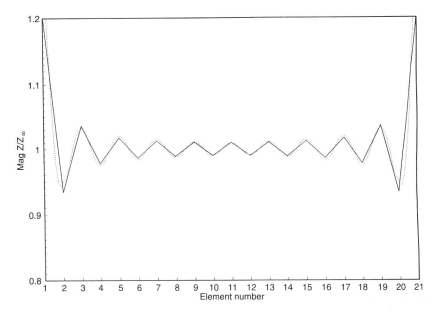

Figure 8.24 21 linear infinite arrays of resonant dipoles, *H*-plane scan at 0 deg.

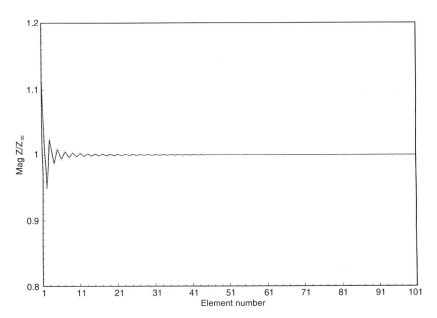

Figure 8.25 201 linear infinite arrays of resonant dipoles/screen, *H*-plane scan at 0 deg.

292 FINITE ARRAYS

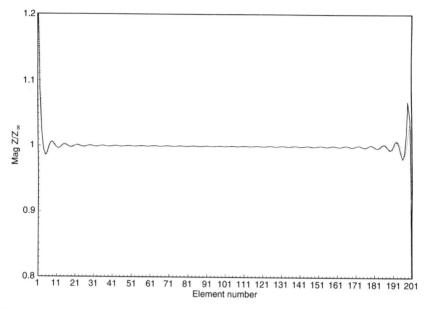

Figure 8.26 201 linear infinite arrays of resonant dipoles/screen, *H*-plane scan at 45 deg.

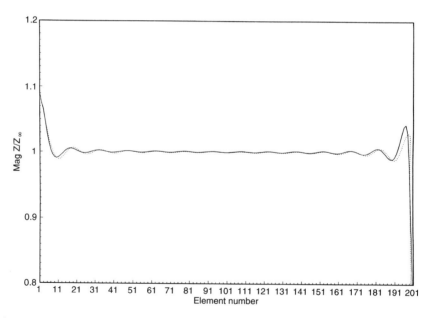

Figure 8.27 201 linear infinite arrays of resonant dipoles/screen, *H*-plane scan at 60 deg.

FINITE-BY-INFINITE ARRAY GIBBSIAN MODEL   293

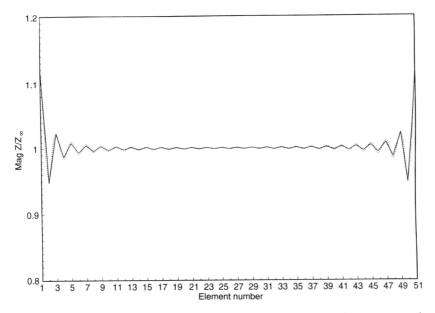

Figure 8.28 51 linear infinite arrays of resonant dipoles/screen, *H*-plane scan at 0 deg.

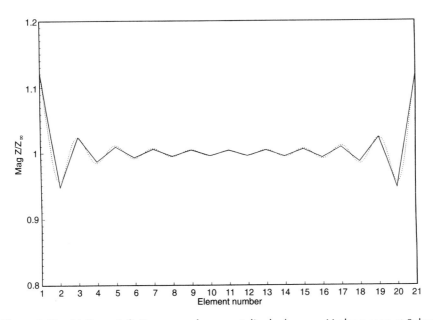

Figure 8.29 21 linear infinite arrays of resonant dipoles/screen, *H*-plane scan at 0 deg.

294 FINITE ARRAYS

From these curves several things are evident. First a pseudo-standing wave in *scan impedance* exists over the face of the array, even for broadside radiation. Second, the period of the pseudo-standing wave is a function of the scan angle as one might expect. And third, the two pseudo-travelling waves that constitute the pseudo-standing wave are attenuated with distance from their respective edges.

The period across the array is almost exactly constant, and has been determined to be given closely by

$$\text{Period} = \frac{0.5}{1 - \sin\theta_0} \quad \text{(in wavelengths)}, \tag{8.9}$$

where θ_0 is the scan angle from broadside. This result may be interpreted as a grating lobe type formula from two pseudo-waves, one at the scan angle and a second along the array toward the edge.

Turning now to *E*-plane scan, Figs. 8.30 and 8.31 show arrays of 201 parallel dipole sticks, again a $\lambda/2 \times \lambda/2$ lattice, for scan angles of 0 and 60 deg. Dipole lengths were again adjusted for resonance in an infinite array. Now the decay of oscillations from edge to center is slower. In general the excursions are larger than those for *H*-plane dipole arrays. At broadside, the *E*-plane scan edge value is lower, while for *H*-plane it is higher than unity; the deviations are comparable (Hansen and Gammon, 1996a). Figure 8.32 gives data for a 51-element array at zero scan.

When a screen was added for *E*-plane scan cases, a new phenomenon appeared. At broadside, the expected half-wave-spaced oscillations appeared

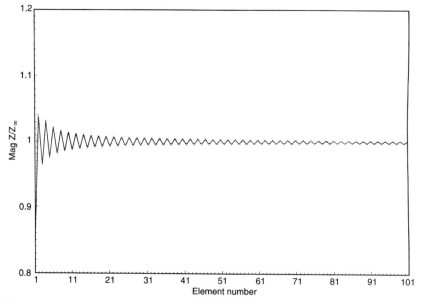

Figure 8.30 201 linear infinite arrays of resonant dipoles, *E*-plane scan at 0 deg.

FINITE-BY-INFINITE ARRAY GIBBSIAN MODEL   **295**

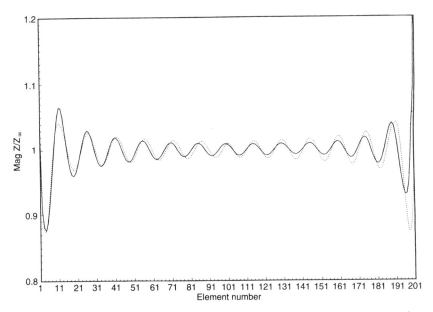

Figure 8.31 201 linear infinite arrays of resonant dipoles, *E*-plane scan at 60 deg.

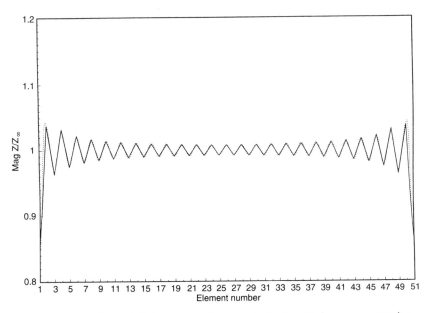

Figure 8.32 51 linear infinite arrays of resonant dipoles, *E*-plane scan at 0 deg.

296   FINITE ARRAYS

to be modulated by an oscillation of larger period, reminiscent of a heterodyne process. Figure 8.33 is for an array of 201 resonant dipole sticks, dipole length 0.4841λ, dipole radius 0.005λ, screen spacing 0.25λ and lattice $\lambda/2 \times \lambda/2$. The previous investigation of H-plane scan of dipoles and dipoles/screen, and of E-plane scan of dipoles, showed no significant change with dipole radius, as long as it was small. In the E-plane dipole/screen case here, the radius affects the modulation period. Assuming a heterodyne process, with the "difference frequency" operating, the period P is simply $1/P = 1/P_1 - 1/P_2$, where P_1 is the basic *scan impedance* period of one wavelength (zero scan), and P_2 is P_1 modified by a factor: $P_2 = P_1(1 + \Delta)$. Figure 8.34 shows the parameter Δ for five radii from 0.001λ to 0.02λ (Hansen and Raudenbush, 1996). This log–linear relationship is due to the change in mutual impedances with change in radius, all reflected through the impedance matrix inverse. Why this appears only for E-plane dipoles with screen is not known.

The obvious question is how this phenomenon changes with scan angle. At 30 deg no modulation is apparent. At 15 deg scan there is modulation present but the pattern of it is not clear. This new phenomenon represents a challenge in array modeling.

An infinite array of half-wave dipoles with element spacing sufficiently large to allow a grating lobe exhibits a blind angle at the grating lobe angle; that is, the *scan resistance* and *scan reactance* become infinite; see Chapter 7. As a result, the magnitude of the reflection coefficient is unity and no power is radiated. The obvious question is: What happens to *scan impedance* in a finite array at GL incidence. The H-plane dipole array simulator was used with element spacing of 0.5858λ, which allowed a grating lobe to appear at

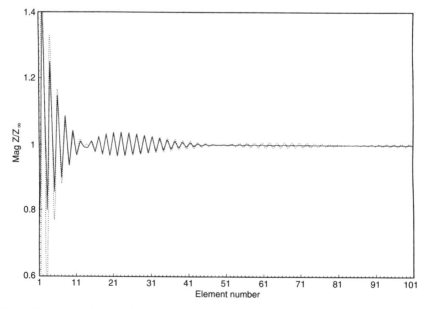

Figure 8.33   201 linear infinite arrays of resonant dipoles/screen, E-plane scan at 0 deg.

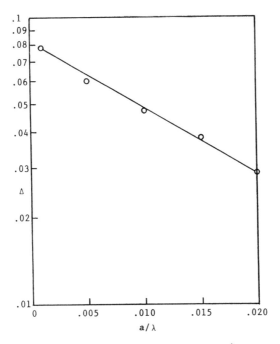

Figure 8.34 Modulation factor Δ vs. radius.

45 deg. Figure 8.35 shows *scan impedance* magnitude for a 201-stick array at 45 deg scan; it increases smoothly from near zero to roughly 1.4, with a few oscillations at the rear. Smaller arrays show closely similar behavior, with of course fewer oscillations. The normalization factor, which is the value in the center of the array, shows a log–log behavior with number of elements, as seen in Fig. 8.36. It can be seen that very large arrays approach a blind angle, with the infinite impedance. Phase of the *scan impedance* is roughly 45 deg; so the phase is approximately that of a conductor.

The same array at the other scan angles rapidly reverts to normal behavior; the smooth *scan impedance* behavior over the array occurs only for a narrow range of angles around the GL angle. At other scan angles, there is minor half-wave modulation.

8.3.2 A Gibbsian Model for Finite Arrays

It is well known that a bandlimited square wave exhibits oscillations that are termed Gibbs's phenomenon. That is, the Fourier transform with finite limits of the sinc spectrum produces the square wave with these oscillations. An examination of various computed results reveals that the peaks and dips of the oscillations are regularly spaced, and that the amplitudes of the first, second, third, etc., peaks are the same regardless of the upper band limit. There the first peak occurs at each edge of the square wave, with the second peak adjacent, and so on. As the upper band limit increases, the total number of peaks increases,

298 FINITE ARRAYS

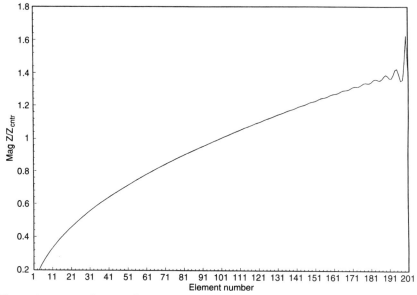

Figure 8.35 201 linear infinite arrays of resonant dipoles, *H*-plane scan at 45 deg.

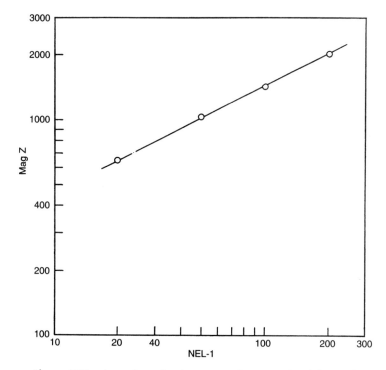

Figure 8.36 Array impedance center value at grating lobe angle.

being roughly the upper limit divided by π, and of course the spacing changes, but the amplitudes remain the same. There is little mention in the literature regarding how to calculate these peaks for a square wave, but Hamming (1972) quotes results derived by Carslaw (1930). A clever technique is used by the latter to determine the extrema positions. The finite series representing the square wave is differentiated, and it is recognized that this sum has a closed-form result. When set to zero, the roots immediately give the extrema locations. These are $m\pi/2n$, where m is the extremum index and n is the upper limit on the summation. The amplitudes at these points are immediately expressed as an integral which, for large n, can be expressed as a Sine Integral; the result is $(2/\pi)\mathrm{Si}(m\pi)$. Thus for large band limit values, the peak amplitudes are readily determined to be 1.1790, 0.9028, 1.0062, 0.9499, and so on.

It might be expected that a single pulse with linear exponential phase across it would better represent the scanned finite array. The aperture distribution should be $\exp(-jtU\sin\theta_0)$ for $|t| \leq 1$, and zero for $|t| > 1$. The transform of this aperture distribution is the spectrum, which is (Sneddon, 1951)

$$f(t) = \mathrm{sinc}\left[(\omega - \alpha)\frac{\tau}{2}\right]. \tag{8.10}$$

Here $\alpha = U\sin\theta_0$. This spectrum is then transformed with finite limits to produce the scanned pulse with Gibbs's type oscillations. The transform is

$$f(t) = \int_{-U}^{U} \tau \,\mathrm{sinc}\left[(\omega - \alpha)\frac{\tau}{2}\right] \exp(j\omega\tau)\, dw. \tag{8.11}$$

The transform integration is complex, and produces a real part involving Sine Integrals, and an imaginary part involving Cosine Integrals. The latter primarily contribute deltalike functions at the pulse edges, while the former constitute the oscillations over the pulse. After various changes of variables the real part becomes (Hansen and Gammon, 1996b)

$$f(t) = \mathrm{Si}\left[(U + \alpha)(1 + T)\right] + \mathrm{Si}\left[(U - \alpha)(1 + T)\right] \\ + \mathrm{Si}\left[(U + \alpha)(1 - T)\right] + \mathrm{Si}\left[(U - \alpha)(1 - T)\right]. \tag{8.12}$$

The extrema locations are found by setting the derivative of f w.r.t. T equal to zero using the previous value for U, and dropping the phase term:

$$f' = \frac{\sin\left[(U + \alpha)(1 + T)\right]}{1 + T} + \frac{\sin\left[(U - \alpha)(1 + T)\right]}{1 + T} \\ - \frac{\sin\left[(U + \alpha)(1 - T)\right]}{1 - T} - \frac{\sin\left[(U - \alpha)(1 - T)\right]}{1 - T} = 0. \tag{8.13}$$

Because of the oscillations of each of these four terms, there are many clustered roots. A useful approximation assumes the first three Si to have large arguments, hence approximately equal to $\pi/2$. Then,

$$f' \simeq \frac{\sin[(U-\alpha)(1-T)]}{1-T} = 0, \qquad (8.14)$$

which gives

$$T \simeq 1 - \frac{n\pi}{U-\alpha} = 1 - \frac{n\pi}{U(1-\sin\theta_0)}. \qquad (8.15)$$

Thus the period is increased by a pseudo-grating lobe factor $1/(1-\sin\theta_0)$. This reduction in number of oscillations with increasing scan angle closely fits the results from the computer simulations. Next a model will be constructed for these *scan impedance* results (Hansen, 1996).

The method used here has been called "observable-based parametrization" (Carin and Felsen, 1993), where the observed data (from the simulators) is fitted with appropriate formulations, in this case the Gibbs's oscillations. These are nominally $(2/\pi)[Si(\zeta\pi) - 1]$, where approximately $\pi L/\lambda$ extrema (at $\zeta = n$) occur from array edge to center. For H-plane scan of a dipole array the $[Si(\zeta\pi) - 1]$ Gibbsian factor applies directly. When a screen is added, the decay from edge to center is much faster; it is fitted until $[Si(\zeta\pi) - 1]^{1.5}$. This is due to the different decay of mutual coupling with screen. E-plane scan of dipoles shows a smaller decay; an exponent of 0.75 is used. These Gibbsian models fit the simulator data well except for the edge element. The fit is fair for the front element when the amplitude is doubled; a transition is used between the edge and the subsequent extrema. For the rear edge element a polynomial fit was made to the simulated data.

Next the Gibbsian model results will be compared with those from the simulators, for 201 elements. Dipole array results, with H-plane scan, are shown in Figs. 8.19 through 8.21, for scan angles of 0, 45, and 60 deg. Model data are dotted lines, while simulator data are solid lines. It can be seen that the agreement is excellent. Note that the match between the discrete array points and the dotted curve, as in Fig. 8.20, is important, rather than the comparison between the dotted peak and the array solid line. Figures 8.22 and 8.23 give the comparison for 51 elements at 0 and 60 deg, while Fig. 8.24 is for 21 elements at broadside. Again excellent agreement.

Scan impedance of dipole with screen arrays of 201 elements for H-plane scan are displayed in Figs. 8.25 through 8.27; again dotted lines are the Gibbsian model, and solid lines are simulator. The more rapid decay of the oscillations is apparent, although the edge values are comparable to those of the preceding set of data. Figures 8.28 and 8.29 show broadside arrays of 51 and 21 elements. Fits are excellent.

E-plane scan produces a much slower decay of oscillations in *scan impedance*, as seen in Figs. 8.30 and 8.31, which are for arrays of 201 dipoles. Figure 8.32 gives *scan impedance* for a 51-element broadside array. Again the match between model data and simulator data is excellent, except for edge elements. A dipole/screen array at broadside shows the modulated oscillations discussed in Section 8.3.1 in Fig. 8.33. A large-period sine factor has been used to model these oscillations. The match is excellent over part of the array, and fair over other parts. For larger scan angles, where the modulation has disappeared, a Gibbsian factor exponent of 2.5 is used, with fair results. However for the broadside case above, the exponent was reduced to 1.5 because of the effect of the heterodyne factor.

REFERENCES

Bailey, M. C., "Mutual Coupling between Circular Waveguide-Fed Apertures in a Rectangular Ground Plane," *Trans. IEEE*, Vol. AP-22, July 1974, pp. 597–599.

Bailey, M. C. and Bostian, C. W., "Mutual Coupling in a Finite Planar Array of Circular Apertures," *Trans. IEEE*, Vol. AP-22, Mar. 1974, pp. 178–184.

Bird, T. S., "Mode Coupling in a Planar Circular Waveguide Array," *IEE J. Microwaves, Opt. Acoust.*, Vol. 3, 1979, pp. 172–180.

Carin, L. and Felsen, L. B., "Time Harmonic and Transient Scattering by Finite Periodic Flat Strip Arrays: Hybrid (Ray)-(Floquet Mode)-(MOM) Algorithm," *Trans. IEEE*, Vol. AP-41, Apr. 1993, pp. 412–421.

Carslaw, H. S., *An Introduction to the Theory of Fourier's Series and Integrals*, Macmillan, 1930 [Dover, 1950].

Cátedra, M. F. et al., *The CG-FFT Method: Applications of Signal Processing Techniques to Electromagnetics*, Artech House, 1995.

Cha, A. G. and Hsiao, J. K., "A Matrix Formulation for Large Scale Numerical Computation of the Finite Planar Waveguide," *Trans. IEEE*, Vol. AP-22, Jan. 1974, pp. 106–108.

Clarricoats, P. J. B., Tun, S. M., and Parini, C. G., "Effects of Mutual Coupling in Conical Horn Arrays," *Proc. IEE*, Vol. 131H, June 1984, pp. 165–171.

Denison, D. R. and Scharstein, R. W., "Decomposition of the Scattering by a Finite Linear Array into Periodic and Edge Components," *Microwave Opt. Tech. Lett.*, Vol. 9, Aug. 1995, pp. 338–343.

Deshpande, M. D. and Bailey, M. C., "Analysis of Finite Phased Arrays of Circular Microstrip Patches," *Trans. IEEE*, Vol. AP-37, Nov. 1989, pp. 1355–1360.

Diamond, B. L., "Small Arrays — Their Analysis and Their Use for the Design of Array Elements," in *Phased Array Antennas, Proceedings of the 1970 Symposium*, A. A. Oliner and G. H. Knittel, Eds., Artech, 1972, pp. 127–131.

Fukao, S. et al., "A Numerical Consideration on Edge Effect of Planar Dipole Phased Arrays," *Radio Sci.*, Vol. 21, Jan.-Feb. 1986, pp. 1–12.

Galindo, V., "Finite Arrays, Edge Effects, and Aperiodic Arrays," in *Theory and Analysis of Phased Array Antennas*, N. Amitay, V. Galindo, and C. P. Wu, Eds., Wiley-Interscience, 1972, Chapter 8.

Gallegro, A. D., Mutual Coupling and Edge Effects in Linear Phased Arrays, MS thesis, Polytechnic Institute of Brooklyn, 1969.

Hamming, R. W., "Numerical Methods for Scientists and Engineers, McGraw-Hill, 1972.

Hansen, R. C., "Formulation of Echelon Dipole Mutual Impedance for Computer," *Trans. IEEE*, Vol. AP-20, Feb. 1972, pp. 780–781.

Hansen, R. C., "Evaluation of the Large Array Method," *Proc. IEE*, Vol. 137, Pt. H, Apr. 1990, pp. 94–98.

Hansen, R. C., "Finite Array Scan Impedance Gibbsian Models," *Radio Sci.*, Vol. 31, Nov.-Dec. 1996, pp. 1631–1637.

Hansen, R. C. and Gammon, D., "Standing Waves in Scan Impedance of Finite Scanned Arrays," *Microwave Opt. Tech. Lett.*, Vol. 8, Mar. 1995, pp. 175–179.

Hansen, R. C. and Gammon, D., "Standing Waves in Scan Impedance: E-Plane Finite Array," *Microwave Opt. Tech. Lett.*, Vol. 11, Jan. 1996a, pp. 26–32.

Hansen, R. C. and Gammon, D., "A Gibbsian Model for Finite Scanned Arrays," *Trans. IEEE*, Vol. AP-44, Feb. 1996b, pp. 243–248.

Hansen, R. C. and Raudenbush, E., "Modulated Oscillations in Finite Array Scan Impedance," *Proceedings IEEE Symposium on Phased Array Systems*, Boston, Mass., 1996.

Ishimaru, A. et al., "Finite Periodic Structure Approach to Large Scanning Array Problems," *Trans. IEEE*, Vol. AP-33, Nov. 1985, pp. 1213–1220.

Levin, D., "Development of Non-Linear Transformations for Improving Convergence of Sequences," *Int. J. Computer Math.*, Sec. B., Vol. 3, 1973, pp. 371–388.

Luzwick, J. and Harrington, R. F., "Mutual Coupling Analysis in a Finite Planar Rectangular Waveguide Antenna Array," *Electromagnetics*, Vol. 2, Jan.-Mar. 1982, pp. 25–42.

Oliner, A. A. and Malech, R. G., "Mutual Coupling in Finite Scanning Arrays," in *Microwave Scanning Antennas*, Vol. II, R. C. Hansen, Ed., Academic Press, 1966a, Chapter 4 [Peninsula Publishing, 1985].

Oliner, A. A. and Malech, R. G., "Mutual Coupling in Infinite Scanning Arrays," in *Microwave Scanning Antennas*, Vol. II, R. C. Hansen, Ed., Academic Press, 1966b, [Peninsula Publishing, 1985], Chapter 3.

Pozar, D. M., "Analysis of Finite Phased Arrays of Printed Dipoles," *Trans. IEEE*, Vol. AP-33, Oct. 1985, pp. 1045–1053.

Pozar, D. M., "Finite Phased Arrays of Rectangular Microstrip Patches," *Trans. IEEE*, Vol. AP-34, May 1986, pp. 658–665.

Roederer, A., "Étude des Reseaux Finis de Guides Rectangulaires à Parois Épaisses," *L'Onde Électrique*, Vol. 51, 1971, pp. 854–861.

Roscoe, A. J. and Perrott, R. A., "Large Finite Array Analysis Using Infinite Array Data," *Trans. IEEE*, Vol. AP-42, July 1994, pp. 983–992.

Shubert, K. A. and Munk, B. A., "Matching Properties of Arbitrarily Large Dielectric Covered Phased Arrays," *Trans. IEEE*, Vol. AP-31, Jan. 1983, pp. 54–59.

Shanks, D., "Non-Linear Transformations of Divergent and Slowly Convergent Sequences," *J. Math. Phys.*, Vol. 34, 1955, pp. 1–42.

Silvestro, J. W., "Mutual Coupling in a Finite Planar Array with Interelement Holes Present," *Trans. IEEE.* Vol. AP-37, June 1989, pp. 791–794.

Singh, S. et al., "Accelerating the Convergence of Series Representing the Free Space Periodic Green's Function," *Trans. IEEE*, Vol. AP-38, Dec. 1990, pp. 1958–1962.

Singh, S. and Singh, R., "On the Use of Levin's T-Transform in Accelerating the Summation of Series Representing the Free-Space Periodic Green's Functions," *Trans. IEEE*, Vol. MTT-41, May 1993, pp. 885–886.

Skinner, J. P. and Munk, B. A., "Mutual Coupling between Parallel Columns of Periodic Slots in a Ground Plane Surrounded by Dielectric Slabs," *Trans. IEEE*, Vol. AP-40, Nov. 1992, pp. 1324–1335.

Skinner, J. P., Whaley, C. C., and Chattoraj, T. K., "Scattering from Finite by Infinite Arrays of Slots in a Thin Conducting Wedge," *Trans. IEEE*, Vol. AP-43, Apr. 1995, pp. 369–375.

Skrivervik, A. K. and Mosig, J. R., "Analysis of Finite Phase Arrays of Microstrip Patches," *Trans. IEEE*, Vol. AP-41, Aug. 1993, pp. 1105–1114.

Sneddon, N., *Fourier Transforms*, McGraw-Hill, 1951.

Steyskal, H., "Mutual Coupling Analysis of a Finite Planar Waveguide Array," *Trans. IEEE*, Vol. AP-22, July 1974, pp. 594–597.

Stroud, A. H. and Secrest, D., *Gaussian Quadrature Formulas*, Prentice-Hall, 1966.

Usoff, J. M. and Munk, B. A., "Edge Effects of Truncated Periodic Surfaces of Thin Wire Elements," *Trans. IEEE*, Vol. AP-42, July 1994, pp. 946–953.

Wasylkiwskyj, W., "Mutual Coupling Effects in Semi-Infinite Arrays," *Trans. IEEE*, Vol. AP-21, May 1973, pp. 277–285.

CHAPTER NINE

Superdirective Arrays

9.1 HISTORICAL NOTES

A useful operational definition of antenna array superdirectivity (formerly called supergain) is directivity (see Chapter 2) higher than that obtained with the same array length and elements uniformly excited (constant amplitude and linear phase). Excessive array superdirectivity inflicts major problems in low radiation resistance (hence low efficiency), sensitive excitation and position tolerances, and narrow bandwidth. Superdirectivity applies in principle to both arrays of isotropic elements, and to actual antenna arrays composed of nonisotropic elements such as dipoles.

Probably the earliest work on the possibility of superdirectivity was by Oseen (1922): see Bloch et al. (1953) for a list of early references. A limited endfire superdirectivity using a monotonic phase function was accomplished by W. W. Hansen and Woodyard (1938). Another early contributor was Franz (1939). Schelkunoff (1943) in a classic paper on linear arrays discussed, among other topics, array spacings less than $\lambda/2$, showing how equal spacing of the array polynomial zeros over that portion of the unit circle represented by the array gives superdirectivity. The field received wide publicity when La Paz and Miller (1943) purported to show that a given aperture would allow a maximum directivity, and when Bouwkamp and De Bruijn (1946) showed that they had made an error and that there was no limit on theoretical directivity. Thus the important theorem: A fixed aperture size can achieve (in theory) any desired directivity value. This theorem is now widely recognized, but the practical implications are less well known. Bloch et al. (1960) say that the theorem has been rediscovered several times; the practical limitations of superdirectivity occur as a surprise to systems engineers and others year after year! In 1946, a burst of wartime research reporting occurred. Reid (1946) generalized the Hansen–Woodyard endfire directivity as $d \to 0$. And Dolph (1946) invented the widely known Dolph–Chebyshev array distribution wherein the equal level oscillations of a Chebyshev polynomial are used to produce an array pattern with equal-level sidelobes. To follow this last development, Riblet (1947) developed Dolph–Chebyshev arrays for spacing below $\lambda/2$, i.e., superdirective.

DuHamel (1953) and Stegen (1953) developed complementary advances in the computation of Dolph–Chebyshev coefficients and directivity. Maximum directivity for an array with fixed spacings was derived, for acoustic arrays, by Pritchard (1953).

Superdirective aperture design thus requires a constraint; arrays with fixed number of elements and spacing, owing to the finite number of variables, do not, as clearly there is an excitation that provides maximum directivity. The Lagrangian process for determining this maximum will be discussed, followed by a discussion of Chebyshev array design and other constrained optimization schemes. Finally, matching problems will be examined.

9.2 MAXIMUM ARRAY DIRECTIVITY

9.2.1 Broadside Directivity for Fixed Spacing

An array with fixed length and number of elements represents a determinate problem. Clearly, a maximum directivity exists. The maximum value can be found by using the Lagrange multiplier method (Sokolnikoff and Sokolnikoff, 1941). The directivity G at broadside can be written as

$$G = \frac{\left(\sum_{n=1}^{N} A_n\right)^2}{\sum_{n=1}^{N}\sum_{m=1}^{N} A_n A_m \operatorname{sinc}\left[(n-m)2\pi d/\lambda\right]}. \tag{9.1}$$

Here it is assumed that N array elements are isotropic and equally spaced by d. In the formulation above, the array amplitudes are A_n and the sinc function $[(\sin x)/x]$ represents the mutual impedance between isotropic elements (Hansen, 1983): $120 \operatorname{sinc} kd$. Although the directivity expression could be maximized directly, it is convenient to constrain the sum of the coefficients to unity, and then to minimize the denominator. The Langrangian equations are

$$2\sum_{n=1}^{N} A_n \operatorname{sinc}\left[(n-i)2\pi d/\lambda\right] - \beta = 0, \qquad i = 1, 2, 3, \ldots, N;$$

$$\sum_{n=1}^{N} A_n = 1. \tag{9.2}$$

β is the Lagrangian multiplier. Solving the first equation for $i=1$ for β and substituting gives N equations in the unknown coefficients for an N-element array. These are easily solved using simultaneous equation computer subroutines with multiple precision as needed. These were solved by hand for $N = 3, 5, 7$ by Pritchard (1953). Hansen (1983) compares the maximum directivity of small arrays with the uniform amplitude directivity versus element spacing; see Fig. 9.1. Above $d/\lambda = 0.5$, the two are very close. Also, some minor oscillations in the directivity curves have been smoothed out, as they are not important here. The coalescing of pairs of curves at zero spacing occurs because arrays of $2N$ and $2N - 1$ elements have the same number of degrees of freedom.

To give an idea of the coefficients and pattern of a small array with modest superdirectivity, Table 9.1 shows the amplitude coefficients for an array of 7 elements with quarter-wave spacing. Directivity is 5.21, and the pattern is the solid line in Fig. 9.2. Also shown (dashed) is the pattern of the same length array with half-wave spacing. Directivity of a corresponding uniform amplitude array is 3.64. The input resistances in the table will be discussed in the next section.

TABLE 9.1 Seven-Element Superdirective Array, $d = \lambda/4$.

Element Number	Amplitude	Resistance (ohms)
1	1.443	0.13299
2	−3.933	−0.04879
3	7.122	0.02694
4	−8.264	−0.02322
5	7.122	0.02694
6	−3.933	−0.04879
7	1.443	0.13299

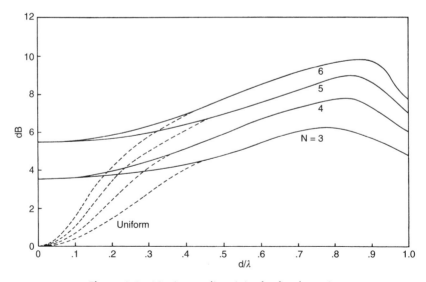

Figure 9.1 Maximum directivity for fixed spacing.

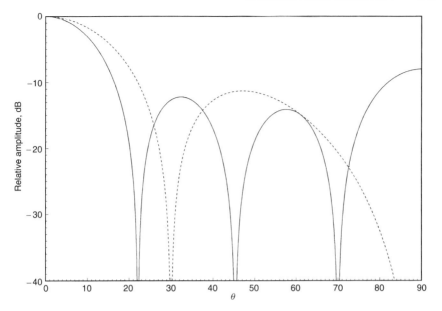

Figure 9.2 Maximum directivity array, $N = 7$, $d = 0.25\,\lambda$.

9.2.2 Directivity as Spacing Approaches Zero

Uzkov (1946) transformed the maximum directivity limit for an array as spacing goes to zero into a sum of Legendre polynomials. His result for an array of N elements is

$$G = \sum_{n=0}^{\infty}(2n+1)[P_n(\cos\theta)]^2, \tag{9.3}$$

where θ is the angle of the main beam from broadside. Tai (1964) independently developed this for broadside arrays, where $\cos\theta = 1$. Since the Legendre polynomial of argument 1 can be written as a product of factors, the result for maximum directivity is simply

$$G \to \left(\frac{1\cdot 3\cdot 5\cdots M}{2\cdot 4\cdot 6\cdots(M-1)}\right), \quad \text{where } M = 2\,AINT\left(\frac{N+1}{2}\right)-1. \tag{9.4}$$

Here M is the number of odd elements or the number of even elements minus 1. Thus the interesting result that 3- and 4-element arrays have the same limiting value, 5- and 6-element arrays have the same value, etc. After reflection this is not surprising since both 3- and 4-element arrays have two variables, both 5- and 6-elements have three variables, etc. This maximum directivity is plotted in Fig. 9.3. The circles show the corresponding limiting directivity for Chebyshev arrays with 10 dB SLR and the squares are for 20 dB SLR. The Chebyshev directivity is less, as the maximum directivity pattern does not have equal-level

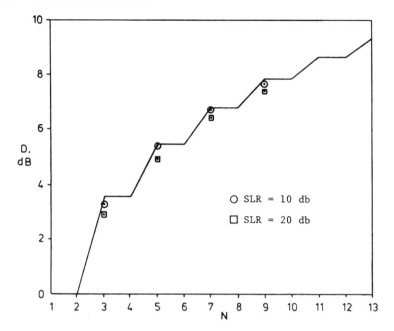

Figure 9.3 Maximum broadside directivity in zero-spacing limit.

sidelobes, and even in the case where there is only one sidelobe (backlobe), the Chebyshev result would be equal to the maximum value only if the sidelobe ratio were properly chosen. See the next section for a discussion of Chebyshev arrays.

For endfire arrays, the maximum directivity in the limit of zero spacing is

$$G \to \sum_{n=0}^{N-1}(2n+1)[P_n(0)]^2 = \sum_{n=0}^{N-1}(2n+1) = N^2. \tag{9.5}$$

9.2.3 Endfire Directivity

W. W. Hansen and Woodyard (1938) developed an endfire line source with modest superdirectivity. This is of interest because the distribution can be sampled to get array excitations and because the amplitude is constant, a feature that is attractive for arrays. They observed that, if the free-space phase progression along the aperture was increased, the space-factor power integral decreased faster than the peak value; thus the directivity increases up to a point. The endfire pattern, for a source of length L, is

$$f(\theta) = \operatorname{sinc}\left(\frac{L}{2}(k\sin\theta - \beta)\right), \tag{9.6}$$

where β is the wavenumber over the aperture. Inverse directivity is proportional to

$$\frac{1}{G} \propto \frac{1}{\text{sinc}^2\left(\frac{L}{2}(k-\beta)\right)} \int \text{sinc}^2\left(\frac{L}{2}(k\sin\theta - \beta)\right) \cos\theta\, d\theta$$

$$= \frac{1}{\text{sinc}^2(\phi/2)} \left(\frac{\pi}{2} + \text{Si}\,\phi + \frac{\cos\phi - 1}{\phi}\right). \tag{9.7}$$

Here Si is the Sine Integral, and $\phi = L(k - \beta)$, the additional phase along the aperture (in addition to the progressive endfire phase). Maximum directivity was determined to occur for $\phi = 2.922$ radians. In many books it is carelessly stated that π extra radians of phase are needed, but there is no physical reason for this; a better approximation to 2.92 is 3. Directivity increase over normal endfire is 2.56 dB, and the sidelobe ratio is 9.92 dB. The distribution is suitable for long arrays; for short arrays a computer optimization of phase is recommended. (See Hansen, 1992).

The Hansen–Woodyard distribution is endfire. In general, the maximum directivity does not occur there. The most general solution for uniform amplitude would allow any element phases needed to maximize directivity. Such a solution could be formally realized for a given number of elements and spacing, but the equations would require a numerical solution. A slightly simpler problem was worked by Bach (1970); he started with a uniform-amplitude array that was phased to produce a main beam at θ_0. The interelement phase factor is $\delta = kd\sin\theta_0$, and the directivity (see Chapter 2) is

$$G = \frac{N}{1 + \frac{2}{N}\sum_{n=1}^{N-1}(N-n)\,\text{sinc}(nkd)\cos(n\delta)}. \tag{9.8}$$

Calculations were made for 2-, 3-, 4-, and 10-element arrays for all beam angles, and for spacings up to λ. Figure 9.4 shows the results for 2- and 3-element arrays, while Fig. 9.5 is for 4- and 10-element arrays. Figure 9.5b is striking in that high directivity occurs along a line roughly for $kd + \delta = 0$, or $\theta_0 = -\pi/2$, with peak directivity near endfire at λ spacing. Along the line roughly for $kd = \delta + 2\pi$ directivity is changing rapidly, perhaps owing to appearance of another lobe. Directivity values are shown at θ_0, but in some cases a "sidelobe" may have higher amplitude. Thus even for uniform amplitude an array is complex.

9.2.4 Bandwidth, Efficiency, and Tolerances

The first of three major difficulties with superdirective arrays is bandwidth, which rapidly becomes a problem as the element spacing decreases below $\lambda/2$. Thus the Q is of concern; for narrowband antennas, bandwidth $\simeq 1/Q$, and the impedance matched bandwidth $\simeq 2/Q$. The ratio of stored to dissipated

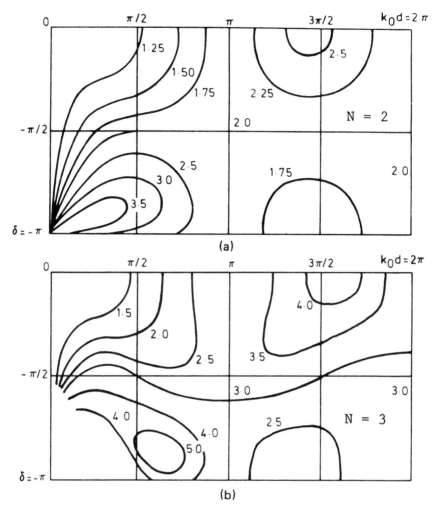

Figure 9.4 Uniform array directivity. (Courtesy Bach, H., "Directivity of Basic Linear Arrays," *Trans. IEEE*, Vol. AP-18, Jan. 1970, pp. 107–110.)

energy, Q, can similarly be written in terms of array coefficients and mutual coupling; which for isotropic elements is

$$Q = \frac{\sum_{n=1}^{N} A_n^2}{\sum_{n=1}^{N} \sum_{m=1}^{N} A_n A_m \operatorname{sinc}\left[(n-m) 2\pi d/\lambda\right]}. \tag{9.9}$$

Calculations have shown (Hansen, 1981b) that, for broadside arrays of fixed length, both directivity and Q increase with the number of elements as

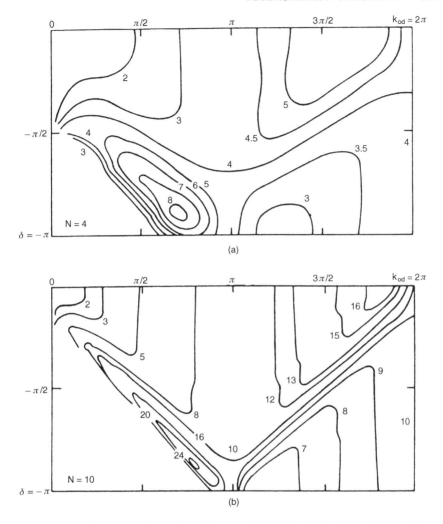

Figure 9.5 Uniform array directivity. (Courtesy Bach, H., "Directivity of Basic Linear Arrays," *Trans. IEEE*, Vol. AP-18, Jan. 1970, pp. 107–110.)

expected. An array with even number of elements has a slightly higher Q than the array with the next larger odd number of elements, possibly because the odd-element sampling is more efficient. For all the cases computed, $\log Q$ varied approximately linearly with directivity. Figure 9.6 shows $\log Q$ versus directivity for odd arrays with lengths 1λ, 2λ, and 5λ. The circles represent points calculated in double precision; extended precision is required for arrays of more elements than those shown in the figure. The straight lines are drawn through the uniform excitation directivity point (equal to G_0) with a slope of π, where G_0 is the gain of the same length array with half-wave spacing: $G_0 = 1 + 2L/\lambda$. There is at this time no physical significance to using this value of slope, but it is suggested by calculations of Rhodes (1974) on super-

312 SUPERDIRECTIVE ARRAYS

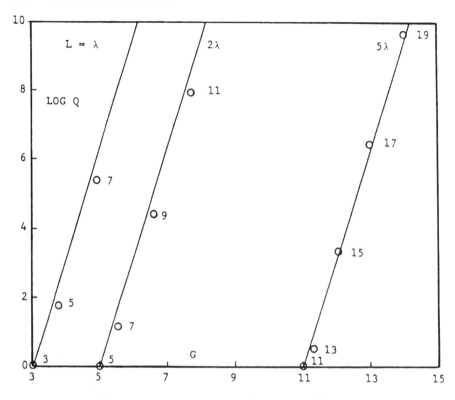

Figure 9.6 Q vs. maximum directivity, odd arrays.

directive line sources. In making a best straight-line fit to the set of points for each of the three arrays, the slopes were in fact clustered around the value of 3.14. However, it is difficult to perform this fit with precision because, as pointed out by Rhodes, the curve of $\log Q$ versus directivity has an oscillatory behavior for low values of Q. If the assumptions above are true, superdirective broadside array behavior is predicted by the equation

$$\log Q = \pi(G - G_0). \tag{9.10}$$

Thus the superdirective array clearly fits into the category of fundamental limitations in antennas (Hansen, 1981a). Whether the assumed slope of π can be physically justified remains an interesting problem at this time.

The Q to be expected from an array of isotropes is approximately a function of the number of elements divided by d/λ. Figure 9.7 shows this relationship for many arrays; each array is represented by a circle. Spacings larger than 0.3λ are not included, as the amount of superdirectivity achieved is small for these. $\log Q$ is approximately linear with $N\lambda/d$; the straight line fit in the figure is

$$\log Q \simeq 0.16043 \left(\frac{N\lambda}{d}\right) - 1.53476. \tag{9.11}$$

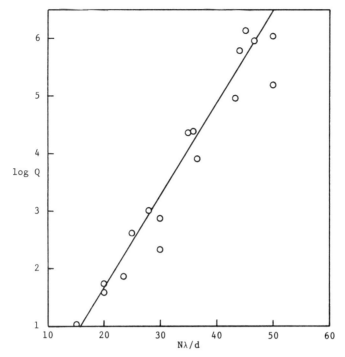

Figure 9.7 Q of broadside linear superdirective arrays.

These data allow the array designer to estimate the degree of superdirectivity (Q) for a given array geometry.

Calculations of performance have been made for superdirective linear arrays of parallel dipoles. Collinear dipoles are not considered as they would overlap neighbors. Figure 9.8 shows $\log Q$ versus directivity for arrays 2 wavelengths long. Both 0.1λ and 0.5λ dipoles are used. Again $\log Q$ versus G is a straight line. As expected, the half-wave dipole array has higher Q than the corresponding isotropic array, owing to energy storage in the dipole near-field. With short (0.1λ) dipoles this energy storage is much larger, resulting in much higher Q.

For endfire linear parallel dipole arrays, the variation of $\log Q$ with directivity is again linear, but the slope changes as the length of the array changes. Figure 9.9 gives data for arrays of lengths 1λ, 2λ, and 5λ; the number of elements in each array is shown in the figure (Hansen, 1990).

A second undesirable feature of superdirective arrays is low efficiency, due both to matching network losses (see Section 9.4.1), and to losses in the antenna elements. Both are caused by the low radiation resistances of these arrays. Typically the elements at the array center show the lowest radiation resistance; calculations for many small arrays of isotropes shows that $R_{\text{rad}} \propto 1/\sqrt{Q}$. Figure 9.10 shows these data, where each circle represents one array. The straight line log–log fit is

$$R_{\text{rad}} \simeq \frac{8.058}{\sqrt{Q}}. \tag{9.12}$$

314   SUPERDIRECTIVE ARRAYS

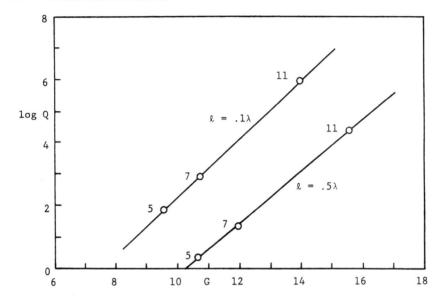

Figure 9.8   Q of 2-wavelength broadside array of parallel dipoles.

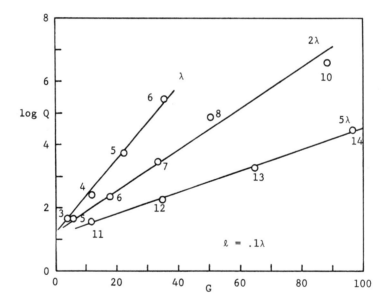

Figure 9.9   Q of endfire arrays of parallel dipoles.

Similar data for endfire arrays are given in Fig. 9.11. A nearly log–log dependence of radiation resistances and Q is observed.

Loss resistance of cylindrical or strip dipoles is easily computed: a dipole of length ℓ, radius a, made of material with surface resistance R_s, has a loss resistance R_{loss} of

MAXIMUM ARRAY DIRECTIVITY 315

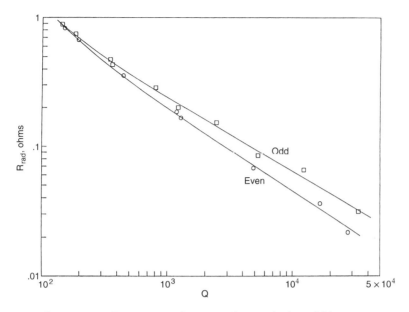

Figure 9.10 Q vs. center element resistance for broadside arrays.

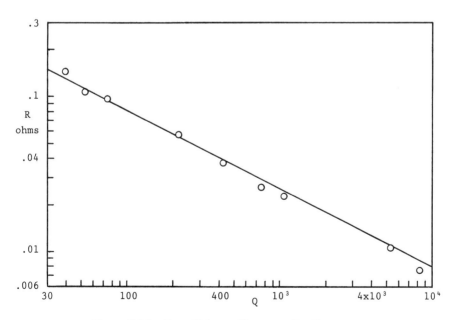

Figure 9.11 R vs. Q for endfire superdirective arrays.

$$R_{\text{loss}} = \frac{R_s \ell (1 - \text{sinc } k\ell)}{4\pi a \sin^2 \tfrac{1}{2} k\ell}. \tag{9.13}$$

For half-wave or resonant dipoles, $R_{\text{loss}} = R_s \ell / 4\pi a$. Strip dipoles are equivalent to cylindrical dipoles, with strip width equal to $4a$. Copper wires have an ideal surface resistance of

$$R_s \simeq 0.000261 \sqrt{f_{MHz}}. \tag{9.14}$$

Over the range of 10 to 1000 mHz, R_s varies from 0.000825 to 0.00825 ohm/□. Using a value of $\ell/a = 25$, a moderately fat dipole, the loss resistance varies from 0.0066 ohm at 10 MHz to 0.066 ohm at 1 GHz. When these typical loss resistance numbers are compared with the radiation resistance values of Figs. 9.10 and 9.11, it is clear that superdirective array efficiency may be a severely limiting consideration. Use of high-temperature superconductor (HTS) materials in the array and feed network can produce high efficiencies, but now the Q is that from Figs. 9.7, 9.8, and 9.9.

The third significant problem with superdirective arrays is tight tolerances. Because superdirectivity involves a partial cancelation of the element contributions at the main beam peak, with more cancelation for more superdirectivity, the tolerance of each element coefficient (excitation) becomes smaller (tighter) with more superdirectivity (Uzsoky and Solymar, 1956). A simple calculation has been made of these effects for Dolph–Chebyshev arrays by perturbing the center element of an odd array, finding the tolerance to reduce the directivity by 0.5 dB. This is not expected to be sensitive to sidelobe ratio and a value of 20 dB was used. Calculations were made for $N = 3$ and 5 as a function of spacing, with the results shown in Fig. 9.12. It was noted that the percentage

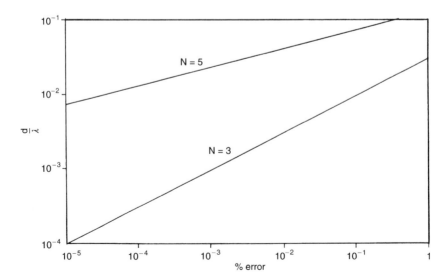

Figure 9.12 Center element tolerance for 0.5 dB directivity drop.

error versus d/λ curve is linear up to spacings of roughly 0.1λ. For $N = 3$ the slope is 2:1, and for $N = 5$ the slope is 4:1. Thus for $N = 3$, if the element spacing is halved, the tolerances must be four times tighter. The percentage tolerance for the center element, to maintain close (0.5 dB) to the expected directivity, is approximately given by

$$\frac{100}{\sqrt{Q}}, \qquad (9.15)$$

but the constant varies with the number of elements.

Thus with $Q = 1000$, for example, the tolerance is 3.2%. Bandwidth and radiation resistance are more serious limitations of superdirective arrays.

9.3 CONSTRAINED OPTIMIZATION

Directivity may be maximized subject to a constraint, which may be on pattern sidelobes, on bandwidth or Q, on allowable tolerances, on S/N, etc. The simplest is sidelobe constraint, after Dolph and Chebyshev.

9.3.1 Dolph–Chebyshev Superdirectivity

The principles and design equations for arrays with equal-level sidelobes, Dolph–Chebyshev arrays, were covered in detail in Chapter 3. However, Dolph's derivation and the formulas of Stegen are limited to $d \geq \lambda/2$. Riblet (1947) showed that this restriction could be removed, but only for N odd. For spacing below half-wave, the space factor is formed by starting at a point near the end of the Chebyshev ± 1 region[1], tracing the oscillatory region to the other end, then retracing back to the start end and up the monotonic portion to form the main beam half. Since the Mth-order Chebyshev has $M - 1$ oscillations, which are traced twice, and since the trace from 0 to 1 and back forms the center sidelobe (in between the trace out and back), the space factor always has an odd number of sidelobes each side, or an even number of zeros. Hence only an odd number of elements can be formed into a Chebyshev array for $d < \lambda/2$. The pattern is given by

$$\begin{aligned} T_M(a\cos\Psi + b), \\ a = \frac{z_0 + 1}{1 - \cos kd}, \\ b = \frac{z_0 \cos kd + 1}{\cos kd - 1}, \end{aligned} \qquad (9.16)$$

where as before $\Psi = kd\sin\theta$. The value of z_0 is:

$$z_0 = \cosh\frac{\operatorname{arccosh} \operatorname{SLR}}{M}. \qquad (9.17)$$

---
[1]The exact starting point depends on N and kd.

The sidelobe ratio is

$$\text{SLR} = T_M(z_0). \tag{9.18}$$

Formulas have been developed by DuHamel (1953), Brown (1957, 1962), Salzer (1975), and Drane (1963, 1964). Those of Drane will be used here as they are suitable for computer calculation of superdirective arrays. The array amplitudes are

$$A_n = \frac{\epsilon_n}{4M} \sum_{m=0}^{M_1} \epsilon_m \epsilon_{M_2-m} T_n(x_n) [T_M(ax_n + b) + (-1)^n T_M(b - ax_n)], \tag{9.19}$$

where $\epsilon_i = 1$ for $i = 0$ and is equal to 2 for $i > 0$; $x_n = \cos n\pi/M$. The integers M_1 and M_2 are, respectively, the integer parts of $M/2$ and $(M+1)/2$. This result is valid for $d \leq \lambda/2$. Small spacings (highly superdirective arrays) may require multiple precision due to the subtraction of terms. Many arrays are half-wave spaced; for these the a and b reduce to

$$a = \tfrac{1}{2}(z_0 + 1), \qquad b = \tfrac{1}{2}(z_0 - 1). \tag{9.20}$$

For half-wave spacing this approach and that of Chapter 2 give identical results! In fact, owing to the properties of the Chebyshev polynomial, the two space factors, in precursor form, are equal:

$$T_{N-1}\left(z_0^{N-1} \cos\frac{\Psi}{2}\right) \equiv T_M\left(\frac{z_0^M(\cos\Psi + 1) + \cos\Psi - 1}{2}\right), \qquad N - 1 = 2M, \tag{9.21}$$

where the superscripts on z_0 indicate that each must be chosen for the proper form. Since many computers have no inverse hyperbolic functions, it is convenient to rewrite the z_0 as

$$z_0 = \tfrac{1}{2}\left[\text{SLR} + \sqrt{\text{SLR}^2 - 1}\right]^{1/M} + \tfrac{1}{2}\left[\text{SLR} - \sqrt{\text{SLR}^2 - 1}\right]^{1/M} \tag{9.22}$$

Directivity for arrays of 3, 5, 7, and 9 elements have been calculated, for sidelobe ratios of 10 and 20 dB. The superdirectivity can be seen in Figs. 9.13 and 9.14 for spacing below 0.5λ, as the ordinary directivity (using the Chebyshev coefficients that are independent of spacing) goes smoothly to 0 dB at zero spacing. The figures display these calculated directivities versus element spacing. Thus a 3-element array offers roughly 3 dB directivity for small spacings, 5 elements offers roughly 5 dB. Next the disadvantages of superdirectivity will be discussed.

Q is given by Lo et al. (1966) as

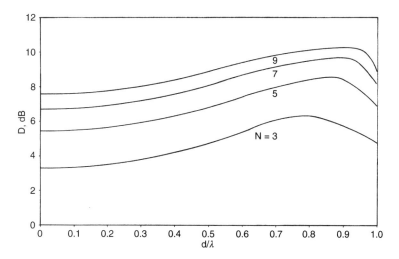

Figure 9.13 Chebyshev array directivity, SLR = 10 dB.

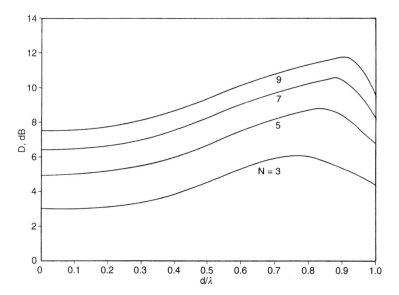

Figure 9.14 Chebyshev array directivity, SLR = 20 dB.

$$Q = \frac{\sum_{n=1}^{N} A_n^2}{\sum_{n=1}^{N}\sum_{m=1}^{N} A_n A_m \operatorname{sinc}[(n-m)kd]}. \quad (9.23)$$

Q, which is the inverse of fractional bandwidth, is plotted for Dolph–Chebyshev arrays versus d/λ in Fig. 9.15. The $N = 3$ array has a log–log slope of Q versus spacing of 4:1, while the $N = 5$ array has a slope of 8:1. The $N = 7$ and 9 arrays have even higher slopes. For small values of Q, the curve is not too accurate.

Bandwidth appears to be more restrictive than tolerances; the $N = 3$ array, to be practical, requires a spacing of the order of 0.1λ or larger. The $N > 3$ arrays are even less forgiving.

Directivity is given by

$$G = \frac{\left(\sum_{n=1}^{N} A_n\right)^2}{\sum_{n=1}^{N}\sum_{m=1}^{N} A_n A_m \operatorname{sinc}[(n-m)kd]}. \tag{9.24}$$

Note the similarity to the result for Q. Two cases of Chebyshev superdirective arrays have been calculated to illustrate the variation of Q with directivity increase. First is a 2-wavelength array of isotropic elements, which with half-wave spacing has a minimum directivity of 5. This 2-wavelength aperture is occupied by 7, 9, 11, and 13 elements. Table 9.2 shows the element spacing, directivity, and Q, for a design SLR of 20 dB. The Q values are shown in Fig. 9.16, where the straight line has been fitted to the data, and is

$$\log Q \simeq 3.3175(G - 5) \tag{9.25}$$

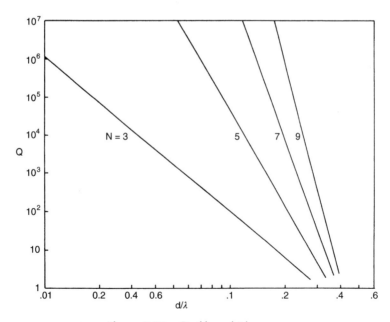

Figure 9.15 Q of broadside arrays.

TABLE 9.2 Two-wavelength Array, SLR = 20 dB

N	d/λ	G	Q
5	0.500	4.69	1.7
7	0.333	5.18	7.0
9	0.250	6.21	1.2×10^4
11	0.200	7.36	5.8×10^7
13	0.167	8.54	5.5×10^{11}

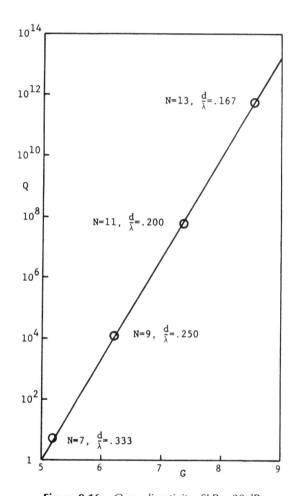

Figure 9.16 Q vs. directivity, SLR = 20 dB.

The form of the equation, with the normal directivity ($G_0 = 5$) subtracted, was suggested by calculations of Rhodes. Rhodes (1971, 1974) maximized directivity of a line source subject to a constraint on Q using the eigenfunctions of the source: prolate spheroidal functions. These are particularly suited to allow a simple derivation as they are doubly orthogonal: over infinite limits, which fits the total energy integral over all angles, and over finite limits, which fits the radiated energy integral over real angles. He found a roughly linear relationship between $\log Q$ and directivity, with a coefficient of roughly 2. This substantiates the suspicion that the Chebyshev design may result in a higher Q than necessary for a given directivity, since the coefficients from Tables 9.2 and 9.3 are in the 3–4 range. To show the effects of SLR, Table 9.3 is a repeat for SLR = 10 dB. And Fig. 9.17 shows a straight line fit, with coefficient 4.1475.

TABLE 9.3 Two-wavelength Array, SLR = 10 dB

N	d/λ	G	Q
5	0.500	4.89	1.8
7	0.333	5.40	37
9	0.250	6.19	7.6×10^4
11	0.200	6.99	3.5×10^8
13	0.167	7.74	3.3×10^{12}

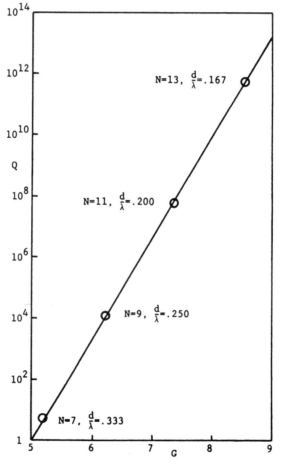

Figure 9.17 Q vs. directivity, SLR = 10 dB.

Higher sidelobes increase directivity and Q. Calculated points for $N = 13$ are inaccurate, double precision on a 32-bit machine is inadequate. Thus the Q increases approximately exponentially with directivity above the normal value. For most increases above normal directivity, it may be inferred from Rhodes that the Q curve has minor oscillations. Note the very rapid increase of Q with directivity; even a 10 percent increase produces a bandwidth limited to a few percent. However, the Chebyshev design may not give the lowest Q; a constrained synthesis is apparently necessary to produce the lowest Q for a given directivity. Newman et al. (1978) have done this for endfire arrays, but only for two cases.

The second case is a half-wavelength aperture, originally computed by Yaru (1951), which with two isotropic elements gives a maximum directivity of 2. Into this small aperture are placed 3, 5, 7, or 9 elements. Table 9.4 gives the pertinent parameters, again for a 20 dB SLR. A straight-line fit of the same type, $\log Q = a(G - 2)$, was tried but the fit is less good. It is not known whether this is due to computing errors or to the small size of the aperture in wavelengths.

DuHamel (1953) extended the Chebyshev design principle to endfire arrays, but only for $d < \lambda/2$. To avoid a back lobe, spacing is customarily made $\leq \lambda/4$. For any scan angle, Ψ is modified as usual to

$$\Psi = kd(\sin\theta - \sin\theta_0), \tag{9.26}$$

where θ_0 is the scan (main beam) angle, and the interelement phase shift is $kd\sin\theta_0$. Coefficients a and b become

$$\begin{aligned} a &= -\frac{3 + z_0 + 2\sqrt{2(z_0 + 1)}\cos kd}{2\sin^2 kd}, \\ b &= \frac{(\sqrt{z_0 + 1} + \sqrt{2}\cos kd)^2}{2\sin^2 kd}. \end{aligned} \tag{9.27}$$

9.3.2 Constraint on Q or Tolerances

Directivity may be maximized subject to constraints on bandwidth (Q) or tolerances. The basic framework of constrained optimization was developed by Gilbert and Morgan (1955) and Uzsoky and Solymar (1956), and was

TABLE 9.4 Half-wavelength Array, SLR = 20 dB

N	d/λ	G	Q
3	0.25	2.11	2.6
5	0.125	3.17	8.7×10^3
7	0.0833	4.40	5.7×10^8
9	0.0625	5.66	1.7×10^{14}

extended by Lo et al. (1966). A review paper is by Cheng (1971). Use of the braket notation introduced by Dirac (Friedman, 1956) allows the formulation to be simplified. Let $\langle A|$ be a row vector and $|A\rangle$ be a column vector. The scalar product then becomes

$$\langle A|A^*\rangle = \sum |A_n|^2, \tag{9.28}$$

where * indicates the complex conjugate. Call the row vector of complex array excitations J and the column vector of path-length phases F:

$$F = \begin{vmatrix} \exp(-jkr_1 \sin\theta) \\ \vdots \\ \exp(-jkr_n \sin\theta) \end{vmatrix}, \tag{9.29}$$

where r_n is the distance from the reference point to the nth element. For a uniformly spaced array, $r_n = (n-1)d$. Now define matrices A and B, where A is

$$|A| = |F^*\rangle\langle F| \tag{9.30}$$

and the elements of B are

$$B_{nm} = \frac{1}{4\pi}\int f_i(\theta,\phi)\exp[-jkd(n-m)\sin\theta]\,d\Omega. \tag{9.31}$$

The pattern of the ith element is f_i. Isotropic elements and a uniformly spaced array allow simplifications of A and B. The elements then become

$$\begin{aligned} A_{nm} &= \exp[-jkd(n-m)\sin\theta], \\ B_{nm} &= \operatorname{sinc}[(n-m)kd]. \end{aligned} \tag{9.32}$$

Now the directivity can be written in abbreviated form:

$$G = \frac{\langle J|A|J^*\rangle}{\langle J|B|J^*\rangle}. \tag{9.33}$$

The tolerance sensitivity S is defined as the ratio of variance of peak field strength produced by errors of variance σ_T^2 (Uzsoky and Solymar, 1956):

$$S = \frac{\sigma_E^2}{\sigma_J^2} = \frac{(\Delta E)^2}{\sigma_J^2} = \frac{\sigma_J^2\langle J|J\rangle}{\langle J|A|J^*\rangle} = \frac{\langle J|J^*\rangle}{\langle J|A|J^*\rangle}. \tag{9.34}$$

Thus sensitivity is also written in abbreviated form. Q is conveniently found from $Q = SG$, or

$$Q = \frac{\langle J|J^*\rangle}{\langle J|B|J^*\rangle}.\qquad(9.35)$$

The directivity is a ratio of two Hermitian quadratic forms, with B positive definite and A at least positive semidefinite. Thus all the eigenvalues of the associated equation

$$|A|J\rangle = G_{\max}|B|J\rangle \qquad(9.36)$$

are zero or positive real. Since A is a single-term dyad there is one nonzero eigenvalue. The eigenvector is found to be

$$|J_{\max}\rangle = |B|^{-1}F\rangle.\qquad(9.37)$$

The corresponding maximum directivity is given by

$$G_{\max} = \langle F|B|^{-1}F^*\rangle.\qquad(9.38)$$

In many applications it is important to maximize G/T, directivity/system noise temperature. This is equivalent to maximizing signal/noise ratio S/N. To do this the element pattern in the integral for B_{nm} is multiplied by the antenna noise temperature $T(\theta, \phi)$. Then the excitation vector that maximizes S/N or G/T is that of Eqn. (9.37).

Another ratio that can be directly minimized is beam efficiency: the fraction of power contained within the main beam, null-to-null. This is

$$\eta_b = \frac{\langle J|A|J^*\rangle}{\langle J|B|J^*\rangle}.\qquad(9.39)$$

Optimization by adjustment of only phases is also possible (Voges and Butler, 1972). This might be attractive for arrays with active modules, where uniform amplitude is desirable. Unfortunately, half of the degrees of freedom are lost, and amplitude plays a much stronger role in pattern shaping.

In practice, because of the limited ensemble of superdirective array design available before unusably high Q occurs, it is simpler to calculate Q, R_{rad}, and G for a few arrays directly. This allows satisfactory design trades to be made, without the complexity of constrained optimization.

9.4 MATCHING OF SUPERDIRECTIVE ARRAYS

Superdirective arrays, like electrically small antennas, tend to have radiation resistances $\ll 50$ ohms, and reactance $\gg 50$ ohms. In both cases the matching network presents a serious problem: a low intrinsic loss is greatly increased, as will be shown.

9.4.1 Network Loss Magnification

A matching network for a superdirective array must cancel the large reactance, and must transform the small radiation plus loss resistance to the nominal value, usually 50 ohms. Generally each element will require a different matching network, although symmetric (broadside) arrays need fewer. If the matching network is composed of discrete L and C components, the low R, high X requirements will produce very large circulating currents in the overall circuit, consisting of the array element, the matching network, and the generator. These circulating currents will magnify the intrinsic loss to a much larger realized loss. Similarly, a distributed matching network under low R, high X conditions will have very large standing waves along the stubs and transformer sections. Again the intrinsic loss is magnified. In both cases the larger loss is due to power being proportional to voltage (or current) squared; the circulating currents or standing waves are large.

A transmission line transformer is typical. Let the matched loss be L, and the antenna voltage standing wave ratio be VSWR. Then the actual loss L_a is (Moreno, 1948)

$$L_a = \frac{(\text{VSWR} + 1)^2 L^2 - (\text{VSWR} - 1)^2}{4 \cdot L \cdot \text{VSWR}}. \tag{9.40}$$

For VSWR $\gg 1$, the actual loss is

$$L_a \simeq \frac{\text{VSWR} \cdot (L^2 - 1)}{4L}. \tag{9.41}$$

Figure 9.18 shows the actual loss versus VSWR for intrinsic (matched) losses of 0.1, 0.2, 0.3, 0.5, 1, 2, 3, and 5 dB. For $R \ll R_0$, the VSWR is VSWR $\simeq (R_0^2 + X^2)/RR_0$, while for $X \gg R_0$, VSWR $\simeq X^2/RR_0$. The latter holds also for $R \ll R_0$ and $X \gg R_0$. Some typical values are given in Table 9.5. For example, a case of $R/R_0 = 0.1$ and $X/R_0 = 10$ produces a VSWR = 1010. A matched loss of 0.01 dB becomes an actual loss of 3.35 dB; a matched loss of 0.1 dB becomes an actual loss of 11.01 dB. A more likely case, since the reactance ratio is usually higher, is $R/R_0 = 0.1$ and $X/R_0 = 30$, giving VSWR = 9010. A matched loss of 0.01 dB now yields an actual loss of 10.56 dB; a matched loss of 0.1 dB gives an actual loss of 20.2 dB. These actual losses of the order of 10 dB are in agreement with measurements on HTS and copper matching networks (Khamas et al., 1988, 1990; Lancaster et al., 1992).

9.4.2 HTS Arrays

Because the radiation resistance of a superdirective array decreases roughly inversely with $\sqrt{Q}$, it is easy to design an array where R_{rad} is comparable to or less than R_{loss}, which makes the efficiency 50% or less. In addition, there is matching network loss, as just described (Hansen, 1991). In practice, the

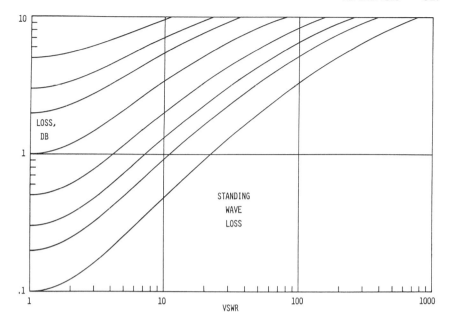

Figure 9.18 Line loss with mismatch.

TABLE 9.5 VSWR from Impedance Mismatches

	X/R_0				
R/R_0	1	3	10	30	100
1	1.000	10.91	102.0	902.0	1.000×10^4
0.3	6.820	33.60	337.0	3004.0	3.334×10^4
0.1	20.05	100.1	1010.0	9010.0	1.000×10^5
0.03	66.68	333.4	3367.0	3.003×10^4	3.334×10^5
0.01	200.0	1000.0	1.010×10^4	9.010×10^4	1.000×10^6

usability of these arrays is limited by the Q or bandwidth; an antenna with Q of several hundreds may be usable, but one with Q of several thousands is generally not. Even if the signal bandwidth is much less than 1%, the environmental instability of very high-Q antennas makes their use difficult. Adaptive tuning is possible but carries its own disadvantages. For these few cases where the antenna Q is manageable but the efficiency suffers, an HTS implementation of the array may be feasible (Dinger et al., 1991).

REFERENCES

Bach, H., "Directivity of Basic Linear Arrays," *Trans. IEEE*, Vol. AP-18, Jan. 1970, pp. 107–110.

Bloch, A. et al., "A New Approach to the Design of Super-Directive Aerial Arrays," *Proc. IEE.*, Vol. 100, Sept. 1953, pp. 303–314.

Bloch, A. et al., "Superdirectivity," *Proc. IRE*, Vol. 48, 1960, p. 1164.

Bouwkamp, C. J. and de Bruijn, N. G., "The Problems of Optimum Antenna Current Distribution," *Philips Res. Rep.*, Vol. 1, 1946, pp. 135–158.

Brown, J. L., "A Simplified Derivation of the Fourier Coefficients for Chebyshev Patterns," *Proc. IEE*, Vol. 105C, Nov. 1957, pp. 167–168.

Brown, J. L., "On the Determination of Excitation Coefficients for a Tchebycheff Pattern," *Trans. IEEE*, Vol. AP-10, Mar. 1962, pp. 215–216.

Cheng, D. K., "Optimization Techniques for Antenna Arrays," *Proc. IEEE*, Vol. 59, Dec. 1971, pp. 1664–1674.

Dinger, R. J., Bowling, D. R., and Martin, A. M., "A Survey of Possible Passive Antenna Applications of High-Temperature Superconductors," *Trans. IEEE*, Vol. MTT-39, Sept. 1991, pp. 1498–1507.

Dolph, C. L., "A Current Distribution for Broadside Arrays which Optimizes the Relationship between Beam Width and Side-Lobe Level," *Proc. IRE*, Vol. 34, June 1946, pp. 335–348.

Drane, C. J., "Derivation of Excitation Coefficients for Chebyshev Arrays," *Proc. IEE*, Vol. 110, Oct. 1963, pp. 1755–1758.

Drane, C. J., "Dolph–Chebyshev Excitation Coefficient Approximation," *Trans. IEEE*, Vol. AP-12, Nov. 1964, pp. 781–782.

DuHamel, R. H., "Optimum Patterns for Endfire Arrays," *Proc. IRE*, Vol. 41, 1953, pp. 652–659.

Franz, K., "The Gain and the (Rudenberg) 'Absorption Surfaces' of Large Directive Arrays," *Hochfreq. Elektroakust.*, Vol. 54, 1939, p. 198.

Friedman, B., *Principles and Techniques of Applied Mathematics*, Wiley, 1956.

Gilbert, E. N. and Morgan, S. P., "Optimum Design of Directive Antenna Arrays Subject to Random Variations," Bell Syst. Tech. J., Vol. 34, May 1955, pp. 637–663.

Hansen, R. C., "Fundamental Limitations in Antennas," *Proc. IEEE*, Vol. 69, Feb. 1981a, pp. 170–182.

Hansen, R. C., "Some New Calculations on Antenna Superdirectivity," *Proc. IEEE*, Vol. 69, Oct. 1981b, pp. 1365–1366.

Hansen, R. C., "Linear Arrays," in *Handbook of Antenna Design*, A. W. Rudge et al., Eds., IEE/Peregrinus, 1983.

Hansen, R. C., "Superconducting Antennas," *Trans. IEEE*, Vol. AES-26, Mar. 1990, pp. 345–355.

Hansen, R. C., "Antenna Applications of Superconductors," *Trans. IEEE*, Vol. MTT-39, Sept. 1991, pp. 1508–1512.

Hansen, R. C., "Hansen-Woodyard Arrays with Few Elements," *Microwave and Opt. Technol. Lett.*, Vol. 5, No. 1, Jan. 1992, pp. 44–46.

Hansen, W. W. and Woodyard, J. R., "A New Principle in Directional Antenna Design," *Proc. IRE*, Vol. 26, Mar. 1938, pp. 333-345.

Khamas, S. K. et al., "A High-T_c Superconducting Short Dipole Antenna," *Electron. Lett.*, Vol. 24, Apr. 14, 1988, pp. 460–461.

Khamas, S. K. et al., "Significance of Matching Networks in Enhanced Performance of Small Antennas When Supercooled," *Electron. Lett.*, Vol. 26, 1990, pp. 654–655.

Lancaster, M. J. et al., "Supercooled and Superconducting Small-Loop and Dipole Antennas," *Proc. IEE*, Vol. 139, Pt. H, June 1992, pp. 264–270.

La Paz, L. and Miller, G. A., "Optimum Current Distributions on Vertical Antennas," *Proc. IRE*, Vol. 31, 1943, pp. 214–232.

Lo, Y. T. et al., "Optimization of Directivity and Signal-to-Noise Ratio of an Arbitrary Antenna Array," *Proc. IEEE*, Vol. 54, Aug. 1966, pp. 1033–1045.

Moreno, T., *Microwave Transmission Design Data*, Dover, 1948, Chapter 2.

Newman, E. H. et al., "Superdirective Receiving Arrays," *Trans. IEEE*, Vol. AP-26, Sept. 1978, pp. 629–635.

Oseen, C. W., "Die Einsteinsche Nadelstichstrahlung und die Maxwellschen Gleichungen," *Ann. Phys.*, Vol. 69, 1922, p. 202.

Pritchard, R. L., "Optimum Directivity Patterns for Linear Point Arrays," *J. Acoust. Soc. Am.*, Vol. 25, Sept. 1953, pp. 879–891.

Reid, D. G., "The Gain of an Idealized Yagi Array," *J. IEE*, Vol. 93, Pt. IIIA, 1946, pp. 564–566.

Rhodes, D. R., "On an Optimum Line Source for Maximum Directivity," *Trans. IEEE*, Vol. AP-19, July 1971, pp. 485–492.

Rhodes, D. R., *Synthesis of Planar Antenna Sources*, Clarendon Press, 1974.

Riblet, H. J., "Discussion on a Current Distribution for Broadside Arrays which Optimizes the Relationship between Beam Width and Side-Lobe Level," *Proc. IRE*, Vol. 35, 1947, pp. 489–492.

Salzer, H. E., "Calculating Fourier Coefficients for Chebyshev Patterns," *Proc. IEEE*, Vol. 63, Jan. 1975, pp. 195–197.

Schelkunoff, S. A., "A Mathematical Theory of Linear Arrays," *Bell Syst. Tech. J.*, Vol. 22, 1943, pp. 80–107.

Sokolnikoff, I. S. and Sokolnikoff, E. S., *Higher Mathematics for Engineers and Physicists*, McGraw-Hill, 1941.

Stegen, R. J., "Excitation Coefficients and Beamwidths of Tschebyscheff Arrays," *Proc. IRE*, Vol. 41, Nov. 1953, pp. 1671–1674.

Tai, C. T., "The Optimum Directivity of Uniformly Spaced Broadside Arrays of Dipoles," *Trans. IEEE*, Vol. AP-12, 1964, pp. 447–454.

Uzkov, A. I., "An Approach to the Problem of Optimum Directive Antennae Design," *Compt. Rend. Acad. Sci. U.R.S.S.*, Vol. 53, 1946, p. 35.

Uzsoky, M. and Solymar, L., "Theory of Super-Directive Linear Arrays," *Acta Phys. Hung.*, Vol. 6, 1956, pp. 185–205.

Voges, R. C. and Butler, J. K., "Phase Optimization of Antenna Array Gain with Constrained Amplitude Excitation," *Trans. IEEE*, Vol. AP-20, 1972, pp. 432–436.

Yaru, N., "A Note on Supergain Antenna Arrays," *Proc. IRE*, Vol. 39, Sept. 1951, pp. 1081–1085.

CHAPTER TEN

Multiple-Beam Antennas

10.1 INTRODUCTION

Multiple simultaneous beams exist when an array of N elements is connected to a beamformer with M beam ports, where N and M may be different. Multiple-beam systems have many uses: in electronic countermeasures, in satellite communications, in multiple-target radars, and in adaptive nulling, for example. The last application uses adaption in beam space as it avoids several serious difficulties that arise with adaption in array space. (Mayhan, 1972).

Not included in this chapter are array feeds for parabolic reflectors, where amplitude and phase control of the elements allows coma correction, beam switching, adaptive nulling, and to a limited degree correction of reflector shape errors (Rudge and Davies, 1970; Blank and Imbriale, 1988; Smith and Stutzman, 1992).

Also not included are horn arrays feeding reflector antennas on communication satellites to provide tailored earth coverage patterns (Clarricoats and Brown, 1984: Richie and Kritikos, 1988; Bird, 1990).

10.2 BEAMFORMERS

This section deals with the methods for forming multiple beams. Multiple beamformers are either networks or quasi-optical lenses. It is customary to call either type a beamforming network (BFN). In general, with a network BFN, beam crossover levels are independent of frequency, while beamwidths and beam angles change with frequency. With a lens BFN, beam angles are fixed, while beamwidths, hence crossover levels, change with frequency. Each class of BFN is discussed.

10.2.1 Networks

10.2.1.1 Power Divider BFN. Probably the simplest BFN uses power dividers to split the received signal from each array element into N outputs; the #1 outputs from the dividers are connected through fixed phasers to a combiner to provide beam #1; etc. Figure 10.1 sketches such a configuration. Since there is a signal decrease to $1/N$, preamplifiers are almost always used at each element to maintain S/N. This type of beamformer is attractive for a few beams, but is in use for as many as 16 beams. The Globalstar satellite S-band transmit array has 91 printed circuit elements, with each connected to an amplifier–isolator–filter chain. Each amplifier is driven by a 16-way power combiner; the #1 inputs to the combiners are connected to a 16-way power divider through fixed phasers; each power divider input represents a transmit beam (Metzen, 1996). The 91-element array is shown in Fig. 10.2.

Another satellite communications system is Iridium; Fig. 10.3 shows an array of 106 patch elements. Each patch is connected to an electronic TR module; the BFN forms 16 shaped beams for earth coverage.

Beamforming networks were originated in the late 1950s by Jesse Butler; thus it is appropriate to discuss the Butler matrix next.

10.2.1.2 Butler Matrix. A Butler matrix BFN connects 2^n array elements to an equal number of beam ports (Butler and Lowe, 1961). It consists of alternate rows of hybrid junctions and fixed phasers (phase shifters), with a typical diagram shown in Fig. 10.4 for an 8-element Butler. This figure exhibits both good and bad features of this BFN: it is a simple network using components easily implemented in stripline or microstrip, but conductor crossovers are required. It has been observed many times that Butler matrix diagrams such as Fig. 10.4 are exactly the same as the calculation flow chart of the FFT. In fact the Butler matrix is the analog circuit equivalent of fast Fourier transform, so we may expect that this BFN has the minimum

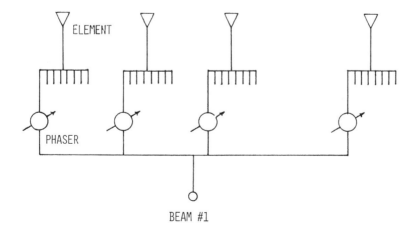

Figure 10.1 BFN using power dividers.

332 MULTIPLE-BEAM ANTENNAS

Figure 10.2 Globalstar multibeam array. (Courtesy Space Systems/Loral.)

Figure 10.3 Iridium multibeam array. (Courtesy Motorola.)

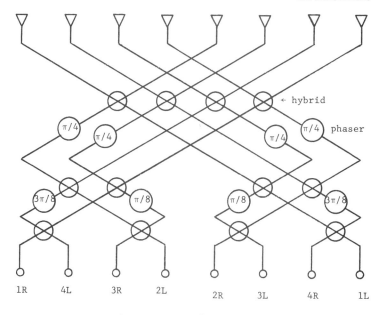

Figure 10.4 Butler matrix BFN.

number of components and minimum path length of all uniform excitation BFNs. Useful references are Butler (1966), Shelton and Kelleher (1961), and Moody (1964).

Using the universal variable $u = (d/\lambda)(\sin\theta - \sin\theta_i)$, where d/λ is the element spacing in wavelengths and θ_i is the axis of the ith beam measured from broadside, the Butler beam patterns are

$$F(u) = \frac{\sin(N\pi u)}{N\sin(\pi u)}. \tag{10.1}$$

All beams have the same shape in u space. For half-wave spacing and large N (number of array elements) the beams are orthogonal. Beam position for any spacing is

$$\sin\theta_i = \pm\frac{i\lambda}{2Nd}, \qquad i = 1, 3, 5, \ldots, (N-1) \tag{10.2}$$

and the beams are spaced apart $1/N$ in u. For half-wave element spacing the beams fill up visible space as sketched in the rosette of beams in Fig. 10.5. The array is uniformly excited and the sidelobe level is -13.2 dB. When the element spacing is increased, the beamwidths become narrower and the beams move closer together. The crossover level, discussed below, does not change, as these effects cancel out. Beam coverage, from the center of the left-most beam to the center of the right-most beam, is

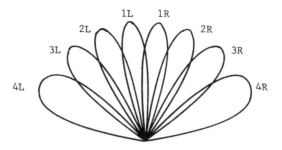

Figure 10.5 Butler BFN beam rosette.

$$\theta_{\text{coverage}} = 2 \arcsin \frac{(N-1)\lambda}{2Nd}. \quad (10.3)$$

This shows the filling of visible space for $d = \lambda/2$, and the smaller coverage for $d > \lambda/2$. As expected, the larger element spacing produces grating lobes, with a grating lobe (GL) corresponding to the right-most beam located just left of the left-most beam, and so on. In other words, a Butler BFN fills visible space with beams, and if $d > \lambda/2$ there will be aliasing and a gain loss due to GLs. When larger element spacing is used, it often is because the elements are larger—subarrays, horns, etc. For this case the GL are replaced by quantization lobes whose levels are reduced by the element pattern; see Chapter 2.

Adjacent beam crossovers occur for $u_x = \frac{1}{2}(u_i + u_{i+1}) = (m+1)\pi/2N$, which gives $u = u_x - u_i = 1/2N$. The crossover level is now

$$F_{xov} = \frac{1}{N \sin(\pi/2N)}. \quad (10.4)$$

Table 10.1 shows this for several powers of 2, along with the beam sidelobe ratios. In both instances the uniform continuous aperture values are obtained for large N; the crossover level approaches $2/\pi$ and the space factor approaches sinc $N\pi u$ for main beam and close-in sidelobes. The nonorthogonal effects for small N are discussed later.

Frequency changes are equivalent to changing d/λ. Increasing frequency narrows the beams and moves them closer together. And grating lobes may

TABLE 10.1 Butler Array Parameters

N	SLR (dB)	Crossover Level (dB)
4	11.30	−3.70
8	12.80	−3.87
16	13.15	−3.91
32	13.23	−3.92
∞	13.26	−3.92

appear if the frequency is raised even a modest amount. As discussed above, these grating lobes are aliased with the primary lobes; the grating lobe beam appearing at −90 deg is associated with the beam port closest to +90 deg, and so on.

10.2.1.3 Blass and Nolen Matrices.

The Blass matrix uses a set of array element transmission lines which intersect a set of beam port lines, with a directional coupler at each intersection (Blass 1960). See the sketch in Fig. 10.6. The top feed line couples out a broadside beam, and is relatively unaffected by the other feed lines. The second feed line tends to couple out a beam at an angle off broadside but the coupling is affected by the top line. A successful analysis must consider each beam port line as a travelling wave (TW) array (of couplers), with interaction between the several TW arrays. See Butler (1966) for an analysis of the Blass matrix. The design can be implemented on a computer, to allow an efficacious trade between network efficiency and beam coupling. However these arrays are difficult to construct as the interaction between couplers makes coupler adjustment and the correlation with measured data difficult. Broadband versions of the Blass matrix incorporating time delay have been discussed by Butler (1966).

The Nolen matrix (Nolen, 1965) is a generalization of both Blass and Butler matrices, and is a canonical BFN configuration. It derives from the fact stated later that orthogonal beams require a unitary coupling matrix, where the matrix elements are the coupling from the nth antenna element to the mth beam port. Unitary matrices can be factored into a product of elementary unitary matrices, of which the type of interest here is

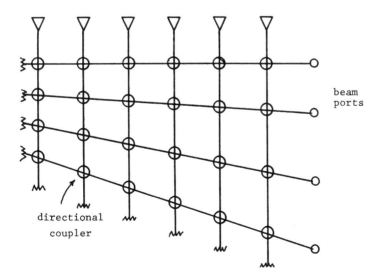

Figure 10.6 Blass matrix BFN.

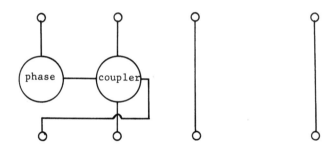

$$\begin{bmatrix} j\cos\alpha \exp j\beta & -j\sin\alpha & & & 0 \\ j\sin\alpha \exp j\beta & j\cos\alpha & & & \\ & & 1 & & \\ 0 & & & 1 & \\ & & & & 1 \end{bmatrix}. \qquad (10.5)$$

This is useful because it represents a simple circuit containing a phaser and a directional coupler, as seen in Fig. 10.7. The concatenation of these prototype network sections produces the Nolen matrix shown in Fig. 10.8. Cummings (1978) has shown that the Nolen matrix can be reduced to the Butler. As in the case of the Blass matrix, the Nolen matrix can have a number of beam ports not equal to the number of antenna elements, although the conceptual development just presented requires them to be equal. The Nolen array is an implementation of the general DFT algorithm and works for any number of

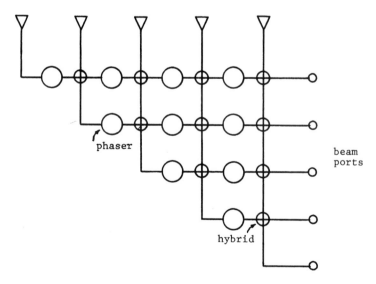

Figure 10.7 Unitary coupling matrix circuit.

Figure 10.8 Nolen matrix BFN.

elements, including prime numbers. As the number of factors in the number of array elements increases, the Nolen matrix can be reduced to simpler forms, culminating in the Butler FFT analog when $N = 2^n$. Owing to the high parts count and the difficulties connected with network adjustments, the Nolen BFN is seldom used.

10.2.1.4 2-D BFN. Multiple beams in both planes can be achieved by cross-connecting two stacks of Butler matrices or of Rotman lenses (see Section 10.2.2.1). Figure 10.9 sketches this arrangement (Sole and Smith, 1987). A stack of Rotman lenses, without the connecting cables, is shown in Fig. 10.10 (Archer, 1975). The hexagonal lattice McFarland matrix of next section is a more efficient (fewer components) matrix BFN. Another 2-D system uses two stacks of 8-port Butler matrices, as sketched in Fig. 10.11 (Detrick and Rosenberg, 1990). The beam cluster from this array is shown in Fig. 10.12. A sophisticated antenna using crossed Butler matrices is for the Iridium satellites. Sixteen-element matrices are connected to the array elements, with 8-element matrices cross-connected (Schuss et al., 1996). Because the coverage area of each array is roughly sector shaped, only half of the normal 256 matrices are needed, and some beam ports are not used. In the next section the more efficient (fewer components) 2-D matrix is described.

10.2.1.5 McFarland 2-D Matrix. The idea of a two-dimensional matrix beamformer utilizing a hexagonal lattice of radiating elements apparently originated at Radiation Systems Inc. in the early 1960s, with work continued at TRW (Williams and Schroeder, 1969). McFarland (1979) discovered how to synthesize a network of power dividers and phase shift units for a two-dimensional array of beams. This beamformer is based on the odd hexagonal lattice where the center is not a single element but rather a triad of elements

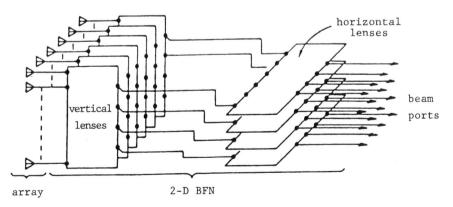

Figure 10.9 2-D Rotman lenses. (Courtesy Sole, G. C. and Smith, M. S., "Multiple Beam Forming for Planar Antenna Arrays Using a Three-Dimensional Rotman Lens," *Proc. IEE*, Vol. 134, Pt. H, Aug. 1987, pp. 375–385.)

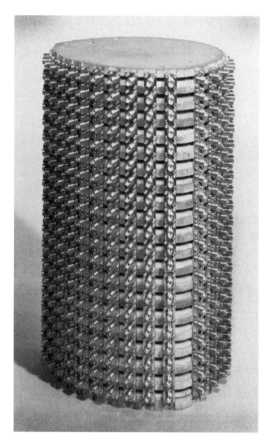

Figure 10.10 Stack of Rotman Lenses. (Courtesy Raytheon Co.)

as seen in Fig. 10.13. This results in every other side of the hexagon of elements having one more element than the adjacent sides. If M is half the number of elements in the largest diagonal, the total number of elements N is equal to $3M^2$. It turns out that this odd lattice can be divided into three symmetric sets of elements as shown in the figure, and each set can be connected to a beamformer. The three interconnected beamformers then provide the hexagonal lattice of beams, and they are connected to a hexagonal lattice of elements. These beamformers are more complex, of course, than a one-dimensional matrix. They contain multiple-arm power dividers, such as three-way dividers, and phase shift sections (Chadwick et al, 1981). Figures 10.14 and 10.15 show two schemes for connecting the elements to the beam ports (McFarland, unpublished notes, circa 1978). In Fig. 10.14 a triad of array elements is connected to a set of divider/phaser boxes, with the outputs providing a triad of beam ports. Figure 10.15 reverses the element triads and the beam port triads. Relatively little work has been done on this type of array, and general synthesis methods are not yet known. In fact, since the

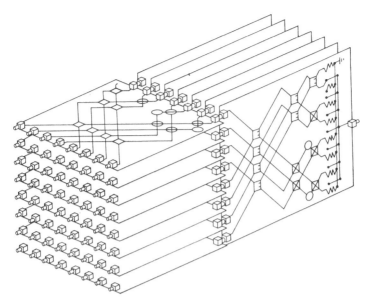

Figure 10.11 2-D Butler matrices. (Courtesy Detrick, D. L. and Rosenberg, T. J., "A Phased-Array Radiowave Imager for Studies of Cosmic Noise Absorption," *Radio Sci.*, Vol. 25, No. 4, July–Aug. 1990, pp. 325–338.)

Figure 10.12 2-D Butler beam rosette. (Courtesy Detrick, D. L. and Rosenberg, T. J., "A Phased-Array Radiowave Imager for Studies of Cosmic Noise Absorption," *Radio Sci.*, Vol. 25, No. 4, July–Aug. 1990, pp. 325–338.)

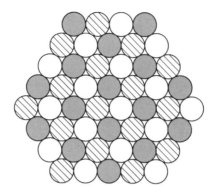

Figure 10.13 $3M^2$ horn cluster triads for $M = 4$, $N = 48$

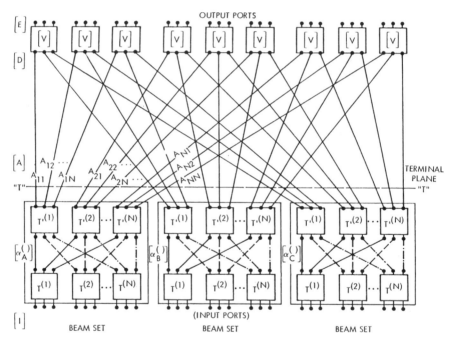

Figure 10.14 Hexagonal lattice BFN triad. (Courtesy J. L. McFarland.)

untimely death of McFarland, the entire subject appears to have languished, in spite of its promise. The two-dimensional beamformer has several advantages. First, it has fewer parts than a cascade of one-dimensional Butlers. Second, a synergistic integration of the beamformer with a switch matrix appears possible. And finally, the geometry fits both the two-dimensional roughly circular array shape and spatial coverage more efficiently.

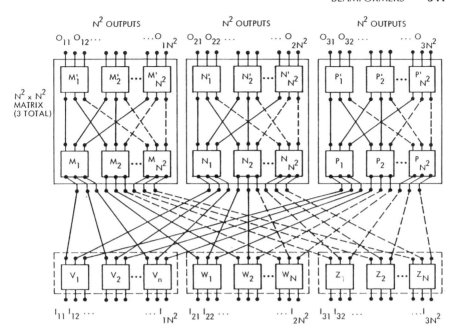

Figure 10.15 Hexagonal lattice BFN triad. (Courtesy J. L. McFarland.)

10.2.2 Lenses

10.2.2.1 Rotman Lens BFN. A bootlace lens is a constrained lens wherein the path lengths from the lens output ports to the radiating element ports are prescribed. An early constrained lens was the R-2R (Brown, 1953, 1969) where the inner and outer lens surfaces are circular arcs, with the outer radius twice the inner. This is the only path length lens with perfect collimation for a feed point anywhere on the focal arc (Kales and Brown, 1965). However, the amplitude asymmetry for beams off-axis raises the sidelobes. This, with the circular array shape, has limited the usefulness of the R-2R. It was originally implemented in geodesic form (deformed parallel plates; Tin Hat). A general theory of bootlace lenses was given by Gent (1957). Another one-dimensional lens (called 1-D because it produces beams in a single plane) designed for a moving feed and a linear radiating array was developed by Rotman and Turner (1963). This soon led to a Rotman lens with an array of fixed feeds for beam ports. Figure 10.16 sketches a Rotman lens; only one beam port is shown. The original lens used parallel plate waveguide for the feed portion, and coax cables for the bootlace portion. Implementation now is typically microstrip or stripline for the feed portion, and either coax cables or printed circuit lines for the bootlace portion (Archer, 1975). Strictly speaking, the lens is the bootlace portion, but it is common practice to refer to the entire assembly as the lens. These lenses have three (perfect) focal

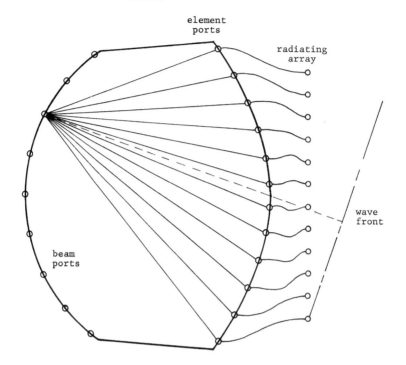

Figure 10.16 Rotman lens beamformer.

points, usually at the center of the feed array, and at two symmetric points between the center and the edge of the feed arc. They are thus a major improvement on the designs of Ruze (1950).

Design of these lenses must involve both geometric trades and mutual coupling effects between the lens ports. The latter is relatively difficult to control, but the former is crucial to the realization of an efficient and compact lens. Thus a careful geometric optics design should be accomplished first; then adjustments must be made to reduce mutual coupling effects. The geometric design trades are described here. The Rotman lens has six basic design parameters: focal angle α, focal ratio β, beam angle to ray angle ratio γ, maximum beam angle Ψ_m, focal length f_1, and array element spacing d. The last two are in wavelengths, and γ is a ratio of sines. A seventh design parameter allows the beam port arc to be elliptical instead of circular. Since the design equations are implicit and transcendental, with only one sequence of solution, the interplay of design parameters is difficult to discern. A series of lens plots is used to show the effects of each parameter. Geometric phase and amplitude errors over the element port arc vary primarily with α and β, and with an implicit parameter which is the normalized element port arc height. Representative plots show how these errors depend upon the parameters.

For lenses where the beam port arc and feed port arc are identical, resulting in a completely symmetric lens, the design equations are greatly simplified

(Shelton, 1978). However, these lenses are seldom used because their design options are much more constrained.

The lens equations equate path lengths from the foci to the array elements; see Rotman and Turner (1963) for a derivation of these. Using the nomenclature of Fig. 10.17, it is convenient to normalize all dimensions by the principal focal length f_1. This is also the lens width at the center. The focal angle α is subtended by the upper and lower foci at the center of the element port curve. It is assumed here that the foci are symmetrically disposed about the axis, and that the lens is also symmetric. Then the parameter β is the ratio[1] of upper (and lower) focal length f_2 to f_1:

$$\beta = \frac{f_2}{f_1}. \tag{10.6}$$

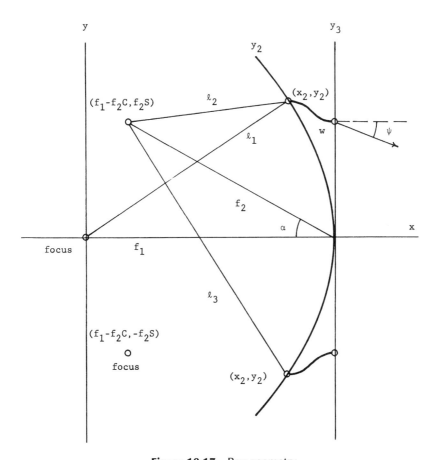

Figure 10.17 Ray geometry.

---

[1] Note that β is the inverse of the ratio g used by Rotman and by McGrath (1988); β is more convenient due to the f_1 normalization.

Clearly the lens width in wavelengths, f_1/λ, is another parameter. Now the angle of the beam radiated by the array is Ψ, and if one of the off-axis foci is excited, then the ratio of lens ray angle α to array beam angle Ψ is γ:

$$\gamma = \frac{\sin \Psi}{\sin \alpha}. \tag{10.7}$$

An indirect parameter of utility is ζ, which relates the distance y_3 of any point on the array from the axis to f_1. This parameter controls the portion of the phase and amplitude error curves that the lens experiences. It is expressed as

$$\zeta = \frac{y_3 \gamma}{f_1}. \tag{10.8}$$

Note that the line lengths w of Fig. 10.16 are an integral and essential part of the lens. The maximum beam angle, Ψ_m, is an important parameter, as is the array element spacing in wavelengths d/λ. The ζ_{max} that corresponds is given by

$$\zeta_{max} = \frac{(NE - 1)\gamma d}{2f_1}, \tag{10.9}$$

where NE is the number of elements in the linear array, since $y_{max} = (NE - 1)/2$. An upper limit on ζ occurs when the tangent to the element port curve is vertical; this also gives $w = 0$. This value of ζ is given by

$$\zeta_{w=0} = \frac{2\sqrt{1 - \beta C}}{S} \sqrt{1 - \frac{1 - \beta C}{S}}. \tag{10.10}$$

Figure 10.18 gives this limiting value versus β, for several values of α. Since the useful range of ζ is roughly from 0.5 to 0.8, a range of β appropriate for a given α may be inferred.

The geometric lens equation is a quadratic in the line length w that connects an element port to the corresponding array element:

$$a\left(\frac{w}{f_1}\right)^2 + \frac{bw}{f_1} + c = 0, \tag{10.11}$$

where the coefficients involve the parameters α, β, and γ:

$$a = 1 - \frac{(1 - \beta)^2}{(1 - \beta C)^2} - \frac{\zeta^2}{\beta^2},$$

$$b = -2 + \frac{2\zeta^2}{\beta} + \frac{2(1 - \beta)}{1 - \beta C} - \frac{\zeta^2 S^2 (1 - \beta)}{(1 - \beta C)^2}, \tag{10.12}$$

$$c = -\zeta^2 + \frac{\zeta^2 S^2}{1 - \beta C} - \frac{\zeta^4 S^4}{4(1 - \beta C)^2}.$$

and $C = \cos \alpha$, $S = \sin \alpha$.

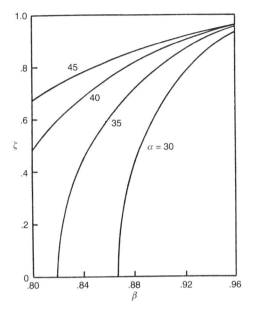

Figure 10.18 Upper Limit on Parameter ζ.

Usually the number of beams and number of elements, the maximum beam angle, and element spacing are specified from the system requirements. Thus, the task is to select the optimum α, β, γ, and f_1/λ.

The lens shape is important, both in conserving space and in reducing loss. For example, a wide lens tends to have path lengths that are more nearly equal, and allows the beam port curve and the element port curve to have different heights, and even different curvatures. As in Fig. 10.18, the width is along the lens axis. Wide lenses have large spillover loss, and higher transmission line loss. A compact lens tends to minimize spillover losses; roughly equal port curve heights now become important, to avoid severe asymmetric amplitude tapers and large phase errors. Curvatures of the two-port curves may be different; use of array element spacing greater than half-wavelength allows more beam ports than element ports to be used. For this case, the beam port curve may be more curved, and the element port curve flatter.

The effects of the seven parameters will be shown through a series of charts (Hansen, 1991). Six charts are shown in Figs. 10.19 through 10.24. Beam ports, which are ticked, are on the left. Element ports, also ticked, are on the right. Foci are indicated by asterisks. The focal length is normalized to unity, so that each tick mark on the axis (and on the ordinate) is 0.05. From the ordinate scale the element positions may be inferred. Each lens curve is extended past the outermost port by half the width of that port. These examples have 9 beam ports and 11 element ports, and of course 11 elements in an equally spaced linear array.

With all other variables fixed, increasing α opens the beam port curve, and closes the element port curve. Port positions are roughly unchanged. But the

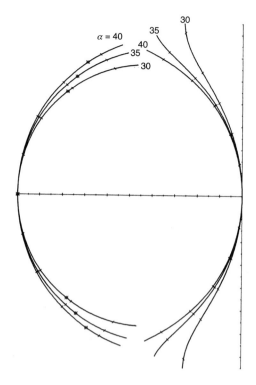

Figure 10.19 Effect of focal angle: $\beta = 0.9$, $\gamma = 1.1$, $\Psi_m = 50$, $f_1 = 4\lambda$, $d = 0.5\lambda$.

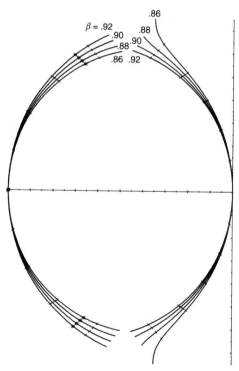

Figure 10.20 Effect of focal ratio: $\alpha = 40$, $\gamma = 1.1$, $\Psi_m = 50$, $f_1 = 4\lambda$, $d = 0.5\lambda$.

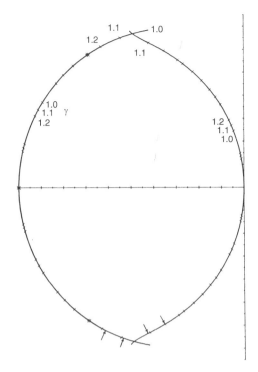

Figure 10.21 Effect of beam angle ratio: $\alpha = 40$, $\beta = 0.9$, $\Psi_m = 50$, $f_1 = 4\lambda$, $d = 0.5\lambda$.

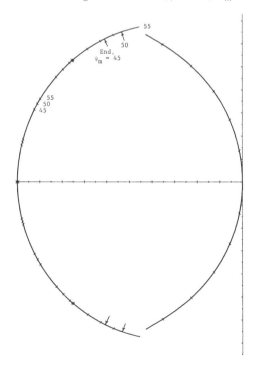

Figure 10.22 Effect of maximum beam angle: $\alpha = 35$, $\beta = 0.92$, $\gamma = 1.1$, $f_1 = 4\lambda$, $d = 0.5\lambda$.

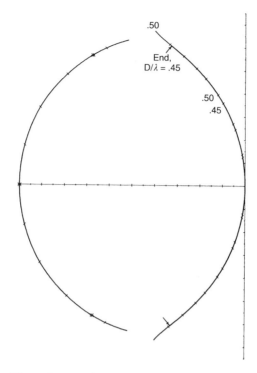

Figure 10.23 Effect of array element spacing: $\alpha = 40$, $\beta = 0.88$, $\gamma = 1.1$, $\Psi_m = 50$, $f_1 = 4\lambda$.

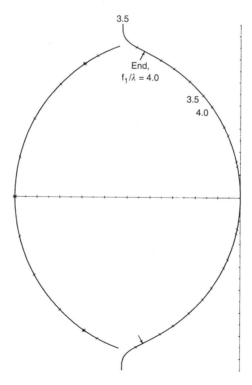

Figure 10.24 Effect of focal length: $\alpha = 40$, $\beta = 0.9$, $\gamma = 1.1$, $\Psi_m = 50$, $d = 0.5\lambda$.

outer foci locations change markedly as expected. The three lens plots of Fig. 10.19 illustrate these effects. It can be seen that a value of α can be selected that roughly equalizes the heights of the two curves. Of course, α must be selected in conjunction with the other variables to minimize phase errors over the aperture. The outer foci should be comfortably inside the beam port curve.

Increasing β has an effect similar to increasing α; the beam port curve opens, and the element port curve closes. Figure 10.20 contains three lens plots to show this. Again, port positions are roughly unchanged. Also the focal locations change relatively little. Again, a value of β can be selected that roughly equalizes the curve heights.

There are pairs of α and β that produce closely the same lens shape, and port positions. However, the foci vary with α, and the connecting lines (from element ports to elements) are different. Table 10.2 shows three α–β lens pairs that have common lens curves and ports, all for $\gamma = 1.1$, $\Psi_m = 50$, $f_1/\lambda = 4$, and $d = \lambda/2$. One may thus infer that the phase error over the aperture for each beam will be different, depending upon α. This will be shown in the next section. For any set of the other four parameters, there are probably some α–β pairs that behave similarly.

Increasing γ leaves both lens curves unchanged, but the beam ports are moved closer together, while the element ports are spread apart. A three-lens set in Fig. 10.21 shows this trend. Although the foci remain fixed, the ends of the curves change, so that the relative position of the foci changes. γ also can affect the relative heights of the two curves.

Values of γ here are 1 or greater, as the cases used are all for large beam angles. When the beam cluster subtends a more modest angle, e.g., 30 deg, values of $\gamma < 1$ are appropriate as they allow a "fat" or curved lens.

When Ψ_m is changed, only the beam port spacings change. Increasing Ψ_m spreads the beam ports and extends the port curve, so that this parameter helps produce a lens with roughly equal heights of beam and element port curves. The three lens plots of Fig. 10.22 depict this behavior.

Element spacing is critical as it controls the appearance of grating lobes (see Chapter 2). For a maximum beam angle of Ψ_m, the spacing that just admits a grating lobe is

$$\frac{d}{\lambda} = \frac{1}{1 + \sin \Psi_m}. \tag{10.13}$$

In general, spacings are kept below this value.

TABLE 10.2 α–β Pairs

Number	α	β
1	30	0.94
2	35	0.92
3	40	0.90

When d is changed, only the element port spacings and the extent of the port curve change, analogous to Ψ_m changing beam port spacing. Figure 10.23 uses two lens plots to show this.

Increasing the lens focal length (width) in general increases the separation between the end ports as well. But changing f_1/λ also changes all spacings, as the lens equations are normalized by f_1. Thus, as shown in the two lens plots of Fig. 10.24, changing f_1/λ also changes the element port arc and element port spacings. The minimum value of f_1 is smaller for Rotman lenses than for other types of lenses (Smith, 1982).

Next, phase and amplitude errors will be examined.

Aperture errors depend upon α and β, and upon eccentricity, but only indirectly on the other parameters. Thus the most insight results from plotting phase and amplitude errors versus the normalized parameter ζ: see Eqn (10.8). Since phase errors are zero at angles corresponding to the three foci, a satisfactory approach uses one beam position midway between the central and edge foci, and a second beam position beyond the edge focus. Amplitude errors occur at all beam ports, so more cases are needed to display amplitude error behavior.

Figures 10.25 through 10.28 show phase error versus ζ for lenses with α of 35, 40, and 45 deg, for various values of β. Note that to get phase error, the values from the figures are to be multiplied by f_1/λ. For the midfoci beams, the phase error is small, except for very large lenses. Phase errors for the wider-angle beams are still modest, and will not be important except for large lenses,

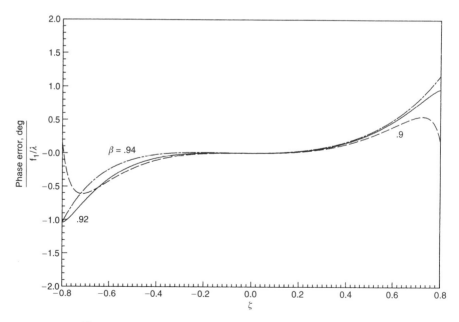

Figure 10.25 Rotman Lens, $\alpha = 35$, ray angle $= 17.5$ deg.

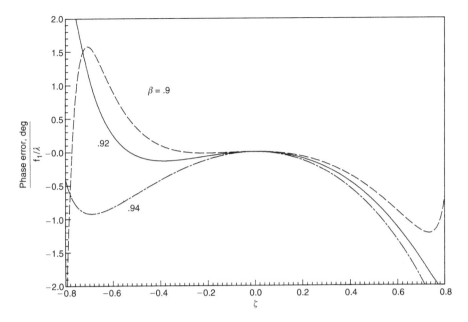

Figure 10.26 Rotman Lens, $\alpha = 35$, ray angle = 45 deg.

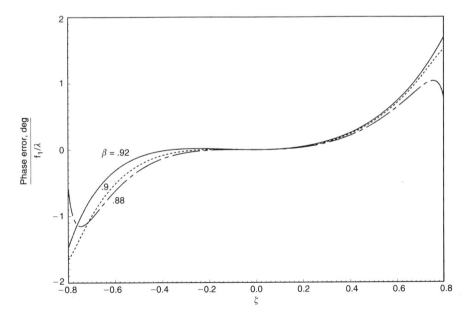

Figure 10.27 Rotman Lens, $\alpha = 40$, ray angle = 20 deg.

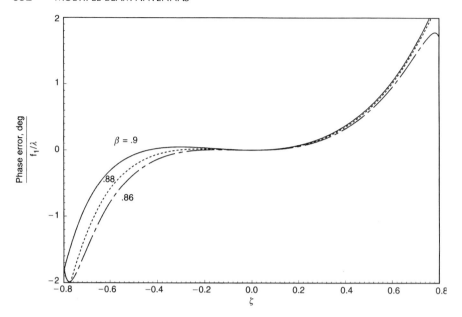

Figure 10.28 Rotman Lens, $\alpha = 45$, ray angle = 22.5 deg.

or designs with $\zeta > 0.75$. In general, the phase errors increase as α is increased, for all beam positions.

Although Rotman and Turner indicated an optimum value of β, which was $2/(2+\alpha)^2$, examination of Figs. 10.25 through 10.28 show that an optimum β exists only for one range of ζ and one ray angle. Best values of β are different for between-foci rays and rays outside the foci. Since the latter usually have larger phase error, the designer could optimize, but the value would vary with both ray angle and ζ_{max}. This old value also gives poor lens shapes when the numbers of beam and element ports are roughly equal.

The results of using an elliptical beam port curve are shown in Fig. 10.29, where the phase errors are equalized at $\zeta = \pm 0.7$ for a ray angle of 45 deg. Note that the elliptical beam port curve is a simple way of realizing the optimum curve of Katagi et al. (1984). This gives a 13% reduction in phase error; the phase errors for the midfoci rays at 17.5 deg become slightly more asymmetric, but are still well below the 45 deg ray errors. Note that the ellipticity of -0.3 only changes the principal radius by 5%, so amplitudes are essentially unchanged. The beam port ellipse major axis is along the lens axis for this ellipticity.

Amplitude errors are calculated using beam port and element port horn patterns of $\operatorname{sinc} \pi u$, where horn widths are all set to a nominal $\lambda/2$. Each port horn has its axis normal to the port curve. Figure 10.30 shows amplitude error, normalized to 0 dB for the axial ray, for a lens with $\alpha = 35$, $\beta = 0.92$. Curves for ray angles of 0 deg, 17.5 deg, and 45 deg, are given. Similarly, Fig. 10.31 is for a lens with $\alpha = 40$, $\beta = 0.9$, for ray angles of 0 deg, 20 deg, and 50 deg. These examples are two of the α–β pairs of Table 10.2, and thus the amplitude errors

BEAMFORMERS **353**

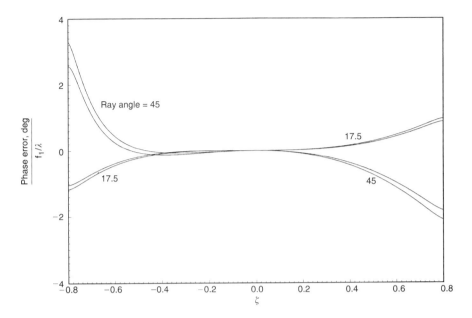

Figure 10.29 Rotman Lens, $\alpha = 35$, $\beta = 0.92$.

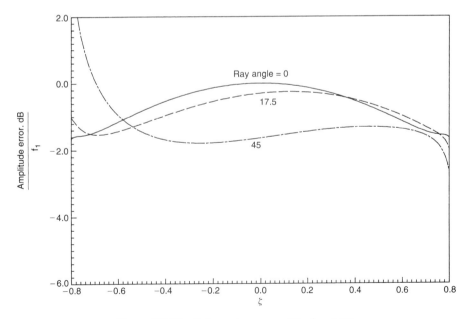

Figure 10.30 Rotman Lens, $\alpha = 35$, $\beta = 0.92$.

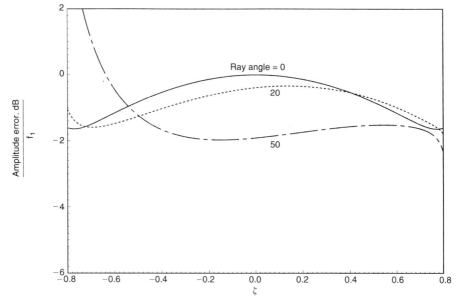

Figure 10.31 Rotman Lens, $\alpha = 40$, $\beta = 0.9$.

are similar. As expected, for wide ray angles the near and far ends of the element port curve experience modest amplitude changes. Compared to the amplitude taper needed to produce 25 dB sidelobes, these amplitude errors are small. Actual lenses may have port widths larger than $\lambda/2$, so the amplitude tapers can be expected to increase, especially for edge beams.

The asymmetry of amplitude for the off-axis beams can be reduced by pointing each port horn at the opposite apex instead of normal to the port curve (Musa and Smith, 1989). For example, a 9-beam, 11-element lens with $\alpha = 40$, $\beta = 0.9$, $\gamma = 1.1$, $\Psi_m = 50$, $d = 0.5\lambda$ and $f_1 = 4\lambda$ has amplitude taper for the outside beam as shown in Table 10.3. Also shown is the taper for apex

TABLE 10.3 Beam One Amplitude Taper; $f_1 = 4\lambda$

Element Number	Axes Normal to Arc (dB)	Axes Through Apex (dB)
1	−8.49	−2.26
2	−7.03	−1.84
3	−5.50	−1.46
4	−4.29	−1.31
5	−3.33	−1.37
6	−2.58	−1.63
7	−2.04	−2.08
8	−1.70	−2.73
9	−1.57	−3.60
10	−1.68	−4.71
11	−2.08	−5.95

pointed horns. Use of apex pointing produces appreciable improvement. Gain is slightly improved.

Element and beam port spillover, phase and amplitude errors, port impedance mismatches, and transmission line loss all contribute to reducing lens gain. Note that, as in the case of a horn feeding a reflector antenna, there is no feed horn spreading loss, owing to the equal-path property through the foci. The small inequality of other paths is subsumed in the path phase and amplitude errors. Gain will be calculated here based on port spillover and on aperture errors. Since the amplitude error calculation includes both beam port and element port horn patterns, spillover is included (Smith and Fong, 1983). The phase and amplitude errors at the element ports are transferred to an array of isotropic elements. Then the problem reduces to that of calculating gain of a symmetric linear array of isotropic elements with complex coefficients, using the formula from Chapter 2. Actual gain is then that of the isotropic array factor multiplied by the element gain, times the impedance mismatch factors.

Variation of gain with parameters is very small. For example, for a typical small lens only 0.2 dB change occurs from the center beam to the edge beam, and the gain values are roughly independent of α, β, γ, etc. With a larger array the gain increases just as expected. Using the same 9-beam, 11-element example of the previous section, gain ranges from 10.2 to 10.4 dB; the latter value, for the center beam, is within 0.1 dB of the gain for a similar uniformly excited array. For this lens the ζ range is ± 0.756.

Effects of feed horn spillover and internal lens reflections can be reduced either by employing dummy (terminated) feed horns adjacent to the edge horns, or through the use of absorber between the ends of the beam port arc and the element port arc (Musa and Smith, 1989).

The design process starts by the selection of a center frequency, at which all dimensions are computed. Then, a starting value of f_1/λ is selected, to keep ζ_{max} well below 0.8. The focal length will be somewhat less than the array length. Next, using the guidelines, α, β, and γ are selected, to (1) locate the outer beam port a modest amount past the outer focus; (2) produce beam port and element port arcs of comparable heights; and (3) yield acceptable phase and amplitude errors at each port. Achieving this may require adjustment of f_1/λ or d/λ, and of α, β, and γ. Use of an elliptical beam port arc is usually not warranted except for large lenses.

When a satisfactory design is realized at the center frequency, phase and amplitude errors at each port are calculated at representative frequencies, to assess performance over the frequency range. And of course the actual beam and element port horn widths are used. At this stage, calculation of a beam rosette (a set of beam patterns) at each frequency is appropriate. Some compromise and iterative adjustment of parameters may be necessary to obtain good wideband results, and to best accommodate mutual coupling effects.

Although the lens width f_1 is less than the array length, the lens height is always greater than the array length. Lens dimensions are reduced by the square root of effective dielectric constant for either stripline or microstrip implementation. See Archer (1984) for examples. Low-sidelobe designs are discussed in Section 10.3. Figure 10.32 shows a typical microstrip Rotman lens.

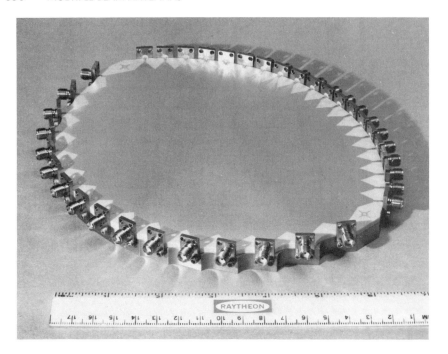

Figure 10.32 Microstrip Rotman Lens. (Courtesy Raytheon Co.)

10.2.2.2 Bootlace Lenses. Although the Rotman lens is a bootlace lens, this section is devoted to two-dimensional bootlace lenses; the multiple beams occur in two dimensions. Unfortunately rotating a Rotman lens to form a symmetric lens does not give good results; a circular focus is not possible. A two-dimensional bootlace lens may have up to four discrete perfect focal points, but they cannot be located with quadrantal symmetry, as the discussion below will show. In general the shape of the lens and array surfaces is affected by the choice of focal points (Cornbleet, 1994).

In practice, bootlace lenses and other lenses are not designed with discrete foci; the optimum design is a MINIMAX solution for circles of least confusion. That is, the aberrations are minimized over the aperture in some sense such as least-mean-squares. Consideration of foci, however, allow lenses to be demarcated into classes, and gives an understanding of how errors depend upon lens geometry.

The simplest bootlace lens has a flat array face and a flat lens face, and has two foci which are symmetrically disposed about the axis in a plane containing the axis. The foci can be coalesced on axis with the result that the lens faces are circular and symmetric (Rao, 1982a). However this arrangement provides the largest errors when scanned.

When the lens surface becomes ellipsoidal, but still with a flat array face, three foci are allowed. These consist of an on-axis focus and two conjugate foci. The three foci are in a plane containing the axis. The lens (and array) outlines are elliptical to allow the cross-plane path lengths to be nearly the same as in-plane path lengths from any focus.

The quadrifocal lens, as expected, has four foci which are disposed in two conjugate pairs about the axis, with all foci on the plane containing the axis. This lens has an ellipsoidal lens surface, as does the trifocal lens, but the array surface is now much more complex. In-plane the cross sections are ellipsoidal, while in cross planes they are circular arcs (Rao, 1982b). This fourth-order surface is easily specified but difficult to construct.

The following is a simple explanation of how a bootlace lens focuses. Consider a bootlace lens with flat array face and flat lens face, with the feed surface a spherical cap. For a focal point off axis, the lace lengths are chosen so that the path length from the focus to the center array element is the same as that for other elements in the plane containing the axis and the focus. The radiated beam angle exactly compensates for the shorter path to the closest element, and the longer path to the farthest element.

However, for elements in the cross plane, if the lens face is symmetric, the path lengths from the focus to elements in the cross plane are always longer than the in-plane path lengths. This is corrected by bringing the lens face (and also array face) elements in the cross plane closer. The result is that the shape of the lens face is now elliptical, with the major axis in-plane and the minor axis cross-plane. Correspondingly, the array face is also elliptical. Thus for a bifocal flat face lens, the flat faces must be elliptical in shape.

The bifocal and trifocal lenses always have the focal points in a plane containing the axis of the lens. It is not possible to design a lens with two or three focal points where one focal point is on the x-axis and a second focal point is on the y-axis, and at the same time have small path length errors for all angles of scan.

Asymmetric phase errors over an array or aperture tend to shift the main beam position. Symmetric errors tend to raise sidelobes, reduce gain, and reduce difference pattern slope at the center. These symmetric error effects are similar to those that are observed in the near-field, where a quadratic error typically occurs. It has been shown (in Chapter 12) that as the phase (or path length) error increases, the first sidelobe rises and with further increases is absorbed into the side of the main beam. Meanwhile, the next null is filling in and the next sidelobe is rising. As sidelobes are shouldered into the main beam, the gain decreases correspondingly. For difference patterns a similar phenomenon occurs. Again the first sidelobe null fills and the sidelobe rises as error increases. The slope changes but only with significant errors.

Bootlace lenses tend to have an error which is maximum at the lens center, and maximum at the lens edge but with opposite sign. This is a result of minimizing the errors over the aperture. This type of error produces similar effects upon the sidelobe and main beam structure. However, the change of difference pattern slope appears to be more significant, probably out of phase with the aperture edge.

Table 10.4 shows typical gain decrease caused by a path length error. Note that these errors are given in wavelengths.

TABLE 10.4 Path Length Error Gain Decrease

Δ/λ	Gain Decrease (dB)
0.05	0.14
0.1	0.50
0.15	1.1
0.2	1.9
0.25	2.8

Lens path length errors of less than roughly 0.1λ have small effect. The small gain decrease is due to the shouldering of the first sidelobes. When the error is about a quarter-wavelength, several sidelobes have been shouldered each side, and the main beam is significantly broadened. The difference pattern sidelobe structure is not much altered, as the sidelobes were originally high. There is not yet significant slope change. Path length errors of the order of a half-wavelength produce a very broad main beam incorporating many sidelobes. The difference pattern now exhibits significant slope change. These give an indication of how path length errors affect lens performance.

The calculation of bootlace lens path length errors is complicated, even when the lenses are designed with discrete foci. However there are some general rules of thumb. The path length error is directly proportional to aperture diameter in wavelength. Error is proportional to the scan angle in the form of $\sin^2 \Psi$. Error is directly proportional to the beam angle ratio γ; this is the ratio of beam angle to feed angle. It is related to the ratio of array height to lens height and to the foci locations and maximum scan angle. In general, γ is around unity and the array lens dimensions are comparable. The path error is roughly inversely proportional to f/D.

In general, having more foci reduces errors when the plane of scan is in the plane of foci. For scan in all planes, the best design appears to be one with all foci on the axis, thereby providing a symmetric lens and array.

A simpler bootlace lens is made with two flat faces, with corresponding elements on the faces connected by phasers. Figure 10.33 sketches such a lens without the feed horns. With a fixed four-horn feed, monopulse tracking is achieved. The phasers provide beam scan. An advantage of this lens is that the aperture amplitude distribution is independent of scan. Figure 10.34 shows an experimental S-band lens with diameter of roughly 8 ft (Wong et al., 1974).

A double lens system can be used to provide wide-angle scanning where the radiating array amplitude distribution does not change with scan angle. Time delay paths are also provided. In one such scheme (Ajioka and McFarland, 1988), a cluster of feeds excites a hemispherical bootlace lens with a hemispherical output face. The outputs in turn excite a transfer array that contains the radiating array; see Fig. 10.35. Figure 10.36 shows part of this wideband, wide angle array system. (Lee, 1988).

BEAMFORMERS 359

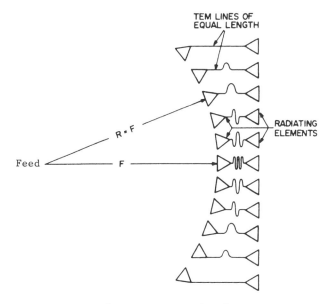

Figure 10.33 Bootlace lens.

Figure 10.34 S-Band bootlace lens. (Courtesy Hughes Aircraft Co.)

360 MULTIPLE-BEAM ANTENNAS

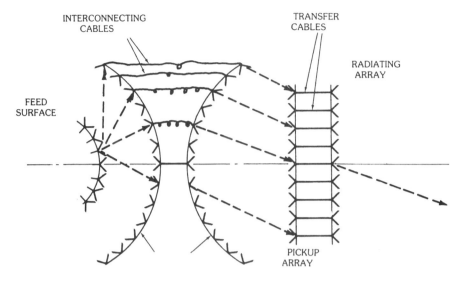

Figure 10.35 Dual hemispherical lens array.

Figure 10.36 Dual hemispherical lens. (Courtesy Hughes Aircraft Co.)

10.2.2.3 Dome Lenses.
A hemispherical, or near-hemispherical, bootlace lens has been used to refract the scanned beam produced by a planar array across the diameter of the lens. That is, if the array radiates at angle θ_0, the lens will radiate the beam at $K\theta_0$, where $K > 1$; see Fig. 10.37. Thus coverage near the horizon is obtained without the problems common to arrays with scan ranges approaching 80 to 90 deg. The dome bootlace lens uses fixed phase or delay "laces," designed to produce a specific K. A $K = 1.5$ dome will provide zenith to horizon coverage with a feed array scanning ± 60 deg. For shipborne applications $K = 2$ gives coverage 30 deg below the horizon (to accommodate roll) with the same array scan range (Schwartzman and Stangel, 1975; Schwartzman and Liebman, 1979). The phase function in the dome is given by

$$\sin\theta_{\text{inc}} + \sin\theta_{\text{refr}} = \frac{\partial\phi}{\partial s} \tag{10.14}$$

where the incident and refracted angles are with respect to the dome normal, and the derivative is that of the insertion phase w.r.t. the arc length (Susman and Mieros, 1979). Because of the circular geometry, a nonlinear phase is required, so that the planar array phase is adjusted with scan angle to provide this correction. An important performance criterion is the area under the gain versus scan angle curve; this is controlled primarily by the planar array and K. However, the maximum value of the curve is controlled by the dome parameters (Steyskal et al., 1979). Figure 10.38 shows a dome bootlace lens

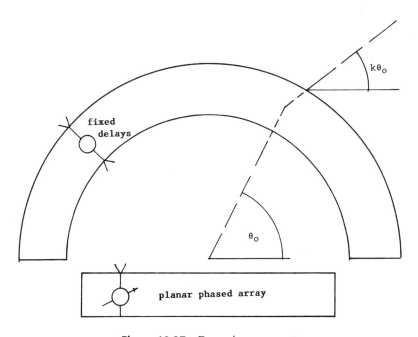

Figure 10.37 Dome lens concept.

Figure 10.38 Dome lens under construction. (Courtesy Sperry Corp.)

antenna. Dielectric dome lenses have also been used; the wall thickness and/or dielectric constant varies with angle (Valentino et al., 1980).

10.2.2.4 Other Lenses Two schemes that provide scanning in one dimension via a variable medium will be described. The first uses a stack of ferroelectric dielectric plates with metal strip in between close to the fixed beam array face. As shown in Fig. 10.39, the stack is the same size as the array, and is roughly one free-space wavelength thick; a voltage gradient is applied across the metal strips to scan the beam in the E-plane. Experimental loss at X-band is about 1 dB (Rao et al., 1996).

In the second scheme a similar stack of metal strips is close to the array face, but the dielectric plates have been replaced by an array of diodes. Because the diode medium produces less phase shift than the ferroelectric medium, the stack is thicker, perhaps $3-4\lambda_0$, with several diode rows normal to the array face. This type of lens is called the radant lens. The diodes are turned on/off to control the phase shift through the medium (Rao et al., 1996). Loss data for this lens were not available.

A completely different type of lens is the constrained lens, where the rays are restricted to specific paths. Waveguide lenses are a good example, where the ray paths are through waveguides whose axes are parallel to the lens axis. Because of the dispersive nature of waveguides, phase shift is a function of length. Thus

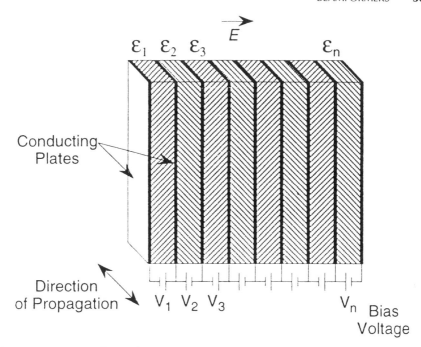

Figure 10.39 Ferroelectric lens. (Courtesy Rao, J. B. L., Trunk, G. V., and Patel, D. P., "Two Low-Cost Phased Arrays," *Proceedings IEEE Symposium Phased Array Systems and Technology*, Boston, Mass., Oct. 1996, pp. 119–124.)

a waveguide lens would have longest guides at the periphery, and shortest guides at the center. Of course such lenses are usually zoned, to reduce thickness and weight, at a cost of reduced bandwidth. A cluster feed provides a multiple beam cluster. Such a lens is shown in Fig. 10.40 (Dion and Ricardi, 1971). Detailed design data are given by Ruze (1950), and Lee (1988).

Another multiple-beam arrangement is that using relatively large subarrays with limited scan; the grating lobes are reduced by using the overlapped subarray technique described for linear arrays in Section 2.3.5. For two-dimensional arrays, this technique can be implemented with networks or with lenses. BFN implementations are discussed by DuFort (1978).

A limited scan array, but not utilizing overlapped subarrays, uses a spherical bootlace lens fed by a corrective lens (DuFort, 1986). In operation, the bootlace lens is excited with phases (and amplitude adjustment if desired) to produce a series of sinc beams that excite the radiating lens or matrix. Each sinc beam excitation produces a flat-top subarray excitation at the radiating array. Together these produce a radiated sinc beam at the desired angle. The configuration is complex, but provides 100% aperture efficiency, with no spillover. It is closely related to the double lens arrangement of Section 10.2.2.2.

364 MULTIPLE-BEAM ANTENNAS

Figure 10.40 Multiple beam waveguide lens. (Courtesy Lincoln Labs.)

10.2.3 Digital Beamforming

Digital beamforming for receive is conceptually the simplest and at the same time the most capable configuration. Each element is connected to a preamp and then to an A/D converter, as shown in Fig. 10.41. The outputs of the A/Ds are connected to a digital data bus, which would probably be a coded serial

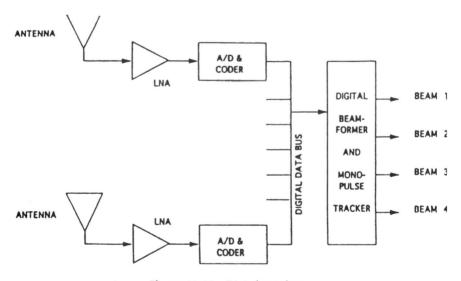

Figure 10.41 Digital topology.

bus. The computer can then form any number of multiple beams, perform rapid beam scan, produce low-sidelobe beams, perform adaptive nulling, implement algorithms for target classification and for multipath discrimination, and do monopulse type tracking with algorithms that simulate monopulse. Because the element combining algorithms can be formulated in true time delay form, there are no bandwidth limitations in the digital beamformer. The serious limitation is, of course, the bit–bandwidth product of the A/D. Table 10.5 gives dynamic range versus number of bits in the A/D converter. At the present time, hardware appears suitable only for low microwave frequencies and below. Advances in superconducting A/D converters are promising, but practical realization of this hardware is years away.

The computer (processor) for a digital array has two speed regimes: the handling of the digitized RF requires coherent beam formation at the A/D converter upper frequency, while the calculation/change of amplitude tapers for sidelobe control, formation of monopulse tracking data, etc., can proceed at a much slower rate. That is, the RF data handling requires a very fast processor; for the array control functions a much slower processor will suffice. Some array functions, such as adaptive nulling, may require a specialized and/or separate processor for such tasks as matrix inversion, singular value decomposition, etc. All of the digital hardware and software is available only the A/D capability is awaited.

10.3 LOW SIDELOBES AND BEAM INTERPOLATION

10.3.1 Low-Sidelobe Techniques

This section discusses various ways of providing decoupled beams so that low sidelobes and high crossover level may both be enjoyed. All of these schemes involve loss. The first uses two separate arrays with beam interlace which could be considered to involve a 3 dB loss. A single array is more practical and the loss can be directly applied as an array taper, it can be applied through a hybrid network, or it can be applied through beam superposition.

10.3.1.1 Interlaced Beams. A direct way of achieving orthogonal beams with low sidelobes and high crossover level employs two sets of interlaced beams. Take a beam set using cosine excitation, where the space factor is

TABLE 10.5 A/D Converter Dynamic Range

Number of Bits	Dynamic Range (dB)
10	60.21
12	72.25
14	84.29
16	96.33

$$F(u) = \frac{\cos \pi u}{1 - 4u^2}. \tag{10.15}$$

The beams are spaced at $u = 0, \pm 2, \pm 4, \ldots$ and the crossover level is $F(1) = 1/3 = -9.54\,\text{dB}$. Now form a second similar beam set spaced at $u = \pm 1, \pm 3, \pm 5, \ldots$. A beam from set one crosses the adjacent beams from set two at $F(0.5) = \pi/4 = -2.10\,\text{dB}$, thus achieving high crossover and modestly low ($-23\,\text{dB}$) sidelobes. Similarly, interlaced cosine square beam sets yield a $-3.27\,\text{dB}$ crossover with $-31.5\,\text{dB}$ sidelobes. Unfortunately, each beam set requires its own array and BFN. Aside from the extra hardware requirements, it may be difficult to get the two array phase centers close together. See sketch in Fig. 10.42.

10.3.1.2 Resistive Tapering. A tapered amplitude, low-sidelobe distribution can be achieved simply by attenuating the array element outputs. For example, a cosine distribution has an aperture efficiency of 0.811. However, to obtain orthogonal beams, the crossover level would be $-9.5\,\text{dB}$ as mentioned. Moving the beams closer to raise the crossover will couple the beams. This type of "loss" is of no help in reducing the beam coupling. Another way of forming a cosine amplitude taper does decouple the beams, but with higher loss as next described.

10.3.1.3 Lower Sidelobes via Lossy Networks. The relationship between sidelobe ratio and crossover level for a lossless BFN does not apply to lossy networks. An important question is how much loss must be accepted to raise

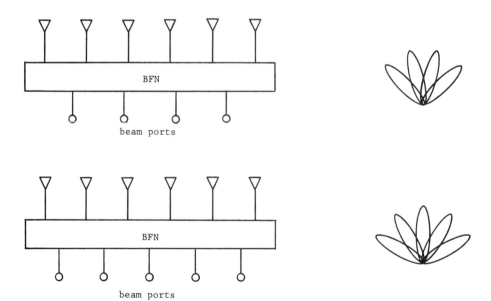

Figure 10.42 Interlaced beams from two multiple-beam antennas.

the crossover level. There is not at this time a definitive answer, but some key results can be given. White (1962) showed that cosine orthogonal beams can be formed from two sets of displaced uniform beams; see Section 10.4. This can be accomplished with an additional network of hybrids as shown in Fig. 10.43. The row of hybrids combines the uniform beams to form the cosine beams, and as shown in the network a BFN output of N beams produces $N-1$ cosine beams. Another way of looking at the operation of the network is that the hybrids isolate the beam ports. Tracing a signal from one beam port shows that half the power goes into hybrid loads, so the efficiency is -3.01 dB and the sidelobe ratio is 23.0 dB. Lower sidelobes can be achieved by combining three sets of uniform beams as shown in Section 10.4 to get cosine squared beams. Figure 10.44 shows this configuration, which yields $N-2$ beams from N BFN outputs. The SLR $= 31.5$ dB and the efficiency is 3/8. Since the only orthogonal aperture distributions appear to be $\cos^M$, performance of several of these is summarized in Table 10.6. Figure 10.45 shows efficiency versus sidelobe level. The network efficiency is

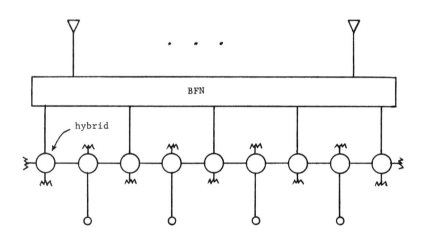

Cosine Beam Ports

Figure 10.43 Formation of orthogonal cosine beams.

TABLE 10.6 Lossy Network Formation of Orthogonal Beams with High Crossovers

Distribution	SLR (dB)	Crossover Level (dB)	η (dB)
Uniform	13.26	-3.92	0
Cosine	23.00	-2.10	-3.01
Cosine Square	31.47	-3.27	-4.26
Cosine Cubed	39.30	-4.44	-5.05

368 MULTIPLE-BEAM ANTENNAS

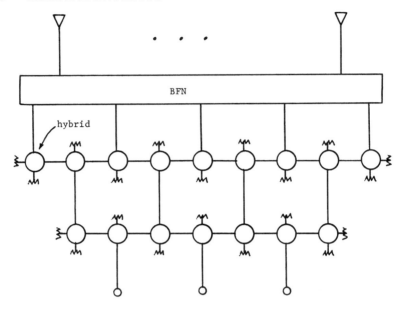

Figure 10.44 Formation of orthogonal cosine square beams.

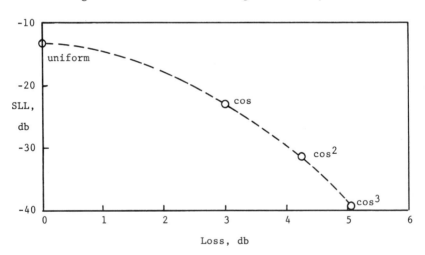

Figure 10.45 Orthogonal distribution sidelobe levels.

$$\eta = \int_0^1 \cos^{2M}(\pi p/2)\, dp = \frac{1 \cdot 3 \cdot 5 \cdots (2M-1)}{2^M M!}. \qquad (10.16)$$

Although the network in the figures will produce the desired orthogonal low-sidelobe, high-crossover beams, it is not clear whether more efficient networks are feasible. Early work by Stein (1962), Allen (1961), and Shelton (1965) have given only general results. Thomas (1978) uses a Rotman lens BFN but the "network" losses are inseparable from the lens losses. See also Winter (1979, 1980). Thus at this time there appears to be no data that indicates that lower losses can be achieved, but also no proof that lower losses are not possible.

10.3.1.4 Beam Superposition. Low sidelobes can be produced by appropriate beam superposition using the Taylor n̄ distribution of Chapter 3. This distribution provides near-optimum directivity and beamwidth for a given SLR by making the first n̄ sidelobes of equal height, with remaining sidelobes decaying as $1/u$, where

$$u = \frac{L}{\lambda}(\sin\theta - \sin\theta_m), \quad (10.17)$$

with θ_m the direction of the mth beam. A transition region exists around n̄, so the n̄ sidelobes are not precisely at the design SLR but taper smoothly into the $1/u$ envelope. The $1/u$ envelope taper is necessary to assure a robust (low-Q) aperture. The Taylor n̄ distribution is given by

$$g(p) = 1 + 2\sum_{n=1}^{\bar{n}-1} F(n, A, \bar{n})\cos n\pi p, \quad (10.18)$$

where the coefficients and terms are defined in Chapter 3. The Taylor n̄ distribution is interesting because of the possibility of approximating $g(p)$ with a few terms.

Except for very small n̄, the only significant term in the $g(p)$ series is the first: $n = 1$. For example, consider the 30 dB space factor with the largest n̄ resulting in a monotonic distribution. The coefficients are given in Table 10.7, where it can be seen that only two coefficients are necessary. This distribution is a "cosine-on-a-pedestal," but not the usual one since here the cosine has value -1 at the aperture ends. n̄ can be increased to the value giving maximum aperture efficiency (n̄ = 23 for this case) without degrading this two-coefficient

TABLE 10.7 Distribution Coefficients for SLR = 30 dB, n̄ = 7

$F_0 =$	1
$F_1 =$	0.28266
$F_2 =$	−0.01313
$F_3 =$	−0.00164
$F_4 =$	0.00468
$F_5 =$	−0.00414
$F_6 =$	0.00225

approximation. Since the change of efficiency from max monotonic $\bar{n}$ to max efficiency $\bar{n}$ is less than 2% for SLR ≤ 40 dB, the monotonic $\bar{n}$ is a good choice.

Taylor showed that the space factor can be written as (see Chapter 3)

$$F(u) = \sum_{n=-\bar{n}-1}^{\bar{n}+1} \epsilon_n F(n, A, \bar{n}) \operatorname{sinc}[\pi(u+n)]. \tag{10.19}$$

Thus the space factor consists of a central beam with $\bar{n}+1$ beams on either side, all spaced at integer values of u. And these constituent beams are each the result of uniform excitation. With only $n = \pm 1$ needed to approximate the Taylor $\bar{n}$ distribution, only three beams are needed to approximate the Taylor $\bar{n}$ space factor. The BFN then must add three contiguous beams with the proper weighting to produce a single low-sidelobe beam. To have N low-sidelobe beams the BFN must have $N+2$ uniform excitation beams of course. A simple implementation would utilize a 3-way power divider at each uniform beamport, with two arms at $2F(1, A, \bar{n})$ and one arm at 1. Two arms of each outside beamport are terminated, and one arm of each of the adjacent beamports. The efficiency is then a function of the number of beamports:

$$\eta = \frac{N-2}{N} = \frac{M}{M+2}, \tag{10.20}$$

where there are N uniform beamports, and M low-sidelobe beams. For 8 beams the loss is 0.97 dB and for 16 beams it is 0.51 dB, for example. This dissipation loss is separate from the aperture efficiency that pertains to the low-sidelobe beams of course. Sidelobes below -32 dB have been achieved by Thomas (1978) using a Rotman lens BFN, with -28 dB sidelobes over a 10% bandwidth. When the BFN is a Rotman or bootlace lens, beam superposition is sometimes called "block feeding."

10.3.2 Beam Interpolation Circuits

With beams having low crossover levels, or for accurate angle of arrival determination, it is desirable to be able to move the beam cluster in space, or to form interpolated beams that can be moved. One simple method of performing the latter will be described. Consider two adjacent beams from a uniformly excited half-wave spaced multiple-beam antenna (MBA). Beam peak angles occur at

$$\sin \theta_i = \frac{m_i}{N}, \tag{10.21}$$

where m_i is an odd integer $\leq$ half the number of array elements N. Each beam has phase of $(\omega t - kd \sin \theta_i)$, which gives $m_i \pi / N$ with ωt suppressed. Now two adjacent beam ports are connected through phasers to a hybrid junction as shown in Fig. 10.46. In this figure the beam phases are $\phi_i = m_i \pi / N$, while the phasers are ϕ_1 and ϕ_2. The sum and difference outputs from the junction are

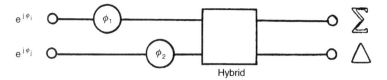

Figure 10.46 Beam interpolation circuit.

$$\Sigma = 2\cos\left[\tfrac{1}{2}(\phi_i - \phi_j + \phi_1 - \phi_2)\right] \exp\left[\frac{j}{2}(\phi_i + \phi_j + \phi_1 + \phi_2)\right],$$

$$\Delta = 2\sin\left[\tfrac{1}{2}(\phi_i - \phi_j + \phi_1 - \phi_2)\right] \exp\left[\frac{j}{2}(\phi_i + \phi_j + \phi_1 + \phi_2)\right].$$

(10.22)

Take first the sum output. If only one phaser is used, the beam phaser versus phaser value is shown below, along with the amplitude loss. The lost power goes into the difference arm.

BEAM PHASE	PHASER ϕ_1	AMPLITUDE
ϕ_i	$\phi_i - \phi_j$	$\cos(\phi_i - \phi_j)$
$\tfrac{1}{2}(\phi_i + \phi_j)$	0	$\cos[\tfrac{1}{2}(\phi_i - \phi_j)]$
ϕ_j	$\phi_j - \phi_i$	0

(10.23)

Since $\phi_i - \phi_j = 2/N$ there is an appreciable loss in small arrays when the beam is moved close to position ϕ_i. This loss can be reduced by using both phasers; the cosine factor may be kept at unity. Then

BEAM PHASE	PHASER ϕ_1	PHASER ϕ_2
ϕ_i	0	$\phi_i - \phi_j$
$\tfrac{1}{2}(\phi_i + \phi_j)$	$\tfrac{1}{2}(\phi_i - \phi_j)$	$\tfrac{1}{2}(\phi_i - \phi_j)$
ϕ_j	$\phi_i - \phi_j$	0

(10.24)

Although only three beam positions are shown above, two phasers can be adjusted to position the sum beam anywhere from beam position ϕ_i to beam position ϕ_{i+1} without loss. The actual beam position is found from the beam phase, from

$$\pi \sin\theta_i = \phi_i.$$

(10.25)

The difference pattern which can be used to steer a monopulse type null for accurate angle tracking behaves similarly:

BEAM PHASE	PHASER ϕ_1	PHASER ϕ_2
ϕ_i	$\pi/2$	$\tfrac{\pi}{2} + \phi_i - \phi_j$
$\tfrac{1}{2}(\phi_i + \phi_j)$	$\tfrac{1}{2}(\pi - \phi_i + \phi_j)$	$\tfrac{1}{2}(\pi + \phi_i - \phi_j)$
ϕ_j	$\tfrac{\pi}{2} - \phi_i + \phi_j$	$\pi/2$

(10.26)

Again the amplitude factor is unity but the phaser values are $\pi/2$ larger than those needed for shifting the sum beam peak. Using adequate phasers, a tracking loop can be constructed which will operate as long as the signal is between beam angles ϕ_i and ϕ_{i+1}. (Mailloux, et al, 1970).

Beam shifting and null shifting can also be accompanied at IF using one receiver channel per beam port (Mailloux, 1994).

10.4 BEAM ORTHOGONALITY

10.4.1 Orthogonal Beams

10.4.1.1 Meaning of Orthogonality. When beams in a multiple-beam antenna (MBA) are orthogonal, a signal injected into a given beam port appears only in that beam pattern, with no signal in other beam patterns or other beam ports. Using a scattering matrix representation of the array (of N isotropic elements) and beamforming network, the independence of M ports requires that the submatrix representing the beam ports have zero elements. Reciprocal BFNs imply a symmetric matrix. And a lossless network requires a unity matrix:

$$\sum_{n=1}^{N+M} S_{ni} S_{nj}^* = \delta_j^i. \tag{10.27}$$

Using this relationship Allen (1961) showed that uncoupled beams (and ports) lead to

$$\int_{-1/2}^{1/2} F_i(\pi u) F_j^*(\pi u)\, du = \delta_j^i, \tag{10.28}$$

where $u = (d/\lambda) \sin\theta$ with θ measured from broadside, and the space factor is

$$F_i(\pi u) = \sum S_{ni} \exp(j 2n\pi u) \tag{10.29}$$

Thus the space factors must be orthogonal over an angular range of

$$-\arcsin\frac{\lambda}{2d} \le \theta \le \arcsin\frac{\lambda}{2d} \tag{10.30}$$

in order to yield uncoupled beams and uncoupled ports (Kahn and Kurss, 1962). When $d/\lambda = 1/2$ (or $d/\lambda = 1$ since the integration limits could have been $= 1$) the beams are orthogonal over visible space. Thus for spacings that are multiples of $\lambda/2$ the beams and ports are both uncoupled for proper beam spacing.

In an extension of the network approach to multibeam analysis, Andersen and Rasmussen (1976) showed that decoupling and descattering (among ports) requires that the array mutual resistances be zero. For isotropic elements the mutual resistance is proportional to sinc mkd, which is zero for half-wave spacing. This is necessary but not sufficient as it implies decoupling (orthogonality) for all beam spacings. Beam orthogonality requires either mutual impedance that is zero for the array spacing, and this is never true even for isotropic or minimum scattering antennas (see Wasylkiwskyj and Kahn, 1970), or that the drive vectors of the array be eigenvectors of the mutual impedance matrix inverse. Only for amplitude distributions consisting of $\cos^M x$, with the proper beam spacings, does the eigenvalue relationship appear to hold. Thus patterns and crossover levels are not independent.

10.4.1.2 Orthogonality of Distributions.
While an MBA is either an array or uses an array as primary feed, much can be learned from a study of continuous (line) sources. A uniform line source has the space factor

$$F(u) = \text{sinc}\,[\pi(u - u_1)], \tag{10.31}$$

where $u = (L/\lambda)\sin\theta$, $u_1 = (L/\lambda)\sin\theta_1$. θ_1 is the beam peak. The aperture distribution, from $p = -1$ to $+1$, is

$$g(p) = \exp(-jp\pi u). \tag{10.32}$$

Orthogonality of the space factors can be simply written (Lewin, 1975)

$$\int_{-\infty}^{\infty} \text{sinc}\,[\pi(u - u_1)]\,\text{sinc}\,[\pi(u - u_2)]\,du = \text{sinc}\,[\pi(u_1 - u_2)] = \delta_2^1, \tag{10.33}$$

where the line source length has been assumed long so that the limits can be approximated by ∞. Orthogonal beams then can be spaced apart in units of $u_\Delta = \pm 1, \pm 2, \ldots$, but only $u_\Delta = 1$ allows adjacent main beams to cross over. This occurs at $u = u_1 + \frac{1}{2}$ and gives a crossover level of $2/\pi = -3.92$ dB. Alternatively, the orthogonality integral could be written over the aperture, since the Fourier transforms of orthogonal functions are also orthogonal, with one conjugated (Harmuth, 1970). For the uniform aperture this gives

$$\int_{-1}^{1} \exp[jp\pi(u - u_1)]\exp[jp\pi(u - u_2)]\,du = \text{sinc}\,[\pi(u_1 - u_2)]$$
$$= \text{sinc}\,(\pi u_\Delta). \tag{10.34}$$

The general class of orthogonal amplitude tapered distributions is $\cos^M(p\pi/2)$. Since the space factors and orthogonality integrals can be evaluated in closed form, it is easy to compare these distributions. The cosine distribution will be used as an example. The space factor can be written either of two ways, with $F(0) = 1$ in both cases:

$$F(u) = \frac{\pi}{4} \{\text{sinc}\,[\pi(\tfrac{1}{2} - u)] + \text{sinc}\,[\pi(\tfrac{1}{2} + u)]\} = \frac{\cos \pi u}{1 - 4u^2}. \tag{10.35}$$

Space factor zeros occur for $u = 3/2, 5/2, 7/2, \ldots$. The first sidelobe peak is at $u = 1.8894$ and the sidelobe ratio is 23 dB. Half-beamwidth, in u, is $u_3 = 0.5945$, while aperture efficiency (directivity with respect to uniform excitation) is 0.8106. The orthogonality integral reduces to

$$2\,\text{sinc}\,(\pi u_\Delta) + \text{sinc}\,[\pi(1 - u_\Delta)] + \text{sinc}\,[\pi(1 + u_\Delta)] = 0, \tag{10.36}$$

which can be simplified to

$$\frac{\sin \pi u_\Delta}{u_\Delta (1 - u_\Delta)^2} = 0. \tag{10.37}$$

Values of beam separation that give beam orthogonality then are the solutions of this equation

$$u_\Delta = 2, 3, 4, \ldots. \tag{10.38}$$

In all cases minus u can be substituted for plus u, but this has not been indicated. Table 10.8 gives pertinent parameters for four aperture distributions.

Beamwidth and aperture excitation efficiency are u_3 and η_t, while the sidelobe peak angle is u_{SL}. The crossover angle and level are u_x and F_x.

It is sometimes stated that linear combinations of orthogonal space factors can also be orthogonal, e.g., a cosine-on-a-pedestal distribution. This is not valid, as examination will show. The orthogonality integral over the aperture, where $u_\Delta = u_1 - u_2$, is

$$\tfrac{1}{2} \int_{-1}^{1} (a + b \cos \tfrac{1}{2}\pi p)^2 \exp(j p \pi u_\Delta)\, dp =$$

$$(a^2 + \tfrac{1}{2} b^2)\,\text{sinc}\,(\pi u_\Delta)$$
$$+ \tfrac{1}{2} ab \{\text{sinc}\,[\pi(\tfrac{1}{2} - u_\Delta)] + \text{sinc}\,[\pi(\tfrac{1}{2} + u_\Delta)]\}$$
$$+ \tfrac{1}{4} b^2 \{\text{sinc}\,[\pi(1 - u_\Delta)] + \text{sinc}\,[\pi(1 + u_\Delta)]\}. \tag{10.39}$$

TABLE 10.8 Orthogonal Space Factors

Distribution	Zeros	u_{SL}	SLR (dB)	$\tfrac{1}{2} u_3$	η	u_x	F_x (dB)
Uniform	1, 2, 3	1.4303	13.26	0.4429	1	.5	−3.94
Cosine	1.5, 2.5, 3	1.8894	23.00	0.5945	0.8106	1	−9.54
Cosine Square	2, 3, 4	2.3619	31.47	0.7203	0.6667	1.5	−15.40
Cosine Cubed	2.5, 3.5, 4.5	2.8420	39.30	0.8292	0.5764	2	−21.34

This reduces to

$$\left(a^2 + \tfrac{1}{2}b^2\right)\operatorname{sinc}\pi u_\triangle + \frac{2ab\cos\pi u_\triangle}{\pi(1 - 4u_\triangle)^2} + \frac{b^2 u_\triangle \sin\pi u_\triangle}{2\pi(1 - u_\triangle^2)} = 0. \quad (10.40)$$

For the special cases of uniform excitation ($b = 0$) and cosine excitation ($a = 0$), the correct values of $u_\triangle$ appear. The general case has values of $u_\triangle$ that make the beams orthogonal, but the beam spacings are not uniform. For example, taking a distribution with 10 dB pedestal (edge excitation -10 dB) that has $a = 0.46248$, $b = 1$, the values of $u_\triangle$ are[2]:

$$u_\triangle = 1.648, 2.271, 3.143, 4.097, 5.074, 6.060, \ldots . \quad (10.41)$$

The implication of this nonuniform sequence is that all adjacent beams are orthogonal ($u_\triangle = 1.648$) but none of the beams are orthogonal to any beams past their neighbor, since $2 \times 1.648 = 3.143$, $3 \times 1.648 = 5.074$, etc. In contrast, the cosine alone distribution has $u_\triangle = 2, 3, 4, \ldots$, so that if adjacent beams are 2 apart, all beams are multiples of 2 apart, hence orthogonal. That the cosine-on-a-pedestal is not orthogonal is not surprising, since the orthogonality integral function of u is related to the space factor: If the space factor zeros are evenly spaced, and the beam peaks are located at the zeros of the other beams, the beams can be orthogonal. Only the $\cos^M$ distributions have this characteristic. Unfortunately these distributions are inefficient compared with Taylor space factors. Such modern space factors are designed by locating the close-in zeros to yield the proper sidelobe level, then allowing the zeros to transition to the integer spacing for large u, as required for low aperture Q. The resulting nonuniform spacing makes the space factors not orthogonal.

Orthogonality is then difficult to achieve for several reasons: apertures must be large in wavelengths and the $\cos^M$ distributions are inefficient; arrays must have half-wave spacing. It is possible to modify these constraints by introducing loss into the BFN. This, of course, will also reduce efficiency. Before this is considered it is necessary to determine the effects of nonorthogonal beams.

10.4.1.3 Orthogonality of Arrays.

The preceding work addressed continuous aperture distributions since they are more tractable. It is well known that arrays of many elements can accurately approximate a continuous distribution simply by sampling it at the array element locations. To realize a given space factor with an array of few elements it is usually necessary to adjust the array polynomial zeros (Elliott, 1977). Unfortunately, this adjustment of zeros destroys the beam orthogonality. To see whether orthogonal beams can be formed from small arrays, the orthogonality integral could be examined directly. For example,

---

[2] These are calculated using a Wegstein rooter.

$$\int \frac{\sin(N\pi u_i)}{N\sin(\pi u_i)} \cdot \frac{\sin(N\pi u_j)}{N\sin(\pi u_j)} \, du \qquad (10.42)$$

represents two uniform excitation beams produced by an N-element array. Integrals of this type appear to be intractable. And there is no theorem for discrete functions that indicates orthogonality of the DFT of orthogonal functions. We are then left with the practical solution of evaluating nominally orthogonal arrays as a function of number of elements. Many elements should give the continuous distribution sidelobes, while fewer elements may show a sidelobe rise owing to the less accurate representation. This is calculated below.

10.4.2 Effects of Nonorthogonality

10.4.2.1 Efficiency Loss.
When aperture distributions that are orthogonal are used, but with a nonorthogonal beam spacing, BFN efficiency is lost owing to beam coupling. Stein (1962) derived the limits on beam coupling and the resultant efficiency. Let the BFN beam overlap parameter β be

$$\beta_{ij} = \int E_i(\theta, \phi) E_j(\theta, \phi) \, d\Omega \qquad (10.43)$$

with $B_{ii} = 1$. The maximum efficiency now becomes

$$\eta_{\max} = \frac{1}{1+\beta^2}. \qquad (10.44)$$

The integral allows the BFN efficiency to be determined easily; DuFort (1985) showed that the BFN efficiency is also equal to the ratio of the average to peak aperture (array) power distribution. Thus for uniform, cosine, and cosine square distributions, the maximum BFN efficiency is 0 dB, -3.01 dB, and -4.26 dB, respectively. Discrete BFN efficiency is given by

$$\eta_{\text{BFN}} = \frac{1 + 2\sum B_n^2}{(1 + 2\sum B_n)^2}, \qquad (10.45)$$

where the B_n, for example, are Taylor $\bar{n}$ coefficients. A Taylor distribution with 25 dB sidelobe ratio and $\bar{n} = 5$ has a BFN efficiency of -1.75 dB, much better than that of the cosine distribution, which has a comparable sidelobe level. For large BFN, the beam overlap parameter is given by

$$\beta = \frac{1}{\pi} \int_{-\infty}^{\infty} \operatorname{sinc} x \operatorname{sinc}(x + \pi u_\Delta) \, dx = \operatorname{sinc} \pi u_\Delta. \qquad (10.46)$$

This parameter and the corresponding beam overlap efficiency are plotted in Fig. 10.47. A similar equation for a cosine squared distribution is

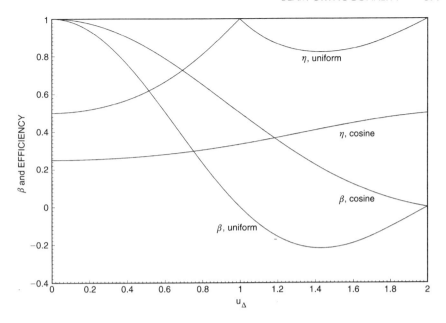

Figure 10.47 Beam overlap coefficient and BFN efficiency.

$$\beta = \frac{1}{\pi} \int_{-\infty}^{\infty} \text{sinc}^2 x \, \text{sinc}^2 (x + \pi u_0) \, dx$$

$$= \frac{4}{\pi} \{2 \, \text{sinc} \, \pi u_\triangle + \text{sinc} \, [\pi(1 + u_\triangle)] + \text{sinc} \, [\pi(1 - u_\triangle)]\}. \qquad (10.47)$$

This parameter and its efficiency are also plotted in Fig. 10.45, where clearly the cosine square efficiency is considerably poorer. Note, however, that the Taylor n̄ distribution can be expected to be much better than even the cosine distribution.

10.4.2.2 Sidelobe Changes. Beam coupling in an actual antenna is produced by mutual coupling changing the array excitation amplitude and phase. The earlier discussions of, and calculations about, orthogonality all assumed isotropic array elements. And for these hypothetical elements certain amplitude tapers and corresponding proper beam locations produced orthogonal beams. But actual arrays use slots, dipoles, loops, patches, etc., when low-gain elements are desired, or horns, flat spirals, conical spirals, log-periodics, etc., when moderate-gain elements are needed. None of these approximate the self- and mutual impedances of isotropic elements, where $Z = 120 \, \text{sinc} \, kd$, and in consequence there are no exactly orthogonal beams from arrays utilizing practical elements. Now the important question becomes: What is the sidelobe degradation (beam coupling) due to beam nonorthogonality? The part of this that is due to mutual coupling affecting the array distribution is answered below.

A dipole array will be used as an example, but results for slot arrays are similar: a dipole array without ground screen is equivalent, in terms of reflection coefficient variation, to a slot array. Call the applied drive voltages V_n and the resulting currents I_n. These are related through the impedance matrix:

$$[V] = [I][Z], \qquad (10.48)$$

where an element Z_{ij} of the latter is the mutual impedance between the ith and jth elements. The array pattern of the mth beam is given by

$$F_m(\theta) = \sum_{n=1}^{N} I_n \exp\{j[nkd(\sin\theta - \sin\theta_m)]\}, \qquad (10.49)$$

where d is the element spacing. Thus sidelobe degradation is computed by placing a beam at the desired angle and computing the pattern. To do this, the complex simultaneous equations are first solved, with the voltage drive vector containing both the amplitude taper and the phasing to position the mth beam. The resulting solution is the current vector which is used in the sum above to calculate the embedded pattern of the mth beam. Mutual impedances are computed from Carter's zero-order theory, using a version arranged for efficient computation (Hansen and Brunner, 1979). As shown in Section 3.1, the beams, for uniform excitation, are centered at angles given by

$$\sin\theta_i = \frac{2i-1}{2Nd/\lambda}, \qquad i = 1, 2, 3, \ldots, N/2. \qquad (10.50)$$

For the uniform case all $I_n = 1$. Low sidelobes are produced by use of a Taylor $\bar{n}$ distribution; see Chapter 3. This highly efficient distribution is particularly suitable for multiple-beam antennas. All patterns are normalized to unity at the nominal beam angle.

Calculations were made for half-wave dipoles, in both parallel and collinear configurations. The highest sidelobe is adjacent to the main beam, but can be on either side; the patterns are asymmetric. To determine any effect of placing the beams at the prescribed locations (for a uniform excitation array), beams were placed at broadside, next to broadside, at the outside position, and next to the outside. For the wider element spacings grating lobes will exist, but in all cases the outside beam is part of the original set of beams. No grating lobe beams were evaluated. For the four beams the highest sidelobe level was found, as shown in Table 10.9 for parallel dipole elements. Sidelobe level varies cyclically with element spacing, increasing slowly then rapidly falling, in a manner similar to that of array directivity with spacing. This is not surprising since both effects are caused by mutual coupling. As the beam position moves out, the sidelobe level break point moves to a smaller d/λ, and as expected the

TABLE 10.9 Sidelobe Level of 8-Element Array of Parallel Dipoles

Beam	d/λ, Sidelobe Level (dB)					
	0.5	0.6	0.7	0.8	0.9	1.0
Broadside	−13.3	−13.3	−13.5	−15.0	−12.8	−11.5
Next Broadside	−12.9	−12.9	−13.0	−13.2	−12.3	−10.9
Next Outside	−10.4	−10.0	−9.8	−12.6	−12.7	−12.8
Outside	−10.0	−9.4	−11.1	−12.7	−12.8	−12.8

performance of the far-out beams is poorer. There is no apparent difference in performance as the beam moves through the prescribed positions; apparently half-wave dipoles are sufficiently different from minimum scattering antennas to overcome any beam position effects. Table 10.10 gives results for collinear dipoles but only for the center beam and outside beam. Except for the strong coupling effects at half-wave spacing for the outside beam, the collinear arrays show less sidelobe level variation with spacing and from beam to beam. For the collinear arrays all sidelobe levels are with respect to the peak of that beam. These calculations have utilized 8 elements as this is a frequently used size of multiple-beam array. Table 10.11 gives sidelobe results for arrays of 16 parallel dipoles. Only the outside beam position was calculated as it represents the worst case. It can be noted that there is somewhat less variation for these larger arrays. And, as expected for 32-element arrays, the realized sidelobes are very close to the design values.

It was expected that the high sidelobes of uniformly excited arrays would be affected to a minor degree while low-sidelobe arrays are more seriously affected. In calculating low-sidelobe arrays a Taylor $\bar{n}$ distribution was used. The value of $\bar{n}$ for each value of sidelobe level was chosen as the largest which gave a monotonic distribution. Table 10.12 shows sidelobe level for Taylor arrays of 25 dB SLR with $\bar{n} = 5$. The poor broadside results for the 8-element array are due to the inadequate sampling of the Taylor distribution that an 8-element array allows[3]. It can be observed that the 16-element array sidelobe level is satisfactory at broadside. As expected there is sidelobe level degradation at large beam angles. This degradation can be reduced by impedance matching

TABLE 10.10 Sidelobe Level of 8-Element Array of Collinear Dipoles

Beam	d/λ, Sidelobe Level (dB)					
	0.5	0.6	0.7	0.8	0.9	1.0
Central	−12.8	−12.9	−12.8	−12.9	−13.0	−13.1
Outside	−17.5	−13.7	−12.4	−12.4	−12.5	−12.6

---

[3]For N so small, a Villeneuve $\bar{n}$ distribution should be used instead of the Taylor $\bar{n}$; see Chapter 3.

TABLE 10.11 Sidelobe Level of 16-Element Array of Parallel Dipoles

Beam	d/λ, Sidelobe Level (dB)					
	0.5	0.6	0.7	0.8	0.9	1.0
Outside	−10.8	−11.6	−13.1	−13.1	−13.1	−13.2

TABLE 10.12 Sidelobe Level of Array of Parallel Dipoles, SLR = 25 dB

Beam	Sidelobe Levels (dB)	
	$d/\lambda = 0.5$	$d/\lambda = 1.0$
	$N=8$	$N=8$
Broadside	−24.0	−20.2
Outside	−20.3	−22.7
	$N=16$	
Broadside	−25.1	
Outside	−21.1	

the elements at an intermediate angle so that the broadside and outside beams are equally mismatched. From these calculations two conclusions can be drawn. First, small arrays are inefficient producers of low-sidelobe patterns; 32 elements closely approximate the efficient continuous distribution, while 16 elements are a fair approximation. Second, sidelobe levels will degrade at extreme beam positions, so the array must be overdesigned; i.e., the design SLR must be larger than the expected value.

Degradation of sidelobe level for outside beams can be improved by compensating the *scan impedance* versus scan angle; see Chapter 7.

ACKNOWLEDGMENT

Photographs courtesy of Phil Metzen, Dr. William Kreutel, Dr. Dave Thomas, Dr. J. J. Lee, Pat Valentino, and Dr. A. Dion.

REFERENCES

Ajioka, J. S. and McFarland, J. L., "Beam-Forming Feeds," in *Antenna Handbook: Theory, Applications, and Design*, Y. T. Lo and S. W. Lee, Eds., Van Nostrand Reinhold, 1988, Chapter 19.

Andersen, J. B. and Rasmussen, H. H., "Decoupling and Descattering Networks for Antennas," *Trans. IEEE*, Vol. AP-24, Nov. 1976, pp. 841–846.

Archer, D., "Lens-Fed Multiple-Beam Arrays," *Microwave J.*, Vol. 18, Oct. 1975, pp. 37–42.

Archer, D., "Lens-Fed Multiple Beam Arrays," *Microwave J.*, Vol. 27, Sept. 1984, pp. 171–195.

Bird, T. S., "Analysis of Mutual Coupling in Finite Arrays of Different-Sized Rectangular Waveguides," *Trans. IEEE*, Vol. AP-38, Feb. 1990; correction p. 1727.

Blank, S. J. and Imbriale, W. A., "Array Feed Synthesis for Correction of Reflector Distortion and Verier Beamsteering," *Trans. IEEE*, Vol. AP-36, Oct. 1988, pp. 1351–1358.

Blass, J., "Multi-Directional Antenna—A New Approach to Stacked Beams," *IRE Nat. Conv. Record*, 1960, Pt. 1, pp. 48–50.

Brown, J., *Microwave Lenses*, Methuen, 1953.

Brown, J., "Lens Antennas," in *Antenna Theory*, Part 2, R. E. Collin and F. J. Zucker, Eds., McGraw-Hill, 1969, Chapter 18.

Butler, J. L., "Digital, Matrix, and Intermediate-Frequency Scanning," in *Microwave Scanning Antennas*, Vol. III, R. C. Hansen, Ed., Academic Press, 1966, Chapter 3 [Peninsula Publishing, 1985].

Butler, J. L. and Lowe, R., "Beam Forming Matrix Simplifies Design of Electronically Scanned Antennas," *Electronic Design*, Vol. 9, Apr. 12, 1961, pp. 170–173.

Chadwick, G. G., Gee, W., Lam P. T., and McFarland, J. L, "An Algebraic Synthesis Method for RN^2 Multibeam Matrix Metwork," *Proc. Antenna Applications Symposium*, Allerton, IL, Sept. 1981, AD-A205 816.

Clarricoats, P. J. B. and Brown, R. C., "Performance of Offset Reflector Antennas with Array Feeds," *Proc. IEE*, Vol. 131, Pt. H, June 1984, pp. 172–178.

Cornbleet, S., *Microwave and Geometrical Optics*, Academic Press, 1994.

Cummings, W. C., "Multiple Beam Forming Networks," Technical Note 1978–1979, Lincoln Laboratory, 1978.

Detrick, D. L. and Rosenberg, T. J., "A Phased-Array Radiowave Imager for Studies of Cosmic Noise Absorption," *Radio Sci.*, Vol. 25, No. 4, July–Aug. 1990, pp. 325–338.

Dion, A. R. and Ricardi, L. J., "A Variable-Coverage Satellite Antenna System," *Proc. IEEE*, Vol. 59, Feb. 1971, pp. 252–262.

DuFort, E. C., "Constrained Feeds for Limited Scan Arrays," *Trans. IEEE*, Vol. AP-26, May 1978, pp. 407–413.

DuFort, E. C., "Optimum Low Sidelobe High Crossover Multiple Beam Antennas," *Trans. IEEE*, Vol. AP-33, Sept. 1985, pp. 946–954.

Dufort, E. C., "Optimum Optical Limited Scan Antenna," *Trans. IEEE*, Vol. AP-34, Sept. 1986, pp. 1133–1142.

Elliott, R. S., "On Discretizing Continuous Aperture Distributions," *Trans. IEEE*, Vol. AP-25, Sept. 1977, pp. 617–621.

Gent, H., "The Bootlace Aerial," *Royal Radar Establishment J.*, Oct. 1957, pp. 47–57.

Hansen, R. C., "Design Trades for Rotman Lenses," *Trans. IEEE*, Vol. AP-39, Apr. 1991, pp. 464–472.

Hansen, R. C. and Brunner, G., "Dipole Mutual Impedance for Design of Slot Arrays," *Microwave J.*, Vol. 22, Dec. 1979, pp. 54–56.

Harmuth, H. F., *Transmission of Information by Orthogonal Functions*, Springer, 1970.

Kahn, W. K. and Kurss, H., "The Uniqueness of the Lossless Feed Network for a Multibeam Array," *Trans. IEEE*, Vol. AP-10, Jan. 1962, pp. 100–101.

Kales, M. L. and Brown, R. M., "Design Considerations for Two-Dimensional Symmetric Bootlace Lenses," *Trans. IEEE*, Vol. AP-13, July 1965, pp. 521–528.

Katagi, T., et al., "An Improved Design Method of Rotman Lens Antennas," *Trans. IEEE*, Vol. AP-32, May 1984.

Lee, J. J., "Lens Antennas," in *Antenna Handbook: Theory, Applications, and Design*, Y. T. Lo and S. W. Lee, Eds., Van Nostrand Reinhold, 1988, Chapter 16.

Lewin, L., "Expression for Decoupling in Multiple-Beam Antennas," *Electron. Lett.*, Vol. 11, Aug. 21, 1975, pp. 420–421.

Mailloux, R. J., *Phased Array Antenna Handbook*, Artech House, 1994.

Mailloux, R. J., Caron, P. R., LaRussa, F. J., and Dunne, C. L., "Phase Interpolation Circuits for Scanning Phased Arrays," report TN-D-5865, NASA/ERC, Cambridge, MA, July 1970.

Mayhan, J. T., "Adaptive Nulling with Multiple-Beam Antennas," *Trans. IEEE*, Vol. AP-26, March 1978, pp. 267–273.

McFarland, J. L., "The RN^2 Multiple Beam-Array Family and the Beam-Forming Matrix," *IEEE APS Symposium Digest*, June 1979, Seattle, Wash., pp. 728–731.

McGrath, D. T., "Constrained Lenses," in *Reflector and Lens Antennas*, C. J. Sletten, Ed., Artech House, 1988, Chapter 6.

Metzen, P., "Satellite Communication Antennas for Globalstar," *Proceedings International Symposium On Antennas* (JINA), Nice, Nov. 1996.

Moody, H. J., "The Systematic Design of the Butler Matrix," *Trans. IEEE*, Vol. AP-12, Nov. 1964, pp. 786–788.

Musa, L. and Smith, M. S., "Microstrip Port Design and Sidewall Absorption for Printed Rotman Lenses," Proc. Inst. Elec. Eng., Vol. 136, Pt. H, Feb. 1989, pp. 53–58.

Nolen, J. N., Synthesis of Multiple Beam Networks for Arbitrary Illuminations, PhD thesis, April 1965.

Rao, J. B. L., "Bispherical Constrained Lens Antennas," *Trans. IEEE*, Vol. AP-30, Nov. 1982a, pp. 1224–1228.

Rao, J. B. L., "Multifocal Three-Dimensional Bootlace Lenses," *Trans. IEEE*, Vol. AP-30, Nov. 1982b, pp. 1050–1056; correction, 1983, p. 541.

Rao, J. B. L., Trunk, G. V., and Patel, D. P., "Two Low-Cost Phased Arrays," *Proceedings IEEE Symposium Phased Array Systems and Technology*, Boston, Mass., Oct. 1996, pp. 119–124.

Richie, J. E. and Kritikos, H. N., "Linear Program Synthesis for Direct Broadcast Satellite Phased Arrays," *Trans. IEEE*, Vol. AP-36, Mar. 1988, pp. 345–348.

Rotman, W. and Turner, R. F., "Wide-Angle Microwave Lens for Line Source Applications," *Trans. IEEE*, Vol. AP-11, Nov. 1963, pp. 623–632.

Rudge, A. W. and Davies, D. E. N., "Electronically Controllable Primary Feed for Profile-Error Compensation of Large Parabolic Reflectors," *Proc. IEE*, Vol. 117, Feb. 1970, pp. 351–357.

Ruze, J., "Wide-Angle Metal-Plate Optics," *Proc. IRE*, Vol. 38, Jan. 1950, pp. 53–59.

Schuss, J. J. et al., "The IRIDIUM Main Mission Antenna Concept," *Proceedings IEEE Symposium on Phased Array Systems and Technology*, Boston, Mass., Oct. 1996, pp. 411–415.

Schwartzman, L. and Liebman, P. M., "A Report on the Sperry Dome Radar," *Microwave J.*, Vol. 22, Mar. 1979, pp. 65–69.

Schwartzman, L. and Stangel, J., "The Dome Antenna," *Microwave J.*, Vol. 18, Oct. 1975, pp. 31–34.

Shelton, J. P., "Multiple-Feed Systems for Objectives," *Trans. IEEE*, Vol. AP-13, Nov. 1965, pp. 992–994.

Shelton, J. P., "Focusing Characteristics of Symmetrically Configured Bootlace Lenses," *Trans. IEEE*, Vol. AP-26, July 1978, pp. 513–518.

Shelton, J. P. and Kelleher, K. S., "Multiple Beams from Linear Arrays," *Trans. IRE*, Vol. AP-9, Mar. 1961, pp. 154–161.

Smith, M. S., "Design Considerations for Ruze and Rotman Lenses," *Radio Electron. Eng.*, Vol. 52, Apr. 1982, pp. 181–197.

Smith, M. S. and Fong, A. K. S., "Amplitude Performance of Ruze and Rotman Lenses," *Radio Electron. Eng.*, Vol. 53, Sept. 1983, pp. 328–336.

Smith, W. T. and Stutzman, W. L., "A Pattern Synthesis Technique for Array Feeds to Improve Radiation Performance of Large Distorted Reflector Antennas," *Trans. IEEE*, Vol. AP-40, Jan. 1992, pp. 57–62.

Sole, G. C. and Smith, M. S., "Multiple Beam Forming for Planar Antenna Arrays Using a Three-Dimensional Rotman Lens," *Proc. IEE*, Vol. 134, Pt. H., Aug. 1987, pp. 375–385.

Stein, S., "On Cross Coupling in Multiple-Beam Antennas," *Trans. IRE*, Vol. AP-10, Sept. 1962, pp. 548–557.

Steyskal, H., Hessel, A., and Shmoys, J., "On the Gain-Versus-Scan Trade-Offs and the Phase Gradient Synthesis for a Cylindrical Dome Antenna," *Trans. IEEE*, Vol. AP-27, Nov. 1979, pp. 825–831.

Susman, L. and Mieras, H., "Results of an Exact Dome Antenna Synthesis Procedure," *IEEE AP Symposium Digest*, June 1979, Seattle, Wash., pp. 38–41.

Thomas, D. T., "Multiple Beam Synthesis of Low Sidelobe Patterns in Lens Fed Arrays," *Trans. IEEE*, Vol. AP-26, Nov. 1978, pp. 883–886.

Valentino, P. A., Rothenberg, C., and Stangel, J. J., "Design and Fabrication of Homogeneous Dielectric Lenses for Dome Antennas," *IEEE AP Symposium Digest*, June 1980, Quebec, Canada, pp. 580–583.

Wasylkiwskyj, W. and Kahn, W. K., "Theory of Mutual Coupling Among Minimum-Scattering Antennas," *Trans. IEEE*, Vol. AP-18, Mar. 1970, pp. 204–216.

White, W. D., "Pattern Limitations in Multiple-Beam Antennas," *Trans. IRE*, Vol. AP-10, July 1962, pp. 430–436.

Williams, W. F. and Schroeder, K. G., "Performance Analysis of Planar Hybrid Matrix Arrays," *Trans. IEEE*, Vol. AP-17, July 1969, pp. 526–528.

Winter, C. F., "Lossy-Feed Networks in Multiple-Beam Arrays," *Trans. IEEE*, Vol. AP-27, Jan. 1979, pp. 122–126.

Winter, C. F., "Multiple Beam Feed Networks Using an Even Number of Beam Ports," *Trans. IEEE*, Vol. AP-28, Mar. 1980, pp. 250–253.

Wong, N. S., Tang, R., and Barber, E. E., "A Multielement High Power Monopulse Feed with Low Sidelobe and High Aperture Efficiency," *Trans. IEEE*, Vol. AP-22, May 1974, pp. 402–407.

CHAPTER ELEVEN

Conformal Arrays

11.1 SCOPE

The essential constituent of a conformal array is curvature. Authorities disagree on whether the array must be part of a curved metallic structure; in this chapter curvature alone is sufficient. Arrays of one or more concentric rings of elements, here called "ring arrays," are treated first. The term "circular array" is not used, as it often means a planar array of circular perimeter. The following sections deal with arrays on curved metallic bodies. Most simple is the cylinder; Section 11.3 treats cylindrical arrays with elements around the full circumference. Partial cylindrical arrays, where only part of the perimeter has elements, are the subject of Section 11.4. Finally, arrays on surfaces with two-dimensional curvature, such as spheres and cones, are examined.

Many applications exist where the array occupies only a small part of a curved body, and where the conformal effects are primarily on patterns and excitation rather than on impedance. For these applications the array can often be designed as a planar array (but with proper conformal phasing), with pattern perturbations predicted by techniques such as geometric theory of diffraction (McNamara et al., 1990). "Smart skin" flush-mounted antennas for aircraft and missiles often are of this type.

One commom application of conformal array is an aircraft or helicopter satellite link, where the signal is circularly polarized, and zenith to horizon coverage is desired. Essentially all elements that radiate circular polarization at broadside radiate linear polarization at $\theta = 90$ deg, or 3 dB directivity drop. Array directivity also changes. A rectangular planar array $L \times W$ has broadside directivity of $4\pi LW\eta/\lambda^2$, where η is the area taper efficiency. The same array at endfire, in the L direction, has directivity of $3\pi W/\lambda\sqrt{2L/\lambda}$. Thus at the horizon the data link is worse by roughly $1/\sqrt{L/\lambda}$; for example, for a uniform amplitude array with a zenith directivity of 23 dB ($L = W = 4\lambda$), the horizon value is 20 dB, and 3 dB lower.

11.2 RING ARRAYS

A ring array, where the elements are located in a circle, offers advantages of many fewer elements than a filled planar array, and a geometry that facilitates 360 deg scanning. Obvious applications of the latter are direction finding in the HF–VHF–UHF bands, navigation systems, communications, and military electronic support systems. Of course the directivity is well below that of a filled array, and many grating lobes may appear.

Analytical efforts on ring arrays existed before 1940, but significant progress was made during World War II by Page (1948a,b), who considered a single azimuthal Fourier mode, and by Knudsen (1951, 1953, 1956), who analyzed a ring array as a series of Bessel function terms and showed that the azimuthal mode number and ring diameter must be compatible to avoid superdirectivity. The properties of individual azimuthal modes were pursued by Tillman and colleagues (Hickman et al., 1961; Hilburn and Hickman, 1968; Hilburn, 1969; Tillman, 1968). The last work is a book containing extensive modal data, both pattern and impedance. Beam cophasal excitation (see next section), in which all elements contribute to the main beam in phase, is almost always used. The harmonic (Bessel) series analysis has been used for sidelobe control through excitation tapering, and for pattern synthesis (Fenby, 1965; James, 1965; Royer, 1966; Longstaff et al., 1967; Redlich, 1970; Lim and Davies, 1975). A harmonic analysis without explicit Bessel functions has also been used (Blass, 1974) as has a polynomial approach (Gerlin, 1974). DuHamel developed a procedure for producing Chebyshev patterns for ring arrays (DuHamel, 1951). Effects of ground for HF ring arrays have been reported by Tillman et al. (1955) and by Ma and Walters (1970). An examination of current distribution on dipoles in a ring array via moment method has been made by Sinnott and Harrington (1973).

A ring array can produce a near-omnidirectional pattern, if the element spacing is not too large. Carter (1943) computed the amplitude of pattern oscillations versus element spacing.

Circular arc arrays, which result when tapered excitation is used, are reported by Lo and colleagues (Lee and Lo, 1965). Gobert and Yang (1974) consider a ring array of nonparallel dipoles.

11.2.1 Continuous Ring Antenna

Assume a local "element" pattern G where ϕ is the azimuth angle, and β is the angular location of the local element. A continuous current density I then produces the space factor (pattern)

$$E(\phi, \theta) = \frac{M}{2\pi} \int_0^{2\pi} I(\beta) G(\phi - \beta, \theta) \, d\beta. \tag{11.1}$$

M is the number of elements which the continuous distribution replaces. If $I(\beta)$ is symmetric, it may be written

$$I(\beta) = \sum_0^\infty I_n \cos n\beta. \tag{11.2}$$

The I_n are the complex current mode amplitudes. The element factor $G(\phi - \beta, \theta)$ may also be expanded as

$$G(\phi - \beta, \theta) = f(\theta) \sum_0^\infty F_m(\theta) \cos[m(\phi - \beta)]. \tag{11.3}$$

Substitution of Eqns. (11.2) and (11.3) into Eqn. (11.1) and integrating yields

$$E(\phi, \theta) = Mf(\theta) \sum \frac{1}{\epsilon_n} I_n F_n(\theta) \cos n\phi, \tag{11.4}$$

where $\epsilon_n = 1$ if $n = 0$, $\epsilon_n = 2$ if $n \neq 0$. The I_n are determined from the desired azimuth pattern expanded in a Fourier series. Call the desired pattern $T_n(\phi)$ to indicate an Nth-order Chebyshev pattern, though any pattern may be used if it can be put into the form

$$T_N(\phi) = \sum_0^\infty C_n^N \cos n\phi. \tag{11.5}$$

The computation of coefficients C_n^N was developed by DuHamel (1951) but his algorithm has poor accuracy for large N. Better formulas are given by Hansen (1981).

Equating (11.4) and (11.5) at $\theta = \theta_0$ yields

$$I_n^{\theta_0} = \frac{\epsilon_n C_n^N}{Nf(\theta)F_n(\theta)} \tag{11.6}$$

and

$$I^{\theta_0}(\beta) = \sum_0^\infty I_n^{\theta_0} \cos n\beta. \tag{11.7}$$

The superscript (θ_0) has been introduced to emphasize that I_n and $I(\beta)$ give the optimum azimuth pattern $T_n(\phi)$ only in the cone $\theta = \theta_0$. At a general elevation angle θ the pattern is

$$E^{\theta_0}(\phi, \theta) = Mf(\theta) \sum_0^\infty I_n^{\theta_0} F_n(\theta) \cos n\phi. \tag{11.8}$$

The choice of $N_0 = k\rho \sin\theta_0$ avoids superdirectivity, and allows efficient excitation of azimuthal modes. Thus $N^{\theta_0} = N_0 \cos\theta_0$, where $N_0 \simeq k\rho$ is N at $\theta_0 = 0$. This selection of N^{θ_0} gives approximately the same amplitude distribution for all θ_0, and a cophasal beam. For a given sidelobe level, the beamwidth is approximately proportional to $1/N$. Thus

$$\text{azimuth beamwidth} \simeq \frac{\text{beamwidth at broadside}}{\cos\theta_0}.$$

This broadening of the beamwidth is only apparent, however. It is due to the fact that the azimuth pattern is measured on a cone $\theta = \theta_0$, and the ratio of the perimeter of the cone base at θ_0 to that at $\theta = 0$ is just $(\cos\theta_0)^{-1}$. The actual spatial extent of the beam in the plane perpendicular to the $\phi = 0$ plane remains approximately constant to near zenith.

Figures 11.1 and 11.2 show the behavior of the azimuth main beam for $\theta_0 = 0$ and $\theta_0 = 45$ deg; there are 128 axial slots in a cylinder of diameter 52.6λ (Munger and Gladman, 1970). The distribution is for a $-50\,\text{dB}$ Chebyshev pattern, so only the main beam shape appears in the plot (all sidelobes are at $-50\,\text{dB}$ for $\theta = \theta_0$). The extensive beam broadening (beyond the $(\cos\theta_0)^{-1}$ factor) and gain loss for $\theta \neq \theta_0$ are due to the deviation of the phase from the optimum (which is essentially $-jk\cos\theta_0\cos\beta p$). Figure 11.3 shows elevation patterns through the beam $\phi = 0$ direction with the azimuth pattern optimized at various θ_0. They are compared with the element elevation pattern, since this represents the maximum possible for each curve. For a linear array the curves would all coincide with the elevation element pattern. For the ring array they coincide with the elevation element pattern only at θ_0, where all elements add in phase to form the beam.

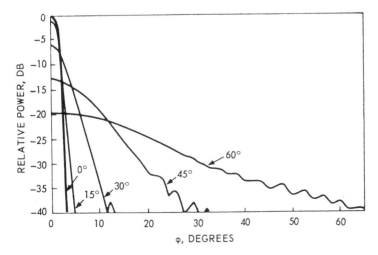

Figure 11.1 Ring-array pattern for $\theta_0 = 0$ deg.

388 CONFORMAL ARRAYS

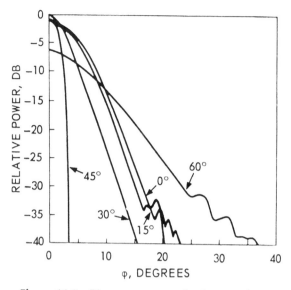

Figure 11.2 Ring-array pattern for $\theta_0 = 45$ deg.

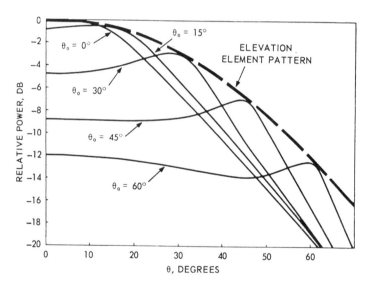

Figure 11.3 Ring-array elevation patterns for azimuth pattern optimized at various θ_0.

Bandwidth of these arrays of isotropic elements is limited by changes in the Bessel coefficients; a change of $\pi/8$ gives a bandwidth of $D/32\lambda$ (Davies, 1983), where D is the ring diameter. However, use of directional elements greatly increases bandwidth (Rahim and Davies, 1982).

11.2.2 Discrete Ring Array

The continuous distribution can be replaced by M elements located at

$$\beta_p = \frac{2\pi(p+f)}{M}, \qquad p = 0, 1, 2, \ldots, M-1, \tag{11.9}$$

where f is a fraction that indicates the position of the beam with respect to the first element. It can be shown (Hansen, 1981) that

$$E(\phi, \theta) = M \sum_{n=0}^{N} \frac{1}{\epsilon_n} I_n^{\theta_0} \left\{ F_n(\theta) \cos n\phi + \sum_{r=1}^{\infty} F_{rM-n}(\theta) \cos\left[(rM-n)\phi - 2\pi r f\right] \right\}. \tag{11.10}$$

Equation (11.10) is the desired pattern plus an error term. For spacing $s = 2\pi p/m$ of less than a half-wavelength, only the $r = 0$ term is significant. For spacing less than one wavelength, only $r = 0$ and $r = 1$ contribute, and so forth. The error term's primary contribution to the pattern is in the form of a grating lobe.

A convenient way of evaluating the grating lobe is to represent the terms in Eqn. (11.9) by integrals and evaluate by the method of stationary phase (Biswell, 1968). An early approximation was given by Walsh (1951). The pattern for an arc array of $2P + 1$ active elements is

$$E(\phi, \theta) = \sum_{p=-P}^{P} I_p G(\phi - \beta, \theta). \tag{11.11}$$

This may be written as a series of integrals according to the Poisson sum formula:

$$E(\phi, \theta) = \sum_{r=-\infty}^{\infty} \int_{-P}^{P} I(\beta_p) G(\phi - \beta, \theta) \exp(j2\pi r p) \, dp \tag{11.12}$$

By change of variable $\beta = 2\pi p/M$,

$$E(\phi, \theta) = \frac{M}{2\pi} \sum_r \int I(\beta) G(\phi - \beta, \theta) \exp(jMr\beta) \, d\beta, \tag{11.13}$$

$$E(\phi, \theta) = \frac{M}{2\pi} \sum_r \epsilon_r \int I(\beta) G(\phi - \beta, \theta) \cos(Mr\beta) \, d\beta, \tag{11.14}$$

where $2\beta(p)$ is the active segment of the ring, M is the number of elements on the full ring, and $I(\beta)$ is a continuous representation of the distribution $I(\beta_p)$.

CONFORMAL ARRAYS

Now substitute Eqns. (11.10) and (11.6) into Eqn. (11.12), and let $\beta(p) = \pi$; it may be shown that the rth term of Eqn. (11.12) is identical to the rth term of Eqn. (11.10). The $r = 0$ term is just the pattern due to the continuous distribution $I(\beta)$, and the higher terms are the error introduced by letting $\beta = 2\pi(p+f)/M$.

The method of stationary phase can be applied to evaluate approximately the integrals in Eqn. (11.12) — in particular to find the contribution to the grating lobe. From Eqns. (11.32) and (11.36) the rth integral of Eqn. (11.12) is

$$I_r(\phi, \theta) = \int_{-\beta(p)}^{\beta(p)} |I(\beta)| \cdot |G(\phi - \beta, \theta)| \exp(jk\rho\beta u) \, d\beta, \qquad (11.15)$$

where

$$u(\beta) = \cos\theta \cos(\phi - \beta) - \cos\theta_0 \cos\beta + \frac{r\beta}{s} \qquad (11.16)$$

and $s = k\rho/M$ is the interelement spacing.

The integral is to be evaluated under the assumption that the largest contribution is from the neighborhood of the point(s) β_0 where the phase term $u(\beta)$ is constant. This assumption is best for $k\rho$ large. The condition for β_0 is now $u'(\beta) = 0$, or

$$\cos\theta \sin(\phi - \beta) + \cos\theta_0 \sin\beta + \frac{r}{s} = 0. \qquad (11.17)$$

In the region around β_0 since $u'(\beta_0) = 0$,

$$u(\beta) \simeq u(\beta_0) + \tfrac{1}{2}(\beta - \beta_0)^2 u''(\beta_0). \qquad (11.18)$$

Furthermore, if β_0 is well inside the range $-\beta(P)$ to $\beta(P)$, the limits may be extended to $-\infty$ and ∞. Finally, assume that $I(\beta)$ and $G(\phi - \beta, \theta)$ are reasonably constant over the region near β_0 so that they may be taken out of the integral. The integral is now in a standard form and becomes

$$I_r(\phi, \theta) = |I(\beta)| \cdot |G(\phi - \beta, \theta)| \sqrt{\frac{2\pi}{k\rho |u''(\beta_0)|}} \qquad (11.19)$$

with

$$u''(\beta_0) = -\cos\theta \cos(\phi - \beta) + \cos\theta_0 \cos\beta. \qquad (11.20)$$

For simplicity, consider $\theta = \theta_0$. Now $u''(\beta_0) = 0$ for $\phi_0 = 2\beta_0$. This gives the maximum of Eqn. (11.16) corresponding to a grating lobe. The position of the grating lobe is now given by Eqn. (11.14), which becomes

$$\sin\frac{\phi_0}{2} = -\frac{r}{2s}\cos\theta_0. \tag{11.21}$$

$r = 0$ gives the main beam at $\phi_0 = 0$, while positive and negative r give grating lobes at positive and negative ϕ_0 angles under the condition

$$\left|\frac{r}{2s}\cos\theta\right| \leq 1. \tag{11.22}$$

However, Eqn. (11.16) is not a valid representation of the integral at $u'(\beta_0) = u''(\beta_0) = 0$, for Eqn. (11.15) must be replaced by

$$u(\beta) = u(\beta_0) + \tfrac{1}{6}(\beta - \beta_0)^3 u'''(\beta_0). \tag{11.23}$$

Now evaluating the integral thus obtained at $\phi_0 = 2\beta_0$,

$$I_n(\phi,\theta) = |I(\beta_0)| \cdot |G(\phi_0 - \beta_0, \theta_0)| \cdot \left(\frac{6}{k\rho|u'''(\beta_0)|}\right)^{1/3} \frac{\Gamma(1/3)\exp(jk\rho\beta_0 u)}{\sqrt{3}}, \tag{11.24}$$

$$|I_r(\phi_0,\theta_0)| = \left|I\left(\frac{\phi_0}{2}\right)G\left(\frac{\phi_0}{2},\theta_0\right)\frac{\Gamma(1/3)}{\sqrt{3}}\left(\frac{12}{k}\right)^{1/3}\left(\frac{s\cos\theta_0}{r\rho}\right)^{1/3}\right|.$$

Equation (11.20) is subject to the restrictions that $I(\beta)$ and $G(\phi - \beta, \theta)$ are slowly varying, $k\rho$ is large, and β_0 is not near 0 but is inside the range $-\beta(P)$ to $+\beta(P)$. To compute the grating lobe height compared with the main beam height, compare the grating lobe computed from Eqn. (11.20) with the main beam computed from

$$I_0 = \int_{-\beta(p)}^{\beta(p)} |I(\beta)| \cdot |G(\beta,\theta)| \, d\beta. \tag{11.25}$$

Figure 11.4 shows the grating lobe height relative to the beam as a function of spacing ($s\cos\theta_0$) for various $\cos^n$ amplitude tapers, with $\rho = 26\lambda$ and $G(\phi - \beta, \theta_0) = \cos(\phi - \beta)$ (Munger, 1969). $\beta(P) = \pi/2$ was used to ensure that $0 < \beta_0 < \beta(P)$. For comparison, actual patterns were computed from Eqn. (11.10) and the grating lobe heights are shown. The maximum difference between the actual grating lobe and that computed from Eqn. (11.20) is 2 dB. Since the approximation form Eqn. (11.20) represents the worst case, it appears to be a useful guide for controlling the grating lobe as a function of spacing, current distribution, element pattern, radius, and arc length. The position of the grating lobe (for $r = -1$ and $\theta_0 = 0$) predicted by Eqn. (11.18) is shown in Fig. 11.5 in comparison with the actual position, with good agreement.

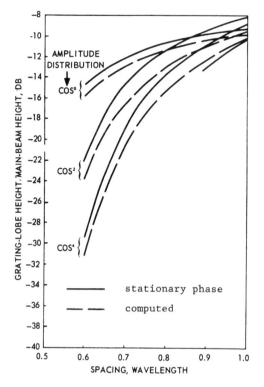

Figure 11.4 Grating lobe height vs. spacing.

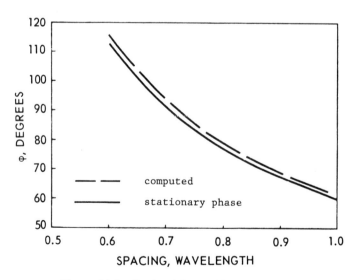

Figure 11.5 Grating lobe position vs. spacing.

11.2.3 Beam Cophasal Excitation

The direct way of achieving a narrow beam from a ring array is to phase all elements so that the element contributions are in phase at the far-field angle ϕ_0. This is called beam cophasal excitation, and requires phases around a ring of radius r_0 of

$$exp\,[-jkr_0 \cos(\phi - \phi_0)] \quad (11.26)$$

The discrete array samples this continuous phase distribution. If the element spacing is half-wave or less, and the number of elements is large, the pattern is approximately that of a continuous ring source, Eqn. (11.1).

The application of azimuthal modal theory to ring arrays was pioneered by Tillman, Hickman, and Neff (1961), and extended by Davies (1965). In concept the phase and amplitude distribution around a ring array can be expressed in a series of azimuthal modes, where the DFT of each mode gives the mode pattern. One can also obtain the self- and mutual impedances for each mode (Tillman, 1968). These modes allow synthesis of a desired pattern, and allow the impedances, directivity, etc., to be determined. They are analogous to the symmetrical components used in analysis of multiphase power systems.

Let the excitation of the mth mode on the ring array of diameter D be

$$g_m(\beta) = \sum_{n=1}^{N} A_n \exp(jkdm \cos \beta), \quad (11.27)$$

where $d = 2\pi D/N$ and β is the angle around the ring array. The pattern of this mode is given by the DFT:

$$f_m(\phi) = J_0\left(2kr_0 \sin\frac{\phi}{2}\right) J_0[kr_0(1 - \cos \theta)], \quad (11.28)$$

where the first Bessel function gives the azimuth pattern, and the second provides the elevation pattern. The J_0 pattern has a poor sidelobe ratio of 7.9 dB, and a slowly decreasing sidelobe envelope, as shown in Fig. 11.6. The abscissa is a universal variable, in this case $u = 2kr_0 \sin(\phi/2)$. Because of the higher sidelobes one might expect the directivity to be higher than that of a linear array across the ring array diameter, and this proves to be true.

It is obvious that sidelobe reduction requires turning off the rear elements, and deweighting the side elements. An amplitude taper is thus introduced such that the elements are excited over an angle less than 180 deg, typically 120 deg. To scan the beam, this amplitude taper (and the corresponding cophasal phase) must be rotated. Wullenweber arrays for HF–DF have been used for many decades; typically element cables are brought to a central control point where a rotating goniometer provides the desired phase and amplitude (Gething, 1991). Large Wullenwebers have been constructed: the University of Illinois antenna covers 4–16 MHz and uses 120 monopole elements (40 at a time) on a ring of diameter 955 ft (Hayden, 1961). Since ring arrays tend to be used at long

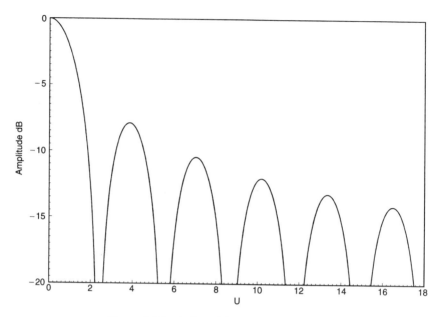

Figure 11.6 J_0 Bessel function vs. argument.

wavelengths, with corresponding relatively narrow bandwidths, the DFT processing can sometimes be performed digitally for receive arrays, using A/D converters. Electronic commutators have been described by Fenby and Davies (1968), Longstaff and Davies (1968), Bogner (1974), and Irzinski (1981). Ring arrays for TACAN also use a rotating distribution (Christopher, 1974; Shestag, 1974). Electronic despin of a spacecraft antenna is described by Gregorwich (1974). The azimuthal amplitude taper will increase the elevation plane beamwidth somewhat. Although the goniometer has been replaced by electronic phase and amplitude control, modern ring arrays are more likely to use modal beamformers. Use of directional elements instead of isotropes (for example vertical HF monopoles) can improve performance if the excited azimuthal sector is not too large. For a 90 deg sector some improvement may be expected, but less for a 120 deg sector. The prinicipal improvement from directional elements is a potential reduction in mutual coupling effects, but this too is nicely used advantageously with modal beamforming.

From Chapter 10 it is recalled that the Butler matrix generates the DFT of the input, so it is logical that the Butler matrix can be used with a ring array to produce a narrow beam, and to rotate it. Here an $N \times N$ Butler matrix is connected with equal-length lines to an N-element ring array; Fig. 11.7 shows this for $N = 16$. Each Butler port will excite one azimuthal mode, so the proper selection of amplitudes and phases on the ports will synthesize the desired pattern using the modes. A further advantage is that adding a progressive phase of $n\phi_0$ to the ports, where $n = 0, 1, \ldots, N - 1$, will rotate the beam to azimuth angle ϕ_0 (Davies, 1965, 1983; Sheleg, 1968, 1973; Skahill and White, 1975). A further advantage of the Butler matrix as a ring array beamformer is

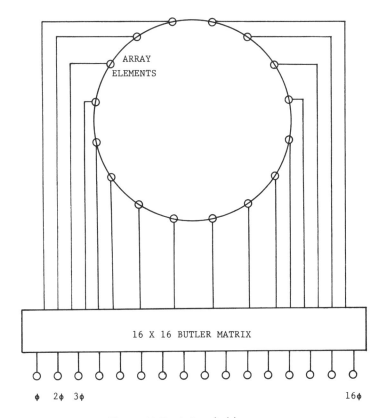

Figure 11.7 Azimuthal beam scan.

that while the modal mutual impedances are different, the effects of each mode appear only at the corresponding Butler port, and hence may be compensated there.

Arc arrays, where the radiating elements are fixed, also suffer from a higher sidelobe envelope at wide angles. Figure 11.8 shows the complete pattern of an arc array covering a subtended angle of 63 deg, with a radius of 12.7λ, and scanned to 15 deg. Note that the wide-angle sidelobe level is roughly −30 dB. When the element spacing is adjusted to make the projected virtual array equally spaced, the sidelobes at −90 deg are now only −20 dB, as shown in Fig. 11.9 (Hansen, 1981). These are due to phase asymmetries that appear on beam scan; they can be reduced by combining a small amount of difference pattern with the sum pattern (Antonucci and Franchi, 1985).

11.3 ARRAYS ON CYLINDERS

In this section cylinders with arrays covering the circumference, and extending a distance L axially are treated. The modal analysis of Section 11.3.3.1 is

396 CONFORMAL ARRAYS

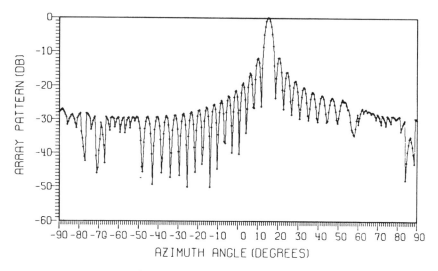

Figure 11.8 Pattern of arc array.

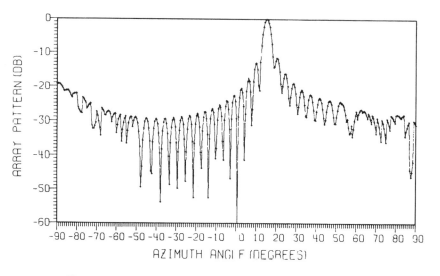

Figure 11.9 Pattern of arc array with equal projected spacing.

applicable, and the use of Butler matrix beamformers is also. However, there are several crucial differences. The axial extent of the array allows elevation pattern control, and higher directivity. Because of the metallic cylinder the element patterns are, of necessity, more direct. A major difficulty occurs in the calculation of mutual impedances. In a ring array simple methods such as the Carter approach of Chapter 7 are used. However, with curved metal surfaces, the mutual impedance expressions produce integrals that are difficult to evaluate, as will be seen below. First the influence of the curved surface on element pattern is examined.

11.3.1 Slot Patterns

The boundary value problem of radiation from a slot (with a sinusoidal distribution) on a metallic circular cylinder was solved exactly several decades ago (Carter, 1943; Harrington and Lepage, 1952; Knudsen, 1959). The solution is a harmonic series of functions appropriate for the cylinder: Bessel and Hankel functions. A standard cylindrical coordinate system is used, where z, θ, ϕ are the axial, polar, and circumferential coordinates. Finite width of the slot has a small effect upon the patterns, but the appropriate factor can be included if desired (Compton and Collin, 1969). With slot voltage V and cylinder radius a, the field for an axial half-wave slot (Wait, 1959) are $E_\theta = 0$ and

$$E_\theta = \frac{V \cos\left(\tfrac{1}{2}\pi \cos\theta\right)}{2\pi^2 ka \sin^2\theta} \sum_{m=0} \frac{\epsilon_m j^m \cos m\phi}{H'_m(ka \sin\theta)}. \tag{11.29}$$

This pattern is that of a half-wave dipole modified by the summation. For a circumferential half-wave slot, again thin, the fields are

$$E_\theta = \frac{2kaV}{\pi^2 \sin\theta \cos\theta} \sum_{m=0} \frac{\epsilon_m j^{m+1} \cos m\phi \cos(m\pi/2ka)}{(k^2 a^2 - m^2) H_m(ka \sin\theta)}, \tag{11.30}$$

$$E_\phi = \frac{2V \cos\theta}{\pi^2 \sin^2\theta} \sum_{m=0}^{\infty} \frac{m j^{m+1} \sin m\phi \cos(m\pi/2ka)}{(k^2 a^2 - m^2) H'_m(ka \sin\theta)}. \tag{11.31}$$

Convergence of these series is controlled by ka. Between ka and $2ka$ terms are usually needed. Large cylinders thus require many terms, but with current computer capability many tens of terms can be included. However, for such large cylinders the Watson transformation has been used to convert the harmonic series above, which results from an evaluation of an integral at real poles, to a residue series of complex-order Hankel functions, evaluated at the complex poles. For large cylinders the residue series, which is analogous to the Regge pole series in quantum mechanics, converges with only a few terms (Compton and Collin, 1969; K. S. Lee and Eichmann, 1970). Returning to the harmonic series, Wait (1959) has computed the sum for the azimuth patterns of primary interest: E_ϕ for an axial slot, and E_θ for a circumferential slot. His results are given in Figs. 11.10 and 11.11. It is interesting to note the axial slot electric field produces diffraction interference behind the cylinder, whereas the circumferential slot pattern exhibits a smooth decay. These patterns are used with the array factor and *scan reflection coefficient* to produce the complete pattern.

11.3.2 Array Pattern

The cylindrical array can be considered to consist of a stack of identical ring arrays. Denote the complex excitation of the pth element in the qth ring by $I_{pq} = I(\beta_p, z_q)$, where β_p is the angular location of the pth and z_q is the z-axis location of the qth ring. The coordinate system is shown in Fig. 11.12. The

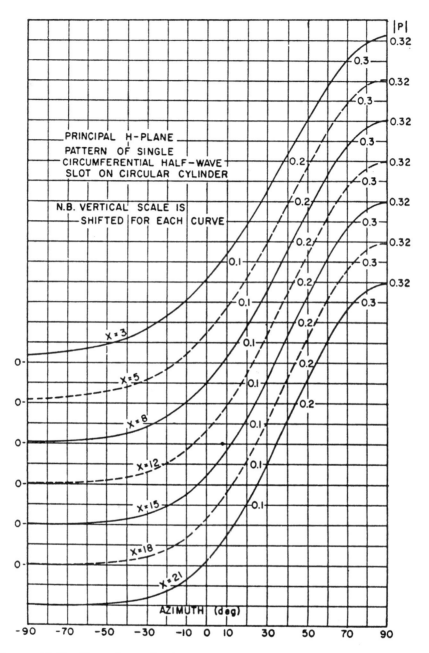

Figure 11.10 Circumferential slot patterns; $x = ka$. (Courtesy Wait, J. R., *Electromagnetic Radiation from Cylindrical Structures*, Pergamon Press, 1959, p. 45.)

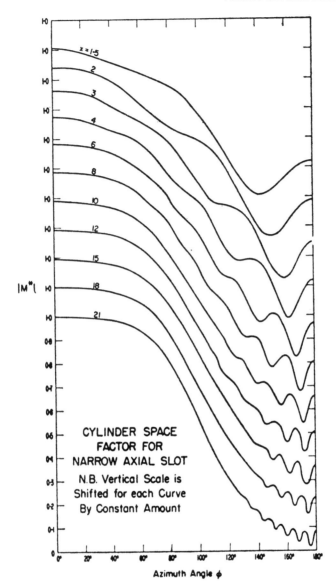

Figure 11.11 Axial slot patterns; $x = ka$. (Courtesy Wait, J. R., *Electromagnetic Radiation from Cylindrical Structures*, Pergamon Press, 1959, p. 30.)

beam is assumed to be pointed in the $\phi = 0$ reference point of the element location. The beam is stepped around the cylinder by redefining the $\beta = 0$ reference to the desired position.

All elements are assumed identical, symmetrical, equally spaced, and pointed along the radius vector. Thus, the azimuth element pattern can be expressed as a function of $|\phi - \beta|$. In general, the azimuth pattern depends

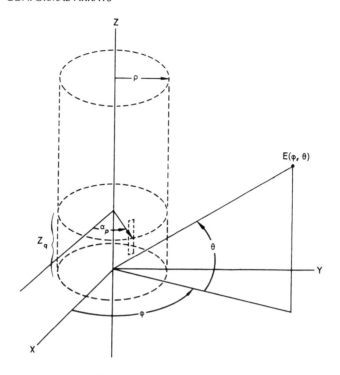

Figure 11.12 Coordinate system.

on the elevation angle θ. The complex element pattern is denoted by $G(\phi - \beta, \theta)$, with the phase referenced to the center of the ring in which it lies. Thus, if it is assumed that the phase center is at the element,

$$G(\phi - \beta, \theta) = |G(\phi - \beta, \theta)| \exp[jk\rho \cos\theta \cos(\phi - \beta)]. \tag{11.32}$$

The far field is

$$E(\phi, \theta) = \sum_p \sum_q I_{pq} G(\phi - \beta_p, \theta) \exp(jqu), \tag{11.33}$$

where $u = kd \sin\theta$ and $d =$ spacing between elements in the axial direction.

A beam can be formed in the direction $\phi = 0, \theta = \theta_0$ by exciting all elements to add in phase in that direction (beam cophasal excitation). Thus, in view of Eqns. (11.32) and (11.33):

$$I_{pq} = |I_{pq}| \exp[jk\rho \cos\theta_0 \cos\alpha_p - jqu_0], \tag{11.34}$$

where $u_0 = kd \sin\theta_0$.

In Eqn. (11.34) the phase terms are separated in β_p and z_q, where $z_q = qd$. This allows the assumption of a current distribution of the form

$$I(\beta_p, z_q) = I^a(\beta_p)I^e(z_q) = I_p^a I_q^e \qquad (11.35)$$

with

$$I_p^a = |I_p^a| \exp(-jk\rho \cos\theta_0 \cos\beta_p) \qquad (11.36)$$

and

$$I_q^e = |I_q^e| \exp(-jqu_0). \qquad (11.37)$$

The superscripts a and e indicate azimuth and elevation distributions, respectively. Note that the azimuth distribution depends on the beam-pointing angle in both azimuth and elevation, whereas I_q^e depends only on θ_0. Writing the current distribution in the form of Eqn. (11.35) allows the pattern (11.33) to be written in the form

$$E(\phi, \theta) = E^a(\phi, \theta)E^e(\theta), \qquad (11.38)$$

where

$$E^a(\phi, \theta) = \sum_p I_p^a G(\phi - \beta_p, \theta) \qquad (11.39)$$

and

$$E^e(\theta) = \sum_q I_q^e \exp(jqu). \qquad (11.40)$$

$E^a(\phi, \theta)$ is just the pattern of a single ring excited by I_p^a; $E^e(\theta)$ is the space factor of a single vertical column of elements excited by I_q^e.

Thus, the analysis can be simplified by assuming the separable aperture distribution and considering the cylindrical-array pattern to be the product of a ring array pattern and a linear array pattern. A pencil beam can be formed by selecting I_p^a to form a beam at $\phi = 0$ in the azimuth cone $\theta = \theta_0$, and by selecting I_q^e to form a beam at θ_0 in the elevation plane $\phi = 0$. Since the pencil beam is the product of two fan beams, the principal sidelobes will lie on the cone and plane in which the fan beams were shaped.

The patterns do not include the effects of mutual coupling. *Scan element patterns*, which do, and the corresponding element efficiencies have been calculated by Schwartzman and Kahn (1964) and by Kahn (1971).

11.3.2.1 Grating Lobes. A cylinder can be covered with a regular lattice, but the projection in any direction produces unequal spacing in azimuth. In elevation, however, the projected element spacings are uniform, so that conventional grating lobe theory can be used for elevation. There are no sharply defined, high-amplitude grating lobes due to azimuth spacing, but the sidelobes of a low sidelobe design may be raised if the element spacing is too large. Since the cylinder must have elements "all around" to allow 360 deg azimuth scanning, there are two variables involved in grating lobe type calculations; element spacing and cylinder radius. In addition, each element pattern usually has its axis in a radial direction so that the element pattern cannot be factored out. Amplitude tapering is typically used to produce moderately low sidelobes, so that any meaningful pattern calculation must include element spacing, cylinder radius, element pattern, and amplitude taper. For this reason there are no simple grating lobe type results for the cylindrical array. Cylindrical arrays that are phased to produce narrow beams tend to be more susceptible to grating lobe problems than do comparable planar arrays because of two factors: (1) On the sides of the active portion of the array the element patterns do not point in the direction of the beam; and (2) in this same region it is necessary to introduce a large interelement phase shift into the excitation to compensate for the curvature of the cylinder. The latter factor is equivalent to scanning the side portion of the array to some angle off the normal to the cylinder at that point; hence, the interelement spacing must be kept correspondingly small to prevent the formation of grating lobes.

Elevation scan to θ_0 is achieved by adding the phase $-qkd \sin \theta_0$ to the qth ring. The performance of the array pattern in the plane $\phi = 0$ is now the same as for a linear array with ring array "elements" with patterns such as those in Fig. 11.3. In particular, a grating lobe appears at θ when

$$\frac{d}{\lambda}(\sin \theta - \sin \theta_0) = \pm m, \, m = 1, 2, \ldots \qquad (11.41)$$

The lobe is at $\phi = 0$ in azimuth because $E^a(\phi, \theta)$ has its maximum at $\phi = 0$ (or nearly so) for all θ. The grating lobe, as it arises from $E^e(\theta)$ has unit magnitude (equal to main beam), but is reduced by $E^a(\phi = 0, \theta)$.

Staggering alternate columns of elements on the cylinder is an effective means of extending the elevation scanning angle for a given ring-to-ring spacing d and maintaining a small grating lobe. Consider the staggered array as a superposition of two regular arrays, each with the normal number of rings Q but only half the number of elements in each ring, $M/2$. The subarrays are identical except one is rotated by half a spacing in azimuth and is raised by half a spacing in the vertical direction, with the phase compensating for the dislocation. The ring-array patterns for the subarrays can be written as follows.

Array I. The beam in the direction of the first element; that is, $f = 0$:

$$E_\mathrm{I} = f(\theta) \sum_n I_n(\theta_0) F_n(\theta) \cos n\phi + f(\theta) \sum_n I_n(\theta_0) F_{M/2-n}(\theta)$$
$$\times \cos[M/2 - n)\phi] + f(\theta) \sum_n I_n(\theta_0) F_{M-n}(\theta) \cos[M - n)\phi]. \qquad (11.42)$$

The $r = 2$ term, which ordinarily would not contribute (for $s < \lambda$) is included because the spacing is now double the normal spacing.

Array II. Beam is in a direction halfway between two elements; that is, $f = 0.5$:

$$E_{II} = f(\theta) \sum_n I_n(\theta_0) F_n(\theta) \cos(n\phi)$$
$$- f(\theta) \sum_n I_n(\theta_0) F_{M/2-n}(\theta) \cos[(M/2 - n)\phi]$$
$$+ f(\theta) \sum_n I_n(\theta_0) F_{M-n}(\theta) \cos[(M - n)\phi]. \quad (11.43)$$

If the array were not staggered, each ring would have M elements and a pattern

$$E^a(\phi, \theta) = f(\theta) \sum_n I_n(\theta_0) F_n(\theta) \cos(n\phi)$$
$$+ f(\theta) \sum_n I_n(\theta_0) F_{M-n}(\theta) \cos[(M - n)\phi]. \quad (11.44)$$

The grating lobe term that arises because of the doubling of the spacing is

$$E_{gl}(\phi, \theta) = \sum_n I_n(\theta_0) F_{M/2-n}(\theta) \cos[(M/2 - n)\phi]. \quad (11.45)$$

Assume array II is raised with respect to array I by $d/2$ and multiplied by $\exp\{jk(d/2)(\sin\theta - \sin\theta_0)\}$ to account for the phase. The patterns of cylindrical arrays I and II are then

$$E_I(\phi, \theta) = \sum_{q=1}^{Q} |I_q|[E^a(\phi, \theta) + E_{gl}(\phi, \theta)] \exp[jqkd(\sin\theta - \sin\theta_0)] \quad (11.46)$$

and

$$E_{II}(\phi, \theta) = \exp\left(\frac{jkd}{2}(\sin\theta - \sin\theta_0)\right)$$
$$\times \sum_{q=1}^{Q} |I_q|[E^a(\phi, \theta) - E_{gl}(\phi, \theta)] \exp[jqkd(\sin\theta - \sin\theta_0)]. \quad (11.47)$$

The sum of E_I and E_{II} is the pattern of the staggered array. It may be put in the form:

$$E(\phi, \theta) = E^a(\phi, \theta) \sum_{q=1}^{2Q} |I_q| \exp\left(\frac{jqkd}{2}(\sin\theta - \sin\theta_0)\right)$$

$$+ E_{gl}(\phi, \theta)\left[1 - \exp\left(\frac{jkd}{2}(\sin\theta - \sin\theta_0)\right)\right]$$

$$\times \sum_{q-1}^{Q} |I_q| \exp[jqkd(\sin\theta - \sin\theta_0)]. \quad (11.48)$$

The first term of Eqn. (11.48) is the pattern of an array of $2Q$ rings spaced at half the normal spacing with M elements on each ring; that is, the staggered array with the "holes" filled in. This term should give no grating lobe because of the half spacing in elevation and the normal spacing in azimuth.

The second term accounts for the grating lobe. The factor

$$\sum_{q=1}^{Q} |I_q| \exp[jqkd(\sin\theta - \sin\theta_0)] \quad (11.49)$$

is the linear array pattern for normal elevation spacing, and gives a grating lobe when

$$\frac{d}{\lambda}(\sin\theta - \sin\theta_0) = \pm 1. \quad (11.50)$$

This gives

$$1 - \exp\left(\frac{jkd}{2}(\sin\theta - \sin\theta_0)\right) = 2. \quad (11.51)$$

Thus, the grating lobe of the cylindrical staggered array is equal to the grating lobe of the linear array, with spacing d, times the grating lobe of a ring array, with spacing $2s$. The elevation and azimuth positions of the product lobe are the positions of the linear and ring array lobes, respectively. The staggered-array lobe appears at the same elevation angle as the lobe of the regular array but is removed from $\phi = 0$ to $\phi = \phi_{gl}$ as determined from a ring array with every other element removed. The advantage gained is the amount the grating lobe of the ring subarrays is down from the main beam.

Figure 11.13 shows the grating lobe height as a function of scan angle for regular and staggered configuration, and the position in elevation and azimuth for the staggered array lobe. The parameters are $N_0 = 128$, $\theta_0 = 0$ deg and a -50 dB Chebyshev distribution (Hansen, 1981).

In obtaining Eqn. (11.48), Eqn. (11.9) was the start. However, the start could equally well have been Eqn. (11.13), using $\beta = 2\pi(p+f)/M$. Also, it was assumed that $f = 0$ for array I and $f = 0.5$ for array II. Equation (11.48) depends only on the fact that $E_{gl}(\phi, \theta)$ for array I is equal to $-E_{gl}(\phi, \theta)$ for array II, which is true for any orientation of the first element

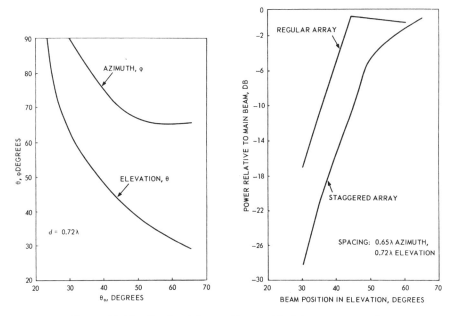

Figure 11.13 Grating lobe position and height vs. scan angle.

with respect to the main beam. Thus Eqn. (11.48) is valid for cylindrical arc arrays in general. Furthermore, $E_{gl}(\phi, \theta)$ may easily be identified from patterns computed directly from Eqn. (11.10) with alternate elements excited.

Figure 11.14 shows $E^a(\phi, \theta) + E_{gl}(\phi, \theta)$ for various θ_0. These patterns are computed from Eqn. (11.11) and the contribution from $E_{gl}(\phi, \theta)$ is easily identified. The effect of staggering is easily seen. For example, if the rings are

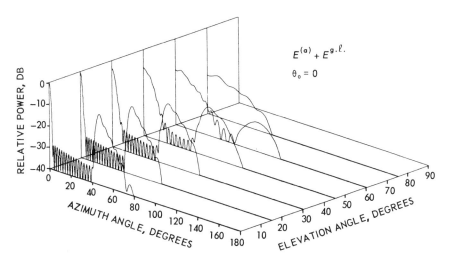

Figure 11.14a 30 dB Chebyshev patterns for $\theta = 0$.

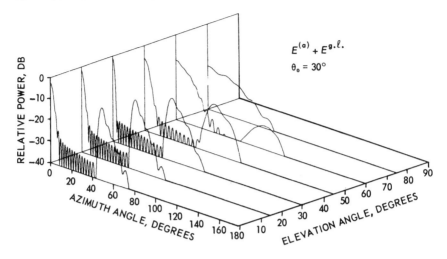

Figure 11.14b 30 dB Chebyshev patterns for $\theta = 30$ deg.

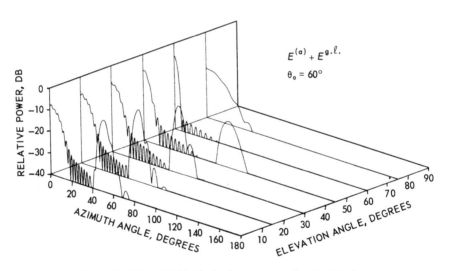

Figure 11.14c 30 dB Chebyshev patterns for $\theta = 60$ deg.

spaced at $d = 0.72\lambda$, a beam at $\theta_0 = 30$ deg gives a grating lobe at about -60 deg. A regular array gives the grating lobe height at -11 dB, which is the difference between the beams at $\theta = 30$ deg and $\theta = 60$ deg. For a staggered array, the grating lobe height is the difference between the beam at $\theta = 30$ deg and the grating lobe at $\theta = 60$ deg, or about -28 dB.

11.3.2.2 Principal Sidelobes. For the regular array the principal sidelobes will lie on the plane $\phi = 0$ and the cone $\theta = \theta_0$, because the regular-array pattern, for the separable distribution, can be thought of as the product of two fan beams.

For the staggered array we can consider Eqn. (11.48). The first term is the product of a linear array fan beam and a ring array fan beam, giving principal sidelobes as a regular array on the plane $\phi = 0$ and the cone $\theta = \theta_0$. The second term gives another set of axes, however. The first two factors of the second term give the linear array grating lobe (without the main beam); the third factor gives the ring array grating lobe (with double spacing). Thus, another set of principal sidelobes lies on the cone $\theta = \theta_{gl}$ and the warped plane $\phi = \theta_{gl}(\theta)$ where θ_{gl} is given by Eqn. (11.41) and $\theta_{gl}(\theta)$ gives a maximum to $E_{gl}(\phi, \theta)$.

For the regular array, then, the full cylindrical array pattern is well represented by two contours through ϕ, θ space: (ϕ, θ_0) and $(0, \theta)$. For the staggered array, the full pattern is well represented by the four contours (ϕ, θ_0), $(0, \theta)$, (ϕ, θ_{gl}), and $(\phi_{gl}(\theta), \theta)$. Figure 11.15 presents the patterns for regular and staggered arrays at various θ_0. The same single ring parameters are assumed as for Fig. 11.14; in addition, 32 rings spaced at 0.72λ are used with a 30 dB Chebyshev distribution for $I_q(e)$. The contour $(\phi_{gl}(\theta), \theta)$ was determined from the patterns of Fig. 11.14 in interpolation between the maximum point on the grating lobe.

The grating lobe can be reduced and elevation scan extended by reducing the azimuth and/or elevation spacing of the staggered array. For example, reducing the azimuth spacing from 0.65λ to 0.5λ (with $d = 0.72\lambda$) increases the scan-angle limit from 30 deg to about 40 deg to maintain a grating lobe of 30 dB, and further reduction to 0.4λ allows scanning to above 75 deg with the grating lobe below 40 dB. Reduction of elevation spacing (with $s = 0.65\lambda$) from 0.72 to 0.6λ allows scanning to above 50 deg for a grating lobe below 40 dB. In the array being implemented, however, the elevation spacing is restricted to a minimum of 0.72λ because of the element size.

11.3.2.3 Cylindrical Depolarization. The cylindrical surface, like any curved surface, depolarizes an incident wave. For example, if a linearly polarized wave is incident in the plane of incidence (the plane containing the cylinder axis and the direction of incidence), polarizations that are parallel or normal to this plane behave differently. With electric field parallel to the plane of incidence, the axial component of field on the cylindrical surface is in the same direction but the circumferential components oppose. Of course, at normal incidence the latter are zero. For normal polarization, the axial components are in the same direction; and the circumferential components are also in the circumferential direction. Thus incident circular polarization produces fields that have the polarization axes oriented right on one side, and left on the other side, with all polarization elliptical except at points of incidence. Hence a cylindrical array should use elements that radiate both axial and circumferential components. Further, the ratio of these components in general for any given element will change as the angle of incidence changes. Arrays on large (in wavelengths) cylinders with tapered circumferential distributions will experience a lower and perhaps negligible level of depolarization as the amplitude at elements significantly away from a

408 CONFORMAL ARRAYS

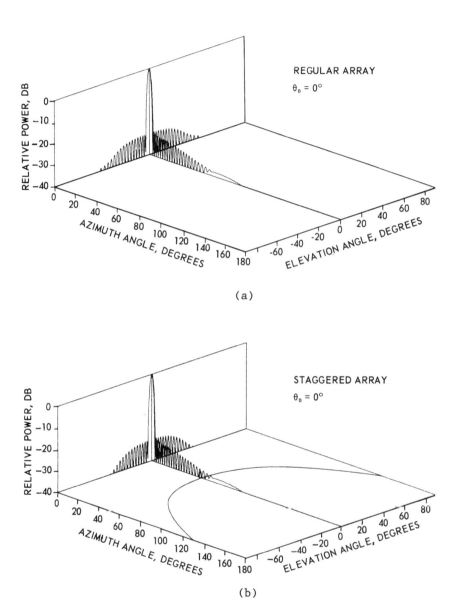

Figure 11.15 30 dB Chebyshev patterns.

ARRAYS ON CYLINDERS   409

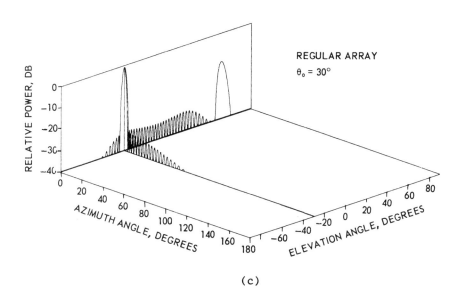

(c)

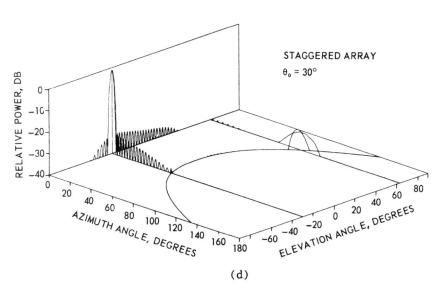

(d)

Figure 11.15   (*Continued*)

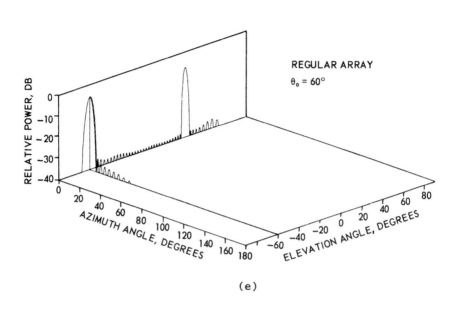

(e)

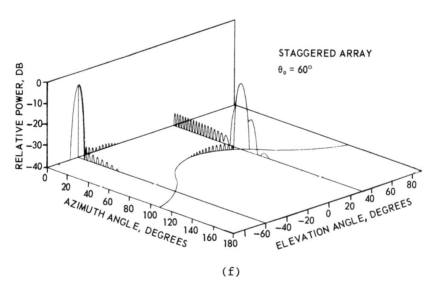

(f)

Figure 11.15 (*Continued*)

projected aperture plane may be small. Conical arrays do not share this advantage unless only the large-diameter portion of the surface is utilized. The general conical case will be discussed later.

11.3.3 Slot Mutual Admittance

In the design of a conformal slot array on the surface of a conducting cylinder, the calculation of the mutual admittance Y_{12} is a crucial step, which has been studied extensively in recent years. Referring to Fig. 11.16, two identical slots, circumferential or axial, are located on the surface of an infinitely long cylinder. The geometrical parameters are

$R =$ radius of the cylinder
$(a, b) =$ dimensions of the slot along (ϕ, z) directions (a is the arc length along the cylinder)
$(z_0, R\phi_0) =$ center-to-center distances between slots
$$s_0 = \sqrt{z_0^2 + (R\phi_0)^2}$$
$$\theta_0 = \arctan(z_0/R\phi_0)$$

The problem is to determine the mutual admittance between these two slots when kR is large. Throught this work it is assumed that (i) the slots are thin, and (ii) their length is roughly a half-wavelength. Then the aperture field in

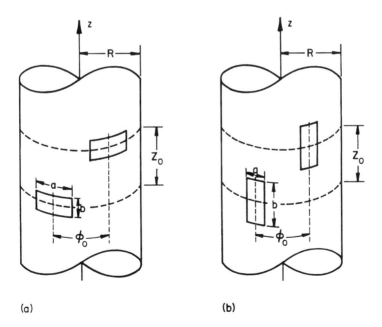

Figure 11.16 Slot geometry: (a) circumferential slots; (b) axial slots.

each slot can be adequately approximated by a simple cosine distribution, which is the so-called "one-mode" approximation. For example, if slot 1 is circumferential, its aperture field under the "one-mode" approximation is given by

$$\bar{E} = V_1 \bar{e}_1, \qquad \bar{H} = I_1 \bar{h}_1, \qquad (11.52)$$

where

$$\bar{e}_1 = \hat{z} \frac{2}{ab} \cos \frac{\pi y}{a}, \qquad y = R\phi, \qquad (11.53)$$
$$\bar{h}_1 = \hat{x} \cdot \bar{e}_1.$$

(V_1, I_1) are respectively the modal (voltage, current) of slot 1. The mutual admittance Y_{12} is defined by

$$Y_{12} = Y_{21} = \frac{I_{21}}{V_1}, \qquad (11.54)$$

where I_{21} is the induced current in slot 2 when slot 1 is excited by a voltage V_1 and slot 2 is short-circuited.

There is an alternative definition of mutual admittance. Instead of Eqn. (11.52), a modal voltage $\underline{V}_1$ (with a bar) may be defined through the expression for the aperture field of slot 1 as follows:

$$\bar{E} = \hat{z} \frac{\underline{V}_1}{b} \cos \frac{\pi y}{a}, \qquad (11.55)$$

or equivalently

$$\underline{V}_1 = \int_{-b/2}^{b/2} (\hat{z} \cdot \bar{E})_{y=0} \, dz. \qquad (11.56)$$

Then a different mutual admittance $\underline{Y}_{12}$ is defined after replacing (V_1, V_2) by $\underline{V}_1 \underline{V}_2)$. It can easily be shown that

$$\underline{Y}_{12} = \frac{a}{2b} Y_{12}. \qquad (11.57)$$

Three remarks are in order: (i) In the limiting case that $b \to 0$, Y_{12} goes to zero as b, whereas $\underline{Y}_{12}$ approaches a constant independent of b. (ii) For the special case $a = \lambda/2$ and $R \to \infty$, it is $\underline{Y}_{12}$ that is identical to the mutual impedance Z_{12} between two corresponding dipoles calculated by the classical Carter's method (Hansen, 1972). (iii) When the slots are excited by waveguides (transmission lines), one often uses Y_{12}. From here on, attention will be given to Y_{12} instead of $\underline{Y}_{12}$.

The mutual admittance defined in Eqn. (11.54) includes the self-admittance Y_{11} as a special case which occurs when two slots coincide. (All the formulas of Y_{12} given in this section, except for the one developed by Lee in Section 11.3.3.1, can be used for calculating Y_{11} by setting $\phi_0 \to 0$ and $z_0 \to 0$.)

11.3.3.1 Modal Series. A modal series was developed by Golden, Stewart, and Pridmore-Brown (1974), utilizing a series for the azimuth poles in the Green's function, and an integral for the continuous spectrum in z. See also Golden and Stewart (1971) and Stewart and Golden (1971). This series has been used extensively in obtaining Y_{12} for slots on a cylinder (Bargeliotes et al., 1976, 1977). The infinite integral must be approximated, and this is facilitated by assumption of a small loss in the medium via a small negative imaginary part of k. This becomes less satisfactory as z becomes larger, but a modified solution for that regime will be discussed later. See also the modal approach of Balzano and Dowling (1974).

The mutual admittance formulas are as follows.

Circumferential Slots

$$Y_{12} = \int_{-\infty}^{\infty} dk_z \sum_{m=-\infty}^{\infty} \psi(m, k_z) G(m, k_z) \exp[-j(m\phi_0 + k_z z_0)], \qquad (11.58)$$

where

$$\psi(m, k_z) = \frac{ab}{8\pi^2 R} \operatorname{sinc}^2 \frac{k_z b}{2} \left[\operatorname{sinc}\left(m\phi_a + \frac{\pi}{2}\right) + \operatorname{sinc}\left(m\phi_a - \frac{\pi}{2}\right) \right]^2,$$

$$\phi_a = \frac{a}{2R},$$

$$G(m, k_z) = Y_0 \frac{jk H_m^{(2)\prime}(k_t R)}{k_t H_m^{(2)}(k_t R)} + \left(\frac{mk_z}{k_t^2 R}\right)^2 \frac{k_t H_m^{(2)}(k_t R)}{jk H_m^{(2)\prime}(k_t R)}, \qquad (11.59)$$

$$k_t = \begin{cases} \sqrt{k^2 - k_z^2}, & k \geq k_z, \\ -j\sqrt{k_z^2 - k^2}, & k \leq k_z. \end{cases}$$

Axial Slots

$$Y_{12} = \int_{-\infty}^{\infty} dk_z \sum_{m=-\infty}^{\infty} \phi(m, k_z) F(m, k_z) \exp[-j(m\phi_0 + k_z z_0)], \qquad (11.60)$$

where

$$\phi(m, k_z) = \frac{ab}{8R}\left[\operatorname{sinc} m\phi_a \frac{\cos(k_z b/2)}{\left(\frac{k_z b}{2}\right)^2 - \left(\frac{\pi}{2}\right)^2}\right]^2, \tag{11.61}$$

$$F(m, k_z) = \frac{Y_0 k_t H_m^{(2)}(k_t R)}{jk H_m^{(2)\prime}(k_t R)}.$$

Lee and Mittra (see Hansen, 1981) developed an alternate modal series, suitable for large z. Consider first circumferential slots. Rewrite Y_{12} in terms of its real and imaginary parts: $Y_{12} = G + jB$. It can be shown that G is given by

$$G = \int_0^{k_0} dk_z \sum_{m=0}^{\infty} \phi \frac{1}{\epsilon_m} \cos(m\phi_0) \cos(k_z z_0) \psi(m, k_z) R(m, k_z), \tag{11.62}$$

where

$$\begin{aligned}
R(m, k_z) &= \frac{2k_0}{\pi k_t^2 R}\left[\frac{1}{M_m^2(k_t R)} + \left(\frac{mk_z}{k_0 k_t R}\right)^2 \frac{1}{N_m^2(k_t R)}\right], \\
M_m^2(x) &= J_m^2(x) + Y_m^2(x), \\
N_m^2(x) &= J_m^{\prime 2}(x) + Y_m^{\prime 2}(x).
\end{aligned} \tag{11.63}$$

Note that G contains a *finite* integral and can be evaluated in a straightforward manner by standard numerical integration techniques. The imaginary part of Y_{12} is given by

$$B = \int_{C_1} dk_z \sum_{m=0}^{\infty} \frac{1}{\epsilon_m} \cos(m\phi_0) \cos(k_z z_0) \psi(m, k_z) W(m, k_z), \tag{11.64}$$

where the integration contour C_1 is shown in Fig. 11.17 and

$$W_m(m, k_z) = \begin{cases} \dfrac{k_0}{k_t}(J_m J_m' + Y_m Y_m')\left[\dfrac{1}{M_m^2(k_t R)} - \left(\dfrac{mk_z}{k_0 k_t R}\right)^2 \dfrac{1}{N_m^2(k_t R)}\right], & k_0 > k_z, \\ \dfrac{-k_0}{|k_t|}\left[\dfrac{K_m'(|k_t|R)}{K_m(|k_t|R)} - \left(\dfrac{mk_z}{k_0|k_t|R}\right)^2 \dfrac{K_m(|k_t|R)}{K_m'(|k_t|R)}\right], & k_0 < k_z. \end{cases}$$

$$\tag{11.65}$$

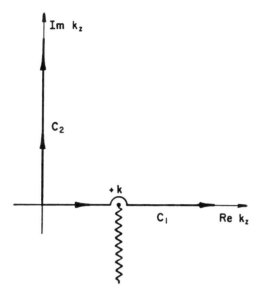

Figure 11.17 Contours in the complex k_z plane.

The computation of B is given in Eqn. (11.64) can be quite laborious because (i) the integration with respect to k_z is of infinite range, and the factor $\cos k_z z_0$ is highly oscillatory for large $k_0 z_0$; (ii) $W(m, k_z)$ has nonintegrable singularities of opposite sign on both sides of $k_z = k_0$; (iii) $W(m, k_z)$ decays slowly with respect to m and k_z.

To circumvent the above difficulties in evaluating B, a method introduced by Duncan (1962) in the study of cylindrical antenna problems is adopted. Rewrite Eqn. (11.64) as

$$B = \text{Im}\left[\sum_{m=0}^{\infty} \frac{1}{\epsilon_m} \cos m\phi_0 \left(-j \int F(m, k_z) \sin(k_z z_0)\, dk_z \right.\right.$$
$$\left.\left. + \int_{C_1} F(m, k_z) \exp(jk_z z_0)\, dk_z \right)\right], \quad (11.66)$$

where

$$F(m, k_z) = [R(m, k_z) + jW(m, k_z)]\psi(m, k_z). \quad (11.67)$$

The imaginary part of the first term inside the bracket of Eqn. (11.66) is

416 CONFORMAL ARRAYS

$$\text{Im}\left[-j\int_{C_1} F(m, k_z) \sin(k_z z_0) \, dk_z\right] = -\int_0^{k_0} R(m, k_z)\psi(m, k_z) \sin(k_z z_0) \, dk_z. \tag{11.68}$$

In order to compute the imaginary part of the second term of Eqn. (11.66), the integration contour C_1 is deformed into C_2 (Fig. 11.17) according to the theory of complex variables. This manipulation leads to

$$\text{Im}\int_{C_1} F(m, k_z) \exp(jk_z z_0) \, dk_z = \text{Im}\int_{C_2} F(m, k_z) \exp(jk_z z_0) \, dk_z. \tag{11.69}$$

Make the change of variable $k_z = j\eta$ in Eqn. (11.69). Substitution of the resultant equation and Eqn. (11.68) into Eqn. (11.66) gives

$$B = \sum_{m=0}^{\infty} \frac{1}{\epsilon_m} \cos(m\phi_0) \left(-\int_0^{k_0} R(m, k_z)\psi(m, k_z) \sin(k_z z_0) \, dk_z \right.$$
$$\left. + \int_0^{\infty} R(m, j\eta)\psi(m, j\eta) \exp(-\eta z_0) \, d\eta\right). \tag{11.70}$$

The final expression for Y_{12} is given by the real part G in Eqn. (11.62) and the imaginary part B in Eqn. (11.70). Several remarks are in order: (i) Not only G but also B is determined by $R(m, k_z)$, which is much simpler than $W(m, k_z)$ defined in Eqn. (11.65). (ii) G contains only a finite integral. (iii) The infinite integral in B, i.e., the second integral in Eqn. (11.58), contains an exponentially decaying factor $\exp[-(z_0 - a)\eta]$ in its integrand. Thus the larger z_0, the faster convergence in the evaluation of B. This is in contrast to the original expression for Y_{12}. (iv) There is no nonintegrable singularity in Eqn. (11.62) or (11.70).

The same method applies to the derivation of an alternative expression for Y_{12} for two axial slots ($a < b$ as shown in Fig. 11.16b). Only the final result is given:

$$Y_{12} = -\frac{ab Y_0}{\pi k_0 R^2} \sum_{m=0}^{\infty} \frac{\cos m\phi_0}{\epsilon_m} \left[\int_0^{k_0} \Phi(m, k_z) \frac{\exp(-jk_z z_0)}{N_m^2(k_t R)} \, dk_z \right.$$
$$\left. + j\int_0^{\infty} \Phi(m, j\eta) \frac{\exp(-\eta z_0)}{N_m^2\left(R\sqrt{k_z^2 + k_0^2}\right)} \, dk_z\right], \tag{11.71}$$

where

$$\Phi(m, k_z) = \left[\operatorname{sinc} \tfrac{1}{2} m \phi_a \frac{\cos \dfrac{k_z b}{2}}{\left(\dfrac{k_z b}{2}\right)^2 - \left(\dfrac{\pi}{2}\right)^2} \right]^2, \quad (11.72)$$

$$\phi_a = 2 \arcsin \frac{a}{2R}.$$

Asymptotic approaches, useful for large cylinder diameters in wavelengths, are included in Section 11.3.4.

Calculated patterns of a radial slot in a cylinder of 20 deg total angle, at a location where $ka = 39$, are shown in Fig. 11.18 for 5, 6, and 7 terms in the modal series. These can be compared with the patterns in Fig. 11.36 based on many terms, and on measurements.

For cylindrical arrays where the finite length of the cylinder affects element impedance, conventional analytical techniques are inadequate. A hybrid Moment Method, spectral approach has been used by Rengarajan (1996).

11.3.3.2 Admittance Data. Using the formulas described, slot mutual impedance data have been calculated for both axial and circumferential slots (Hansen, 1981). All data are at 9 GHz, for slot dimensions 0.9×0.4 in. Figure 11.19 shows mutual admittance between two circumferential slots as a function of rotation angle between the slots. Slot offset is 1.524λ, while cylinder radius is 1.517λ. Both the asymptotic result and the exact modal series result are shown. Two circumferential slots along the same generator, with axial separation a variable, provide the results of Fig. 11.20. Cylinder radius and slot dimensions are unchanged for this figure. The mutual admit-

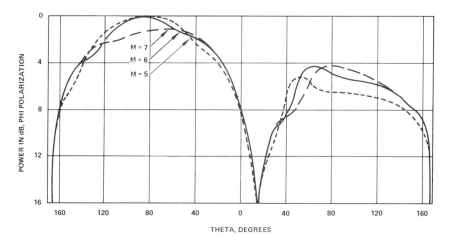

Figure 11.18 Pattern of radial slot on cone. (Courtesy Hughes Aircraft Co.)

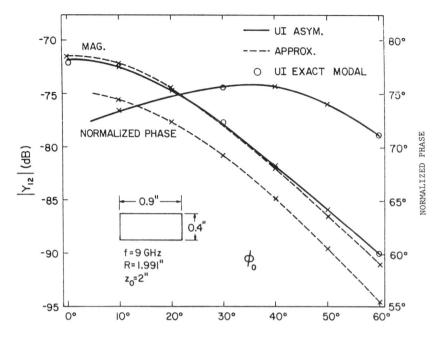

Figure 11.19 Circumferential slot mutual admittance vs. rotation.

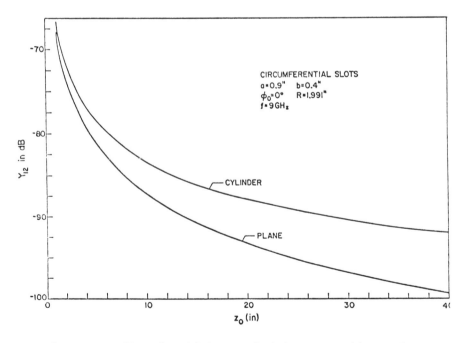

Figure 11.20 Circumferential slot mutual admittance vs. axial separation.

tance magnitude is smaller for slots on a cylinder than for slots in an infinite plane. Figure 11.21 is again for in-line slots, with an axial separation of 6.09λ, as a function of cylinder radius. Y_{12} is normalized by the plane value of $5.37 \times 10^{-5} \exp(j53.55)$ mho.

Axial slot behavior is simpler, and is typified by Fig. 11.22, which gives magnitude of mutual impedance versus axial spacing. Note that the cylindrical

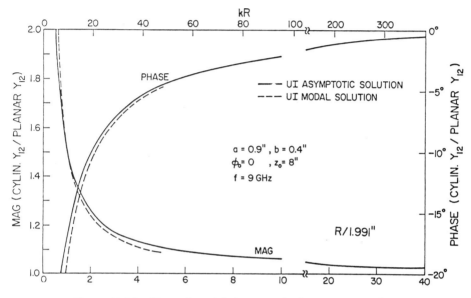

Figure 11.21 Circumferential slot mutual admittance vs. radius.

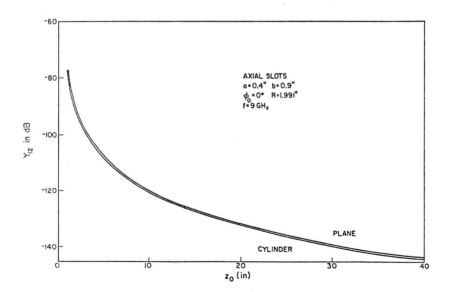

Figure 11.22 Axial slot mutual admittance vs. axial separation.

curvature has little effect, even though the cylinder diameter is only about 3 wavelengths.

11.3.4 Scan Element Pattern

A canonical geometry that yields insight into the behavior of slots upon a cylinder is an array of infinitely long axial thin slots (slits) deployed all around the cylinder. Sureau and Hessel (1969, 1971, 1972), Hessel and Sureau (1971, 1973), and Hessel (1972) derived *scan element patterns* using the unit cell approach of Oliner and Malech (1966), identifying a direct component associated with the integrated saddle point and a set of leaky waves representing complex poles (natural resonances) of the periodic cylinder. Pattern nulls due to external dielectric sheet covers were also exhibited. The unit cell approach was extended by Munger et al. (1971), who considered an infinitely long cylinder covered with a regular array of axial slots. A different approach gives insight into coupling between azimuthal modes. Borgiotti and Balzano (1970, 1972a, 1972b) and Balzano (1974) decomposed a ring of slots into a sum of eigen excitations (modes) and then computed coupling between modes on adjacent rings. If this were generalized to obtain mutual admittance between two slots, each on a different ring, the result would be the modal series admittance, using the cylindrical Green's function (Bowman et al., 1969). This approach, developed by Pridmore-Brown and Stewart, and extended by Lee is the most useful as it gives mutual admittance between slots directly. It was described in detail in Section 11.3.3.1.

Scan element patterns of elements disposed around a cylinder can be obtained by solving an equation for each azimuthal mode separately (Herper et al., 1985); Fig. 11.23 shows SEP for circumferential dipoles around a cylinder of diameter $(120/\pi)\lambda$, with axial dipole spacing of 0.72λ, and circumferential spacings of 0.5, 0.6, and 0.72λ. As expected from the analogous H-plane planar array SEP, the wider spacings show oscillations at broadside, leading to a drop at a pseudo-grating lobe angle. The drops occur at angles smaller than those for the planar case, but are less steep. See also Borgiotti (1983).

An asymptotic approach has been used by Sureau and Hessel (1972) with the SEP calculated from a combination of a space wave and creeping waves. See also S. W. Lee (1978), S. W. Lee and Safavi-Nairi (1978), and Bird (1985). Asymptotic methods for general cylindrical structures have been developed by Shapira et al. (1974a,b). Figure 11.24 shows SEP for an array of rectangular waveguide radiators around a cylinder of diameter $(185/\pi)\lambda$, with axial spacing of 0.8λ and circumferential spacing of 0.6λ. E-field was circumferential, with guide dimensions of 0.32 by 0.75λ. This larger cylinder shows a steeper drop, more like the planar array results of Chapter 7.

Mutual coupling for a concave cylindrical array has been investigated using Floquet theory by Tomasic and Hessel (1989), and for asymptotic theory by Tomasic et al. (1993). This geometry is useful for Rotman and bootlace lenses.

ARRAYS ON CYLINDERS   421

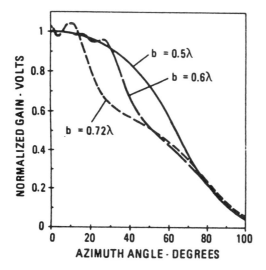

Figure 11.23 Scan element patterns for several spacings. (Courtesy Herper, J. C., Hessel, A., and Tomasic, B., "Element Pattern of an Axial Dipole in a Cylindrical Phased Array — Part I: Theory," "Part II: Element Design and Experiments," *Trans. IEEE*, Vol. AP-33, Mar. 1985, pp. 259–272, 273–278.)

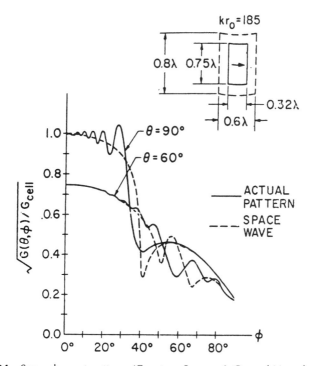

Figure 11.24 Scan element pattern. (Courtesy Sureau, J. C., and Hessel, A., "Realized Gain Function for a Cylindrical Array of Open-Ended Waveguides," in *Phased Array Antennas*, A. A. Oliner and G. H. Knittel, Eds., Proc. of 1970 Symposium at PIB, Artech House, 1972, pp. 315–322.)

422 CONFORMAL ARRAYS

11.4 SECTOR ARRAYS ON CYLINDERS

11.4.1 Patterns and Directivity

In many applications such as missiles and aircraft, full azimuth scanning is not needed. Sector arrays, where the elements occupy a sector of subtended angle ϕ_a, are appropriate. When the sector angle is small, perhaps < 60 deg, the array can be designed as a planar array with minor adjustments. However, large sectors require an examination of all the curvature effects. Figure 11.25 shows relative directivity versus sector included angle for several element pattern variations. The directivity is projected area times the element directivity. Many phased arrays have a $\cos^{1.5}$ behavior in the strong coupling plane, and a cos behavior in the weak coupling plane, owing to *scan impedance* variation. An example of an array with 0.65λ spacing on a 27.4λ diameter cylinder, for sector angles of 60, 90, and 120 deg, is shown in Fig. 11.26 (Hessel, 1972). Larger projected element spacings at large angles allow the grating lobe to increase with larger sector angle. This lobe could be suppressed with closer element spacings of course. However the directivity penalty cannot. From a 60 deg sector to one of 120 deg, the projected area has doubled, but the directivity has increased only 1.0 dB.

When the cylinder radius is large, microwave optics approaches are suitable. The far field is divided into three regions: illuminated, shadow, and a transition region between; see Fig. 11.27 (Pathak and Kouyoumjian, 1974). In the illuminated region GTD is used to obtain the pattern; in the shadow region creeping waves exist; and the transition region is difficult. The latter region

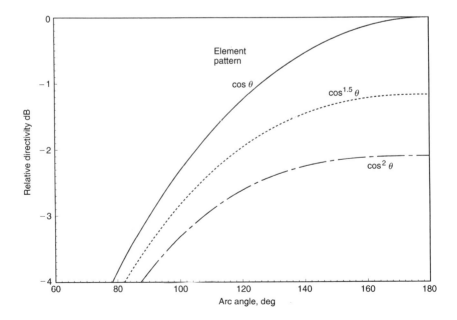

Figure 11.25 Arc array directivity relative to flat array across diameter.

SECTOR ARRAYS ON CYLINDERS   423

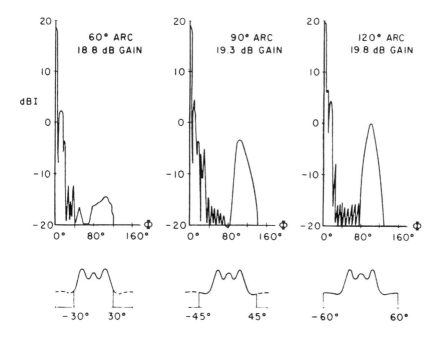

Figure 11.26 Scan element patterns of arc arrays. (Courtesy Hessel, A., "Mutual Coupling Effects in Circular Arrays on Cylindrical Surfaces — Aperture Design Implications and Analysis," in *Phased Array Antennas*, A. A., Oliner and G. H. Knittel, Eds., Proc. of 1970 Symposium at PIB, Artech House, 1972, pp. 273–291.)

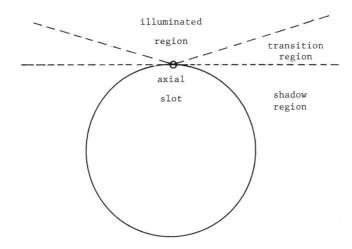

Figure 11.27 Field regions.

covers an angular region each side of the tangent line proportional to $(kr_0)^{-1/3}$. An axial slot (circumferential electric) field is roughly constant, but drops to about 0.7 at $\phi = 90$ deg, independently of cylinder diameter as long as it is large. A circumferential slot (axial electric) field in contrast drops to roughly $0.4(2/kr_0)^{1/3}$. A slow dependence on cylinder diameter exists. A cylinder of 20λ diameter gives field of -17.9 dB, while one of 40λ diameter gives -19.9 dB. The GTD analysis used FOCK functions; see Fock (1965) and Bowman et al. (1969). The complexity of the creeping-wave Watson transformation formulation puts it outside the scope of this book; useful references are Watson (1918), Bremmer (1949), and Felsen and Marcuvitz (1973).

The nonseparability of the elevation and azimuth factors in the pattern expression means that, for a fixed amplitude taper, both the elevation and azimuth patterns change with elevation scan angle θ_0. The practical implications of this are serious: to maintain low sidelobes in azimuth, as the elevation scan angle changes, the array excitation must change.

In addition to the switching and phasing configurations for rotating the beam in azimuth previously mentioned, lenses such as the R-2R have been used for cylindrical array feeds (Boyns et al, 1968; Holley et al, 1974).

11.4.2 Comparison of Planar and Sector Arrays

The question often arises whether the cylindrical array makes efficient use of aperture and hardware — in particular, when compared with the standard planar-array approach. For 360 deg azimuth coverage, four planar arrays, each scanning ±45 deg, are generally used, so the cylinder is compared with the four-sided planar configuration. Identical elements are assumed. For elevation scanning and elevation pattern, the two configurations give nearly identical results, assuming that both use a separable cophasal distribution and identical distributions in elevation. The planar array elevation pattern is the array factor multiplied by the elevation element pattern, while the cylindrical array elevation pattern is the array factor multiplied by the "ring array element" elevation pattern. Figure 11.3, for example, shows the "ring array element" elevation patterns compared with the element elevation pattern. If anything, it is an advantage of the cylinder that the ring array elevation patterns tend to suppress sidelobes more than does the element pattern alone.

The first-order comparison, then, can be reduced to that of the ring array azimuth pattern to the linear array azimuth pattern.

Assume that no more than 180 deg of the arc will be excited on the ring. Then the amplitude excitation of a linear array can be projected[1] onto an arc whose chord is equal to the length of the linear array, with the phase of the elements on the arc corrected to give a linear phase front. For small angles off broadside — that is, the main beam and first few sidelobes, the arc can be expected to give about the same results as the linear array, because for small

---

[1] The increased element density as a function of α on the arc means the amplitude should be reduced by $1/\cos \alpha$; however, this is exactly compensated by the assumed cosine element pattern.

angles the curvature has a negligible effect on the phased contribution from each element. Also, effects of element spacing become apparent only at larger angles. By this reasoning, the projection of, say, a Chebyshev distribution is a convenient means for forming the desired beamwidth and constraining the inner sidelobes of an arc array. Computations bear this out. Furthermore, the farther-out sidelobes tend naturally to be lower, with the exception of the grating lobe. Computations indicate that if the grating lobe is controlled, all sidelobes will be below the inner sidelobes.

The relative performance of the linear and arc arrays can be evaluated in terms of the number of elements and the overall antenna size required. Consider Fig. 11.28. The active apertures are shown in dark lines, and are of the same projected length. The element spacing on the linear array can be fixed at $d = 0.586\lambda$, which is the spacing required to scan ± 45 deg with grating lobe just coming in at ± 90 deg. For the arc, the grating lobe can be controlled by placing the stationary point outside the active arc, say at $\alpha_0 = \alpha(P) + \delta$. Then from Eqn. (11.21) with $\theta_0 = 0$ and $r = -1$, since δ is small,

$$s = \frac{\lambda}{2\sin(\alpha(P) + \delta)} \simeq \frac{\lambda}{2[\sin\alpha(P) + \delta\cos\alpha(P)]}. \tag{11.73}$$

The total number of elements on the four-sided array is about

$$\frac{8\rho \sin\alpha(P)}{0.586\lambda}. \tag{11.74}$$

The total number on the ring is

$$\frac{2\pi\rho}{s} = \frac{4\pi\rho}{\lambda}[\sin\alpha(P) + \delta\cos\alpha(P)]. \tag{11.75}$$

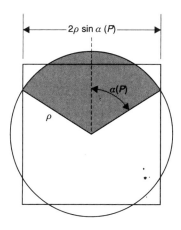

Figure 11.28 Four-sided array vs. circular array.

The ratio is

$$\frac{\text{Number of ring elements}}{\text{Number of linear elements}} = 0.920(1 + \delta \cot \alpha(P)).$$

From this one may conclude that about 92% to 100% of the elements required for a four-sided linear array are required to obtain the equivalent ring array.

Results (Hansen, 1981) indicate that the circular array of size and number of elements equivalent to a four-sided linear array can be made approximately equivalent in broadside performance. Since the linear array beam broadens for scan off broadside, however, the overall performance of the ring array is superior.

In addition, there are some disadvantages of the planar array which the cylindrical array inherently avoids. The ring array beam is identical for all beam positions, while the planar array beam is broader in scanning off broadside. As the ring array beam is scanned by, in effect, commutating the distribution, it is always formed by a distribution which is symmetrical in phase and amplitude. This results in superior beam pointing accuracy independent of frequency change. Finally, the cylindrical array gives 360 deg coverage in azimuth with none of the handover problems associated with the use of several planar arrays. In some applications these advantages can be very important.

The cylindrical array, however, has some disadvantages. For scanning, the amplitude as well as the phase must be switched in azimuth, and the feeding system that results will be more complex than that of a planar array system. (However, computer control is not a problem in view of the separable aperture.) The greatest disadvantage would appear to be that the cylindrical array cannot be physically separated as can the four planar arrays. This means that the cylinder must be in a position to look 360 deg, while each planar array need see only a 90 deg sector. More important, it means that the cylinder cannot be tilted back to increase the elevation coverage, as is common practice with planar arrays. For this reason, a truncated cone might be considered to extend the elevation coverage and still retain the advantages of circular symmetry. Comparison with tilted face planar arrays is slightly more complex; see Knittel (1965) and Corey (1985).

11.4.3 Ring and Cylindrical Array Hardware

A few examples of experimental arrays are given here. Figure 11.29 shows a ring array of 128 sectoral horns, designed to be fed by a parallel plate circular lens. Ring diameter is 26.4λ; horns are $1/2 \times 3/2 \times 3\lambda$ (Boyns *et al.*, 1970).

Figure 11.30 shows an edge slot Doppler scan Microwave Landing System array; each column is excited by a phaser (Hannan, 1976). Another L-band ring array is pictured in Fig. 11.31. It uses two rings of 124 elements each, with 30 excited to form a beam. Elements are probe fed flared horns with chokes to reduce ring-to-ring coupling. This 20 ft diameter array uses an elaborate switching and phasing unit shown in Fig. 11.32. (Gabriel and Cummings, 1970). Provencher (1972) reviews ring array practice. A cylindrical array

Figure 11.29 Ring array. (Courtesy Naval Electronics Lab. Center.)

Figure 11.30 Doppler scan edge slot array. (Courtesy Hazeltine Corp.)

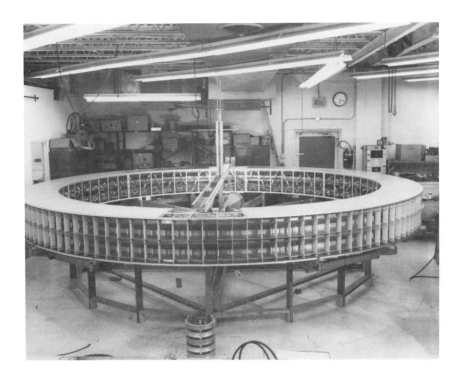

Figure 11.31 Electronic scanned dual beam Wullenweber. (Courtesy Scanwell Labs.)

Figure 11.32 Electronic scanned dual beam ring switch. (Courtesy Scanwell Labs.)

Figure 11.33 Cylindrical frequency scanning array. (Courtesy Naval Research Lab.)

with 256 elements in 32 columns is shown in Fig. 11.33. Each column of 8 dipoles incorporates a serpentine feed for frequency scanning in elevation. The switch matrix utilizes 8 active columns, and provides an amplitude taper for 25 dB sidelobe level (Sheleg, 1975).

11.5 ARRAYS ON CONES AND SPHERES

Cones and spheres are sometimes useful conformal array structures. A sphere, or more usually a hemisphere, can in principle scan in all directions, with only modest degradation at angles near 90 deg where the other hemisphere would have contributed. Conical arrays fit missile front ends, aircraft fuselage noses, smart munitions, etc. The difficulties, however, are enormous. A sphere cannot be populated by a regular lattice smaller than a dodecahedron (Fullerine), nor can a cone, owing to the two-dimensional curvature. When polarization is taken into account, the picture is far worse. An array on a cone, designed for linear circumferential polarization on-axis, requires slot polarizations that rotate as the observer rotates around the cone. This severe distortion is due to the projection of the polarization.

With modern computer control of phase and amplitude, the use of a somewhat irregular lattice on a sphere is acceptable. But the fabrication, assembly, and feed network are more costly due to these irregularities. A special case is the spherical dome array. This is a hemispherical bootlace lens, with a conventional scanned planar array in the diametrical plane. Limited scan of the planar array is magnified to wide-angle scan by the lens; see Chapter 10 for details.

Considerable analytical and experimental work has been done on conical arrays (Hansen, 1981), as will be described below.

11.5.1 Conical Arrays

In the design of electronically scanned conical arrays, problems arise that are different than those for arrays on planar and cylindrical surfaces. The gain of conical arrays changes as a function of the scan angle, and also depends on the part of the radiating structure that is visible from the far field at the beam pointing position. The polarization in the far field changes as a function of scan angle for radiating elements whose polarization is fixed with respect to the surface of the cone. Sum and difference patterns become sensitive to the incident polarization. The manner in which these quantities vary with scan angle will be discussed. The problem of pattern synthesis and analysis is also examined, and several techniques are discussed.

The general configuration for conical arrays consists of a set of radiating elements placed on a conical surface. The far-field patterns to be radiated from this array are pencil beams with suitably controlled sidelobes. The pencil beam is summed from a direction perpendicular to the generatrices through the axis of the cone. The radiating elements comprising the conical array are assumed to have symmetry in the plane perpendicular to the axis of the cone, the ϕ plane, and the array placed on the conical surface is assumed to have symmetry in the ϕ plane. Whatever the shape of the active part of the array, the projected aperture of the active part will be a function of the position angle defining the beam pointing direction in the plane defined by the axis of the cone and a generatrix (the θ plane). The active part is defined as that part that is turned on to receive (or transmit) energy. The projected aperture will be constant for any ϕ scan at constant θ because of the symmetry mentioned above.

The beamwidth of a planar array changes as a function of the beam pointing in direct proportion to the projected aperture perpendicular to the beam pointing direction. This relation is an approximation but is quite accurate for angles between broadside and 50 to 60 deg. In a conformal array the beamwidth is not so easily related to the geometry of the array because both the shape of the active part of the array and its projected area change as a function of the beam

ARRAYS ON CONES AND SPHERES   431

pointing direction. An estimate of the beam and sidelobe shapes will be given below.

For the beam pointed perpendicular to the generatrix of the cone, the radiation characteristics may be approximated using an equivalent planar array whose area is the projected area of the active part of the conformal array. Thus one would expect beamwidths and sidelobe levels commensurate with the projected aperture dimensions. At the other extreme, for a beam pointed along the axis of the cone, a different condition exists. For small angles near the cone axis, the array when projected forward would look approximately like a concentric ring array. The center rings of the array are missing, since the extreme tip of the cone is probably not usable in practice. For a 6-ring by 24-element/ring array the pattern would be somewhat as shown in Fig. 11.34. For comparison, the figure also shows the pattern of an array of the same size on the same cone but with the center rings filled. The latter is the usual pattern for a uniform circular array. It will be noted that the ring array has a narrower beamwidth and higher sidelobes than the filled array. This narrower beamwidth can be explained with reference to an interferometer which has a beamwidth one-half of that of a completely filled linear array of the same length. The

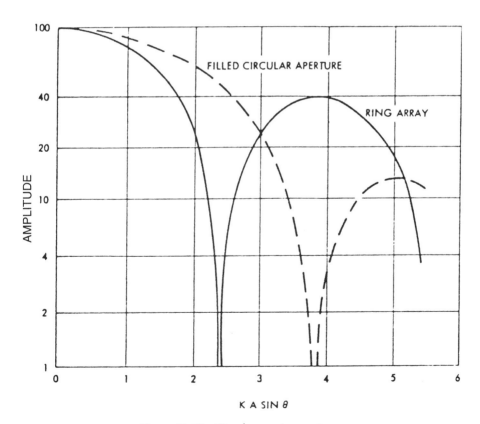

Figure 11.34 Circular aperture patterns.

432 CONFORMAL ARRAYS

"sidelobes" are as high as the main beam. This particular case is an intermediate one. Thus it can be seen that there is a trade-off between array filling, beamwidth, and sidelobe level.

Patterns of individual slots are of interest. Figure 11.35 shows measured and computed elevation patterns of a circumferential $\lambda/2$ slot on a cone of 20 deg total angle; the slot is 6.22λ from the tip. Flattening of the pattern by the cone in the lit region is marked. Azimuthal patterns of a radial slot at the same position are given in Fig. 11.36. The back lobe is -14 dB.

11.5.1.1 Lattices on a Cone. In a conical (and also a cylindrical) array, advantage can be taken of the circular symmetry of the surface to reduce the steering problem essentially to that of scanning in one dimension only with switching in azimuth. Figure 11.37 shows an end-on view of the cone with the shaded portion representing the excited area of the surface. If the beam lies in the plane perpendicular to the cone axis, its position will be as shown in the figure — and symmetrical with respect to the active area of the array. If the beam is scanned toward endfire while ϕ_1 is held fixed, it can be seen that the projection of the beam onto the plane of the paper will still fall at ϕ_1. Hence the active area is still symmetrically located with respect to the beam. The beam can be thought of as being broadside to the cone in the ϕ-plane and capable of being electronically scanned in the plane that passes through the cone axis (the θ-plane).

The beam is now steered around in the ϕ-plane to ϕ_2, by switching the active portion of the array to maintain a symmetrical relationship with the new beam pointing direction. Here also, the beam can be electronically steered toward endfire without changing the angle ϕ_2 or disturbing the symmetry conditions. Hence, it may be concluded that for any angle of ϕ, in the ideal case of continuously illuminated aperture, the beam can be considered to be broadside to

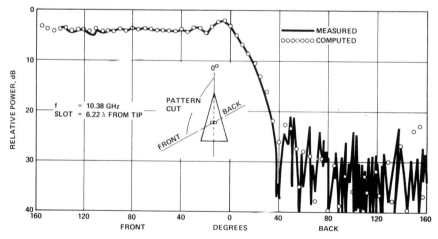

Figure 11.35 Pattern of circumferential slot on cone. (Courtesy Hughes Aircraft Co.)

ARRAYS ON CONES AND SPHERES   433

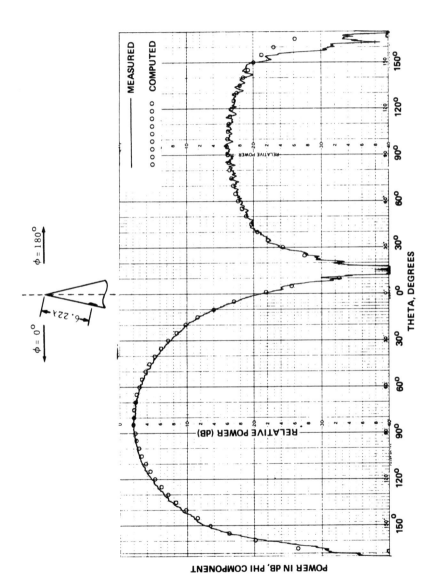

Figure 11.36 Pattern of radial slot on cone. (Courtesy Hughes Aircraft Co.)

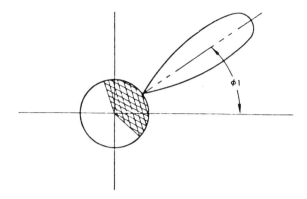

Figure 11.37 Excited part of cone.

the cone in the ϕ-plane and need be electronically scanned only in the θ-plane, with switching in the ϕ-plane.

With a discrete number of elements on the cone, the ideal situation outlined above can only be approximated. However, if a reasonably large number of elements is used, the approximation will be quite close. For example, if 16 elements are used in the larger rings around the cone, the largest portion of the area of excitation can be switched in $22\frac{1}{2}$ deg steps. Thus, it is necessary to electronically scan the beam only $\pm 11\frac{1}{4}$ deg off the perpendicular to the excited area to achieve full coverage around the axis of the cone by a sequence of switching and phasing operations. Since any one active area of the cone scans only $\pm 11\frac{1}{4}$ deg in the ϕ plane, the interelement spacing in that plane need be only slightly less than the spacing required for an array that does not scan at all in that plane. Hence, the interelement spacing problem is reduced approximately to that associated with scanning in the θ-plane only. It may also be attractive to provide amplitude and phase control at each element instead of switching.

The conical array can be thought of as a set of cylindrical arrays of different radii as a first-order approximation. Hence, it is to be expected that it, too, will have a tendency to have grating lobes in the ϕ plane. Therefore, it is anticipated that the interelement spacing in that plane will have to be kept smaller than would normally be necessary for an equivalent planar array. However, the fact that each circle of elements on the cone nearer the tip is smaller than the previous one may tend to alleviate this problem by introducing a modest amount of quasi-randomness into the element placement.

A more nearly continuous illumination in both principal axes of the cone can be achieved by staggering the elements in alternate circles as shown in Fig. 11.38. Thus, although the actual spacing between any two elements on a circle with a 6λ diameter is $1.2\lambda_0$ (assuming 16 elements per circle), the effective spacing is only one-half that value. An interelement spacing of $0.6\lambda_0$ is rather large for a circular array; however, each circle nearer the tip of the cone will bring the elements closer together until the point of physical interference is

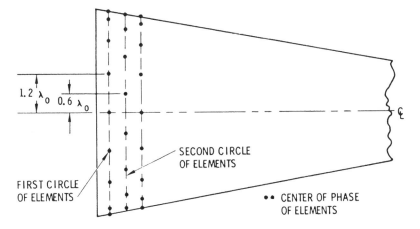

Figure 11.38 Staggered elements on cone.

reached. The average interelement spacing in the ϕ-plane for the section of the cone that has 16 elements per ring should thus be less than $0.5\lambda_0$.

In the smaller regions of the cone, fewer elements will be needed per circle. The type of element to be used will influence the decision on just how many should be used in each circle. The crossed waveguide elements (Kummer, 1972; Kummer et al., 1971, 1973) can be fitted together quite closely on a flat surface provided that they are rotated at an angle of approximately 20 to 25 deg, depending on wall thickness, to the principal axes of the lattice (see Fig. 11.39). They can be brought most closely together on a curved surface if they still maintain that angle. When the elements in alternate circles are staggered in a symmetrical fashion, the lattice is rotated 45 deg as shown in Fig. 11.40.

11.5.1.2 Conical Depolarization and Coordinate Systems.

A cone is less hospitable to an array than a cylinder in that its surface cannot be covered with a regular lattice. The question of orienting the radiators on the cone so that optimum performance in all directions is obtained involves the geometry of the radiator. The case of a beam pointing directly ahead is considered first. Since the array is conformal, some sort of flush-mounted radiator will be required. The slot and the open-ended waveguide are the two most commonly used flush-mounted radiators. The polarizations of these radiators are nearly identical; hence the polarization of the slot radiator is considered.

The E-field in one quadrant of the far field of a slot is shown in Fig. 11.41. In this figure the ground plane is assumed to be in the x, y-plane. It can be seen that the polarization of the field is everywhere perpendicular to the ground plane in the immediate vicinity of that plane. If the slots are imagined to be placed on a 10 deg half-angle cone, then in the forward direction each slot will be viewed from an angle of 10 deg above its nominal ground plane. The polarization at that angle can be visualized by constructing a surface on

436 CONFORMAL ARRAYS

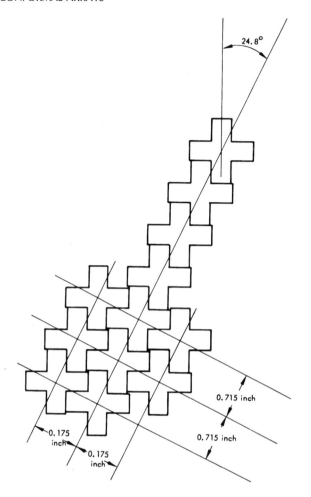

Figure 11.39 Tilted lattice for crossed-guide elements.

Fig. 11.41 that is 10 deg above the x, y-plane. This surface is shown in the figure by dashed lines.

It is instructive to start near the x-axis and note the change in polarization angle as the 10 deg surface is followed around to the y-axis. (The scale of angles in x, y-plane is used for convenience.) At 0 deg the polarization (shown by the dark arrows) is found to be parallel to the ground plane and, hence, is termed horizontal. As the progression continues around the quadrant toward y-axis, the polarization vector at first turns quite rapidly downward, then turns increasingly more slowly in the same direction, until at the y-axis it is vertical. At 45 deg, the polarization is still very nearly vertical instead of nearly 45 deg as might be expected. The result of this unequal rate of rotation of the polarization vector is that, when the slots are placed on a cone, they must rotate at an uneven rate with their position around the cone in order for them to all have the same polarization directly ahead. It can be shown that, in the quadrants,

ARRAYS ON CONES AND SPHERES 437

Figure 11.40 Staggered and tilted elements.

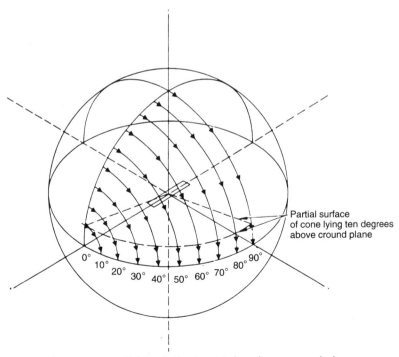

Figure 11.41 Field polarization 10 deg above ground plane.

438 CONFORMAL ARRAYS

the slots must be oriented so that their projections on the endfire view are also parallel with the projections of the principal axis slots. This requirement is due to the fact that the far-field polarization of the E-field of a slot is always perpendicular to the projection of the long dimension of the slot onto a plane perpendicular to the line-of-sight.

To obtain a picture of how the slots oriented to favor endfire operation would look from other angles in space, a paper cone was made with a number of slots drawn on it (Fig. 11.42). When the cone is viewed from endfire, the projections of all these slots are parallel as desired (see Fig. 11.42a). When the slots are viewed from a broadside position nearest the axial slots, they are predominantly lined up in an approximately axial fashion (Fig. 11.42b); hence, cross-polarization would not be too much of a problem. However, when the slots are viewed from the broadside region nearest the transverse slots, they are not properly lined up at all (Fig. 11.42c). With this slot arrangement, large amounts of the available power would go into cross-polarized lobes, and the effective aperture would be much smaller than the projected area of the cone in this direction.

There are a number of ways in which radiators providing variable polarization might be mounted on a cone, but complete symmetry cannot be maintained with a minimum number of elements or with simple control functions for beam formation and steering. An arrangements that simplifies the control task has the slots located on rings at fixed interelement spacings about each ring, and oriented along both generatrices and circumferential circles; see Fig. 11.43. Only the radial and circumferential components of the excitation must be computed, and the phase computation is simple (Kummer, 1972).

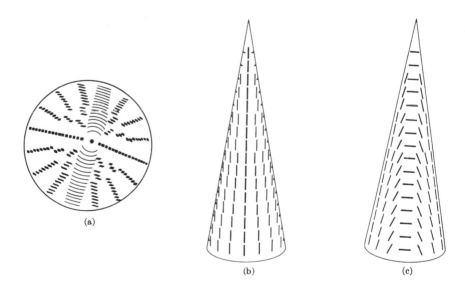

Figure 11.42 Conical array designed for endfire. (Courtesy Hughes Aircraft Co.)

ARRAYS ON CONES AND SPHERES 439

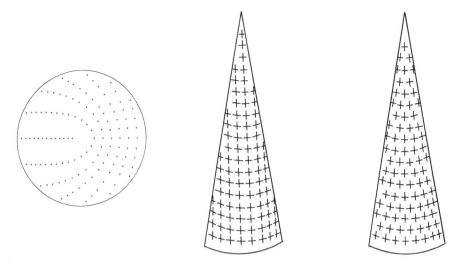

Figure 11.43 Cone with elements on rings.

The unusual geometry of the cone requires special consideration of the coordinate system used to represent the patterns. It has been found that in some coordinate systems it is impossible to properly define the two principal planes of the beam when they are scanned to the nose-fire position (Howard et al., 1969). Hence a special coordinate system was devised that follows the peak of the beam as it is scanned (Howard et al., 1969). The major requirement for the new coordinate system is that two fixed planes of the new system intersect the main beam of the antenna pattern at right angles. This requirement is satisfied by a variable spherical coordinate system, with angular coordinates ϕ' and θ'. The system is positioned so that $\phi' = 0$, $\theta' = 90$ deg corresponds to the main beam pointing direction $\phi = \phi_1$, $\theta = \theta_1$. This variable coordinate system can be related to the fixed conventional coordinate system through the transformations presented in Kummer et al. (1971). In this reference:

$$\cos\theta = \cos\theta' \sin\theta_1 + \sin\theta' \cos\phi' \cos\theta_1, \tag{11.76}$$

$$\sin\theta = \sqrt{\sin^2\theta' \sin^2\phi' + (\sin\theta' \cos\phi' \sin\theta_1 + \cos\theta' \cos\theta_1)^2}, \tag{11.77}$$

$$\sin\phi = \frac{\sin\theta' \sin\phi'}{\sin\theta}, \tag{11.78}$$

$$\cos\phi = \frac{\sin\theta' \cos\phi' \sin\theta_1 - \cos\theta' \cos\phi_1}{\sin\theta}. \tag{11.79}$$

The transformations have little effect on the pattern representations for broadside beam pointing directions (θ_1 near 90 deg), but have considerable effect for endfire beam pointing directions (θ_1 near 0 deg).

Another transformation is necessary to preserve the polarization of the test point used to perform the pattern plots. The test point polarization must be fixed in the Cartesian frame of the physical antenna cone. Again using Kummer et al. (1971), the new polarizations are related to the old polarizations through

$$l_{\theta'} = \frac{l_{\theta}(\cos\theta\cos\phi\cos\theta_1 + \sin\theta\sin\theta_1) - l_{\phi}\sin\phi\cos\theta_1}{\sin\theta'}, \quad (11.80)$$

$$l_{\phi'} = \frac{l_{\theta}\cos\theta_1\sin\phi + l_{\phi}(\cos\theta\cos\phi\cos\theta_1 + \sin\theta\sin\theta_1)}{\sin\theta'}. \quad (11.81)$$

For convenience, the new coordinates were renormalized as $\phi = \phi'$ and $\theta = \theta' - 90$ deg, so that the main beam peak is always centered at $\phi = 0$ deg, $\theta = $ deg.

Using these modifications to the coordinate system, a series of patterns was calculated for a crossed-slot configuration that partially filled a cone. The slot arrangement consisted of six rings of crossed slots with 24 slots per ring. The two arms of each crossed slot were fed in such amplitude and phase that linear polarization of the proper orientation resulted at the peak of the beam. The large end only of the cone was filled with elements, and the ring nearest the tip was 10 wavelengths from it. Total length of the cone was $12.2\lambda_0$ and the base had a diameter of $4.4\lambda_0$. The elements thus extended only about 20% of the way from the base of the cone to its tip, and the computer patterns only present a rough indication of those of a completely covered cone (Bargeliotes and Villeneuve, 1975; Bargeliotes et al., 1974, 1976).

In the calculations each element was weighted by the gain that it has in the direction of the peak of the beam. It was determined that this weighting yields the highest signal-to-noise ratio when the array is operating in the receiving mode. Figure 11.44 presents a series of patterns computed for a beam steered 20 deg off the noise-fire position ($\theta_1 = 20$ deg). These patterns represent only the desired component of polarization. There is a sizable cross-polarized component at some angles. The asymmetry of the conical geometry impedes a straightforward presentation of the antenna patterns, as evidenced by the complexity of the set of pattern plots presented in Fig. 11.44. Pattern computations were next made for elements arranged in a staggered configuration similar to that shown in Fig. 11.45. Sixteen elements per ring were assumed for the initial computation. With that number of elements in each ring, it was estimated that 10 rings would fit on the cone. Thus, a total of 160 elements "filled" the large end of the cone. The small end of the cone was then left "empty." The diameter of the base of the cone was 4.724λ and its total length was 13.335λ. The spacing between rings was $0.45\lambda_0$ to prevent grating lobes. The 10 rings of elements thus covered the lower four wavelengths of the cone. In the nose-fire direction the projected radius of the empty area was over twice as large as the projected thickness of the annulus of the filled area.

The broadside patterns of an array of this type will be quite normal because the visible surface of the array at any one point is approximately rectangular. Further, the majority of the slots are "seen" from favorable angles in regard to polarization.

ARRAYS ON CONES AND SPHERES **441**

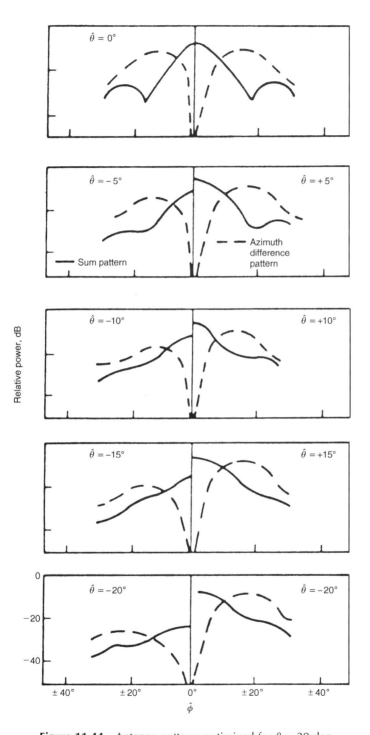

Figure 11.44 Antenna patterns optimized for $\theta = 20$ deg.

442 CONFORMAL ARRAYS

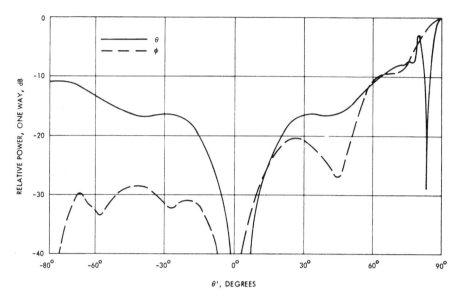

Figure 11.45 Patterns for staggered element array.

In the nose-fire direction, however, a marked difference in beamwidths for the two principal planes appears, as can be seen in Fig. 11.44. Three reasons were determined to explain this difference:

1. The weighting applied to the elements tends to "turn off" the elements that are being viewed from an unfavorable angle. Slots in two areas of the annulus are excited at a much reduced level. This different excitation tends to create an interferometer effect in one principal plane while the effective width of the useful aperture in the other plane is reduced to considerably less than the full diameter of the base.
2. The geometry of the cone is such that when the slots are oriented in a position to optimize the polarization of the signal in the nose-fire position, they are no longer optimum at any other angle. In the process of computing a pattern it is necessary to fix the polarizations of the slots so that there is no cross-polarized component at the peak of the beam. At other angles, more or less energy will go into this component depending on the steepness of the cone and the direction of the cut for the computed pattern. These two factors are related to the rapid change in polarization angle in the radiated pattern of a slot when viewed from a point close to either null in that pattern. This rapid change in polarization further aggravates the interferometer effect.
3. The large empty area in the center of the "working" annulus is in essence aperture blockage which contributes also to the interferometer effect, making it somewhat worse than in either of the other situations alone.

11.5.1.3 Projective Synthesis. Because of the effects of the conical surface upon element patterns, no direct synthesis technique is feasible. The study of depolarization in the previous section utilized projection; this method proved successful in conical array synthesis. A virtual planar array, with the desired size and amplitude tapers, is placed normal to the desired conical array beam. Then each planar array linearly polarized element is projected onto the conical surface including the rotation of polarization, amplitude, and phase (Kummer et al., 1973; Bargeliotes et al., 1977).

The projective synthesis result may, if necessary, be improved by an iterative method that includes a rigorous calculation of element patterns on a cone. These calculations are discussed briefly in Section 11.5.1.4.

To increase array efficiency, the element patterns can be broadened by dielectric loading (Villeneuve et al., 1975, 1982).

Quantization effects of digital phasers and of digital polarization rotators have been studied (Bargeliotes et al., 1977). Results are similar to those obtained for planar arrays with phasers. Four-bit polarization rotators are adequate unless cross-polarization levels or difference pattern null depths must be very low.

11.5.1.4 Patterns and Mutual Coupling. As was the case of arrays of elements on a cylinder, arrays of elements on a cone can be attacked either by harmonic series or by asymptotic methods, the latter for portions of the cone where the radius is large in wavelengths. Early work on calculation of near fields of slots on cones was done by Bailin and Silver (1956); the series include associated Legendre functions and spherical Bessel functions. Considerable effort has been expanded on these calculations, as they are useful for patterns and for mutual admittance (Pridmore-Brown, 1972; Pridmore-Brown and Stewart, 1972). The reader may also wish to refer to work on asymptotic analyses of elements on cones (Chan et al., 1977; Pridmore-Brown, 1973). Damiano uses a computer algebra package to compute zeros of the harmonic functions (Damiano et al., 1996).

11.5.1.5 Conical Array Experiments. Experiments with arrays on cones include a large array on a cone (Munger et al., 1974) and a small array (Thiele and Donn, 1974). The most extensive work was performed at Hughes Aircraft, Culver City. Initial experiments utilized a linear array along a generatrix (Behnke et al., 1972). Following this, an experimental array using crossed slot elements on a cone was constructed and extensively tested (Villeneuve, 1979). The crossed slot element is sketched in Fig. 11.46, and a close-up view of the array is shown in Fig. 11.47. Details of this research program are given in Hansen (1981), Bargeliotes et al. (1974, 1977), Bargeliotes and Villeneuve (1975), Behnke et al. (1972), Howard et al. (1969), Kummer et al. (1971, 1973), Villeneuve et al. (1974, 1975, 1982), and Villeneuve (1979).

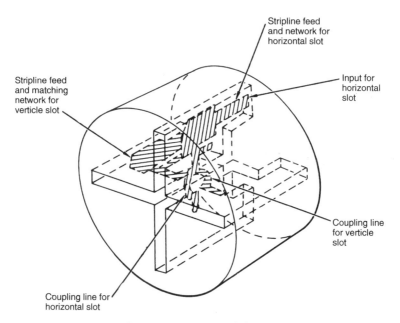

Figure 11.46 Crossed slot element.

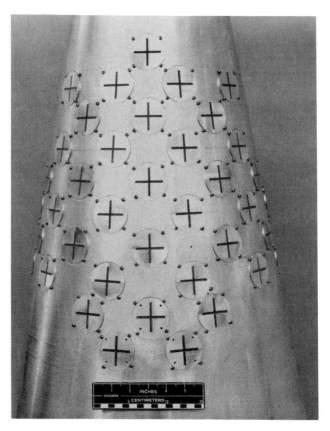

Figure 11.47 Crossed elements on a cone. (Courtesy Hughes Aircraft Co.)

11.5.2 Spherical Arrays

A spherical array behaves much like a cylindrical array scanned only in azimuth, but the lattice and depolarization problems are unique to the sphere. Geometry, pattern coverage, and grating lobes in spherical arrays have been investigated by Hoffman (1963), MacPhie (1968), A. K. Chan et al. (1968), Sengupta et al. (1968a and b), and Schrank (1972). Thinned arrays on a sphere have been probabilistically evaluated by Panicali and Lo (1969). A spherical patch array is shown in Fig. 11.48. (Hockensmith and Stockton, 1977; Stockton and Hockensmith, 1979).

An early work applying the harmonic (modal) series to a slot on a sphere and a cone was given by Bailin and Silver (1956). Much of the mutual coupling work for cones is, with minor changes applicable to coupling of elements on a sphere. Both modal series and asymptotic methods are useful (Hessel et al., 1979).

Figure 11.48 Switched patch spherical array. (Courtesy Ball Aerospace and Communications.)

Acknowledgment

Photographs courtesy of Jerry Boyns, Peter Hannan, Dr. William Gabriel, Boris Sheleg, Dr. Wolfgang Kummer, and Dean Paschen.

REFERENCES

Antonucci, J. and Franchi, P., "A Simple Technique to Correct for Curvature Effects on Conformal Phased Arrays," *Procedural 1985 Antenna Applications Symposium*, Dec. 1985, RADC TR-85-742, Vol. 2, pp. 607–630, AD-A165 535.

Bailin, L. L. and Silver, S., "Exterior Electromagnetic Boundary Value Problems for Spheres and Cones," *Trans. IRE*, Vol. AP-4, Jan. 1956, pp. 5–16; corrections Vol. AP-5, July 1957, p. 313.

Balzano, Q. "Analysis of Periodic Arrays of Waveguide Apertures on Conducting Cylinders Covered by Dielectric," *Trans. IEEE*, Vol. AP-22, Jan. 1974, pp. 25–34.

Balzano, Q. and Dowling, T. B., "Mutual Coupling Analysis of Arrays of Apertures on Cones," *Trans. IEEE*, Vol. AP-22, Jan. 1974, pp. 92–97.

Bargeliotes, P. C. and Villeneuve, A. T., "Pattern Synthesis of Conformal Arrays," Hughes Aircraft, Culver City, Calif., Final Report, Jan. 1975, AD-B004 752.

Bargeliotes, P. C. et al. "Dynamic Impedance Matching in Conformal Arrays", Hughes Aircraft Co., Culver City, Calif., Final Report Jan. 1974.

Bargeliotes, P. C., Villeneuve, A. T., and Kummer, W. H., "Phased Array Antennas Scanned Near Endfire," Hughes Aircraft, Final Report No. P-76/232, Mar. 1976, AD-A028 781.

Bargeliotes, P. C. et al., "Conformal Phased Array Breadboard," Hughes Aircraft Co., Culver City, Calif., Final Report, Jan. 1977, AD-A038 350.

Behnke, M. C., Villeneuve, A. T., and Kummer, W. H., "Advanced Conformal Array Antenna Techniques," Hughes Aircraft Co., Culver City, Calif., Final Report, Jan. 1972, AD-894 772.

Bird, T. S., "Accurate Asymptotic Solution for the Surface Field Due to Apertures in a Conducting Cylinder," *Trans. IEEE*, Vol. AP-33, Oct. 1985, pp. 1108–1117.

Biswell, D. E., *Circular and Cylindrical Array Antenna Investigation*, Texas Instruments, Feb. 1, 1968.

Blass, J., "An Analysis of the Radiation from Circular Arrays of Directive Elements," *Trans. IEEE*, Vol. AP-22, Jan. 1974, pp. 84–87.

Bogner, B. F., "Circularly Symmetric R. F. Commutator for Cylindrical Phased Arrays," *Trans. IEEE*, Vol. AP-22, Jan. 1974, pp. 78–81.

Borgiotti, G. V., "Conformal Arrays," in *The Handbook of Antenna Design*, Vol. 2, A. Rudge et al., Eds., IEE/Peregrinus 1983, Chapter 11.

Borgiotti, G. and Balzano, Q., "Mutual Coupling Analysis of a Conformal Array of Elements on a Cylindrical Surface," *Trans. IEEE*, Vol. AP-18, Jan. 1970, pp. 55–63.

Borgiotti, B. G. and Balzano, Q., "Conformal Arrays on Surfaces with Rotational Symmetry," in *Phased Array Antennas*, A. A. Oliner and G. H. Knittel, Eds., Proc. of 1970 Symposium at PIB, Artech House, 1972a, pp. 301–314.

Borgiotti, G. V. and Balzano, Q., "Analysis and Element Pattern Design of Periodic Arrays of Circular Apertures on Conducting Cylinders," *Trans. IEEE*, Vol. AP-20, Sep. 1972b, pp. 547–555.

Bowman, J. J., Senior, T. B. A., and Uslenghi, P. L. E., *Electromagnetic and Acoustic Scattering by Simple Shapes*, North-Holland, 1969.

Boyns, J. E. et al., "A Lens Feed for a Ring Array," *Trans. IEEE*, Vol. AP-16, Mar. 1968, pp. 264–267.

Boyns, J. E. et al., "Step-Scanned Circular Array", *Procedures Conformal Array Antenna Conference*, Naval Electronics Lab. Center, San Diego, Calif., Jan. 1970, AD-875 378.

Boyns, J. E. et al., "Step-Scanned Circular-Array Antenna," *Trans. IEEE*, Vol. AP-18, Sept. 1970, pp. 590–595.

Bremmer, H., *Terrestrial Radio Waves*, Elsevier, 1949.

Carter, P. S., "Antenna Arrays around Cylinders," *Proc. IRE*, Vol. 31, Dec. 1943, pp. 671–693.

Chan, A. K., Ishimaru, A., and Sigelmann, R. A., "Equally Spaced Spherical Arrays", *Radio Sci.*, Vol. 3, May 1968, pp. 401–404.

Chan, K. K. et al., "Creeping Waves on a Perfectly Conducting Cone," *Trans. IEEE*, Vol. AP-25, 1977, pp. 661–670.

Christopher, E. J., "Electronically Scanned TACAN Antenna," *Trans. IEEE*, Vol. AP-22, Jan. 1974, pp. 12–16.

Compton, R. T. and Collin, R. E., "Slot Antennas," in *Antenna Theory*, Part 1, R. E. Collin and F. J. Zucker, Eds., McGraw-Hill, 1969, Chapter 14.

Corey, L., "A Graphical Technique for Determining Optimal Array Antenna Geometry," *Trans. IEEE*, Vol. AP-33, July 1985, pp. 719–726.

Damiano, J.-P., Scotto, M., and Ribero, J. M., "Application of an Algebraic Tool to Fast Analysis and Synthesis of Conformal Printed Antennas," *Electron. Lett.*, Vol. 32, Oct. 24, 1996, pp. 2033–2035.

Davies, D. E. N., "A Transformation between the Phasing Techniques Required for Linear and Circular Aerial Arrays," *Proc. IEE*, Vol. 112, Nov. 1965, pp. 2041–2045.

Davies, D. E. N., "Circular Arrays," in *The Handbook of Antenna Design*, Vol. 2, A. Rudge et al., Eds., IEE/Peregrinus, 1983, Chapter 12.

DuHamel, R. H., Antenna Pattern Synthesis, PhD thesis, University of Illinois, 1951.

Duncan, R. H., "Theory of the Infinite Cylindrical Antenna Including the Feedpoint Singularity in Antenna Current," *J. Res. N.B.S.*, Vol. 66D, Mar.-Apr. 1962, pp. 181–188.

Felsen, L. B. and Marcuvitz, N., *Radiation and Scattering of Waves*, Prentice-Hall, 1973.

Fenby, R. G., "Limitations on Directional Patterns of Phase-Compensated Circular Arrays," *Radio Electronic Eng.*, Vol. 30, 1965, pp. 206–222.

Fenby, R. G. and Davies, D. E. N., "Circular Array Providing Fast 360° Electronic Beam Rotation," *Proc. IEE*, Vol. 115, Jan. 1968, pp. 78–86.

Fock, V. A., *Electromagnetic Diffraction and Propagation Problems*, Pergamon Press, 1965.

Gabriel, W. F. and Cummings, W. C., "Electronically Scanned Dual-Beam Wullenweber-Antenna," *Proceedings Conformal Array Antenna Conference*, Naval Electronics Lab. Center, San Diego, Calif., Jan. 1970, AD-875 378.

Gerlin, A., "Polynomial Approach to Study of Single-Ring Circular Antenna Array," *Proc. IEE*, Vol. 121, Apr. 1974, pp. 255–256.

Gething, P. J. D., *Direction Finding and Superresolution*, IEE/Peregrinus, 1991.

Gobert, J. F. and Yang, R. F. H., "A Theory of Antenna Array Conformal to Surface of Revolution," *Trans. IEEE*, Vol. AP-22, Jan. 1974, pp. 87–91.

Golden, K. E., Stewart, G. E., and Pridmore-Brown, D. C., "Approximation Techniques for the Mutual Admittance of Slot Antennas on Metallic Cones," *Trans. IEEE*, Vol. AP-22, Jan, 1974, pp. 43–48.

Golden, K. E. and Stewart, G. E., "Self and Mutual Admittances for Axial Rectangular Slots on a Conducting Cylinder in the Presence of an Inhomogeneous Plasma Layer," *Trans. IEEE*, Vol. AP-19, Mar. 1971, pp. 296–299.

Gregorwich, W. S., "An Electronically Despun Array Flush-Mounted on a Cylindrical Spacecraft," *Trans. IEEE*, Vol. AP-22, Jan. 1974, pp. 71–74.

Hannan, P. W., "2-D Coverage with 1-D Phased Arrays," *Microwave J.*, Vol. 19, June 1976, pp. 16–22.

Hansen, R. C., "Formulation of Echelon Dipole Mutual Impedance for Computer," *Trans. IEEE*, Vol. AP-20, 1972, pp. 780–781.

Hansen, R. C., *Conformal Antenna Design Handbook*, Naval Air Systems Command, Sept. 1981, AD-A110 091.

Harrington, R. F. and Lepage, W. R., "Directional Antenna Arrays of Elements Circularly Disposed about a Cylindrical Reflector," *Proc. IRE*, Vol. 40, Jan. 1952, pp. 83–86.

Hayden, E. C., "Propagation Studies Using Direction-Finding Techniques," *J. Res. N.B.S Radio Propagation*, Vol. 65D, May-June 1961, pp. 197–212.

Herper, J. C., Hessel, A., and Tomasic, B., "Element Pattern of an Axial Dipole in a Cylindrical Phased Array — Part I: Theory," "Part II: Element Design and Experiments," *Trans. IEEE*, Vol. AP-33, Mar. 1985, pp. 259–272, 273–278.

Hessel, A., "Mutual Coupling Effects in Circular Arrays on Cylindrical Surfaces — Aperture Design Implications and Analysis," in *Phased Array Antennas*, A. A. Oliner and G. H. Knittel, Eds., Proc. of 1970 Symposium at PIB, Artech House, 1972, pp. 273–291.

Hessel, A. and Sureau, J. C., "On the Realized Gain of Arrays," *Trans. IEEE*, Vol. AP-19, Jan. 1971, pp. 122–124.

Hessel, A. and Sureau, J. C., "Resonances in Circular Arrays with Dielectric Sheet Covers," *Trans. IEEE*, Vol. AP-21, Mar. 1973, pp. 159–164.

Hessel, A. et al., "Mutual Admittance between Circular Apertures on a Large Conducting Sphere," *Radio Sci.*, Vol. 14, Jan.-Feb. 1979, pp. 35–41.

Hickman, C. E., Neff, H. P., and Tillman, J. D., "The Theory of Single-Ring Circular Antenna Array," *Trans. AIEE*, Vol. 80, Part 1, May 1961, pp. 110–114.

Hilburn, J. L., "Circular Arrays of Radial and Tangential Dipoles for Turnstile Antennas," *Trans. IEEE*, Vol. AP-17, Sept, 1969, pp. 658–660.

Hilburn, J. L. and Hickman, C. E., "Circular Arrays of Tangential Dipoles," *J. Appl. Phys.*, Vol. 39, Dec. 1968, pp. 5953–5959.

Hockensmith, R. P. and Stockton, R., "The ESSA Solution," *IEEE National Telecommunications Conference Record*, Dec. 1977, pp. 1–4.

Hoffman, M., "Conventions for the Analysis of Spherical Arrays," *Trans. IEEE*, Vol. AP-11, July 1963, pp. 390–393.

Holley, A. E., DuFort, E. C., and Dell-Imagine, R. A., "An Electronically Scanned Beacon Antenna," *Trans. IEEE*, Vol. AP-22, Jan. 1974, pp. 3–12.

Howard, J. E., Kummer, W. H., and Villeneuve, A. T., "Integrated Conformal Arrays," Final Report on Contract N00019-68-C-0214, Hughes Aircraft Co., Culver City, Calif., 1969.

Irzinski, E. P., "A Coaxial Waveguide Commutator Feed for a Scanning Circular Phased Array," *Trans. IEEE*, Vol. MTT-29, Mar. 1981, pp. 266–270.

James, P. W., "Polar Patterns of Phase-Connected Circular Arrays," *Proc. IEE*, Vol. 112, 1965, pp. 1839, 1947.

Khan, W. K., "Efficiency of Radiating Element in Circular Cylindrical Arrays," *Trans. IEEE*, Vol. AP-19, Jan. 1971, pp. 115–117.

Knittel, G. H., "Choosing the Number of Faces of a Phased Array for Antenna for Hemisphere Scan Coverage," *Trans. IEEE*, Vol. AP-13, Nov. 1965, pp. 878–882.

Knudsen, H., "The Necessary Number of Elements in a Directional Ring Aerial," *J. Appl. Phys.*, Vol. 22, Nov. 1951, pp. 1299–1306.

Knudsen, H. L., "Field Radiated by Ring Quasi-Array of an Infinite Number of Tangential or Radial Dipoles," *Proc. IRE*, Vol. 41, June 1953, pp. 781–789.

Knudsen, H. L., "Radiation from Ring Quasi-Arrays," *Trans. IRE*, Vol. AP-4, July 1956, pp. 452–472.

Knudsen, H., "Antennas on Circular Cylinders", *Trans. IRE*, Vol. AP-7, Dec. 1959, pp. S361–S370.

Kummer, W. H., "Conical Arrays," *Proceedings Array Antenna Conference*, NELC, Feb. 1972, Paper 34, AD-744 629.

Kummer, W. H., Behnke, M. C., and Villeneuve, A. T., "Integrated Conformal Arrays," Hughes Aircraft, Culver City, Calif., Final Report, Feb. 1971, AD-882 641.

Kummer, W. H., Seaton, A. F., and Villeneuve, A. T., "Conformal Antenna Arrays Study," Hughes Aircraft Co., Culver City, Calif., Final Report, Jan. 1973, AD-909 220.

Lee, K. S. and Eichmann, G., "Elementary Patterns for Conformal Dipole Arrays Mounted on Dielectrically Clad Conducting Cylinders," *Trans. IEEE*, Vol. AP-28, Nov. 1980, pp. 811–818.

Lee, S. W., "Mutual Admittance of Slots on a Cone: Solution by Ray Technique," *Trans. IEEE*, Vol. AP-26, 1978, pp. 768–773.

Lee, S. W. and Lo, Y. T., "Pattern Function of Circular Arc Arrays," *Trans. IEEE*, Vol. AP-13, July 1965, pp. 649–650.

Lee, S. W. and Safavi-Naini, S., "Approximate Asymptotic Solution of Surface Field Due to a Magnetic Dipole on a Cylinder," *Trans. IEEE*, Vol. AP-26, July 1978, pp. 593–598.

Lim, J. C. and Davies, D. E. N., "Synthesis of a Single Null Response in an Otherwise Omnidirectional Pattern Using a Circular Array," *Proc. IEE*, Vol. 122, Apr. 1975, pp. 349–352.

Longstaff, I. D. et al., "Directional Properties of Circular Arrays," *Proc. IEE*, Vol. 114, June 1967, pp. 713–718.

Longstaff, I. D. and Davies, D. E. N., "A Wideband Circular Array for H. F. Communications," *Radio Electron. Eng.*, June 1968, Vol. 35, pp. 321–327.

Ma, M. T. and Walters, L. C., "Theoretical Methods for Computing Characteristics of Wullenweber Antennas," *Proc. IEE*, Vol. 117, Nov. 1970, pp. 2095–2101.

MacPhie, R. H., "The Element Density of a Spherical Antenna Array," *Trans. IEEE*, Vol. AP-16, Jan. 1968, pp. 125–127.

McNamara, D. A., Pistorius, C. W. I., and Malherbe, J. A. G., *Introduction to the Uniform Geometrical Theory of Diffraction*, Artech, 1990.

Munger, A. D., "Circular-Array Radar Antenna: Cylindrical-Array Theory," Naval Electronics Lab. Center, Report 1608, Jan. 1969, AD-853 730.

Munger, A. D. and Gladman, B. R., "Pattern Analysis for Cylindrical and Conical Arrays," *Procedures Conformal Array Conference*, Naval Electronics Lab. Center, San Diego, Calif., Jan. 1970, AD-875 378.

Munger, A. D., Provencher, J. H., and Gladman, B. R., "Mutual Coupling on a Cylindrical Array of Waveguide Elements," *Trans. IEEE*, Vol. AP-19, Jan. 1971, pp. 131–134.

Munger, A. D. et al., "Conical Array Studies," *Trans. IEEE*, Vol. AP-22, Jan. 1974, pp. 35–43.

Oliner, A. A. and Malech, R. G., "Mutual Coupling in Infinite Scanning Arrays," in *Microwave Scanning Antennas*, Vol. II, R. C. Hansen, Ed., Academic Press, 1966 [Peninsula Publishing, 1985], Chapter 3.

Page, H., "Radiation Resistance of Ring Aerials," *Wireless Eng.*, Vol. 25, Apr. 1948a, pp. 102–109.

Page, H., "Ring Aerial Systems," *Wireless Eng.*, Vol. 25, Oct. 1946, pp. 308–315; corrections Dec. 1948, p. 402.

Panicali, A. R. and Lo, Y. T., "A Probabilistic Approach to Large Circular and Spherical Arrays," *Trans. IEEE*, Vol. AP-17, July 1969, pp. 514–522.

Pathak, D. C. and Kouyoumjian, R. G., "An Analysis of the Radiation from Apertures in Curved Surfaces by the Geometric Theory of Diffraction," *Proc. IEEE*, Vol. 62, Nov. 1974, pp. 1438–1461.

Pridmore-Brown, D. C., "Diffraction Coefficients for a Slot-Excited Conical Antenna," *Trans. IEEE*, Vol. AP-20, Jan. 1972, pp. 40–49.

Pridmore-Brown, D. C., "The Transition Field on the Surface of a Slot-Excited Conical Antenna," *Trans. IEEE*, Vol. AP-21, Nov. 1973, pp. 889–890.

Pridmore–Brown, D. C. and Stewart, G., "Radiation from Slot Antennas on Cones," *Trans. IEEE*, Vol. AP–20, Jan. 1972, pp. 36–39.

Provencher, J. H., "A Survey of Circular Symmetric Arrays," in *Phased Array Antennas*, A. A. Oliner and G. H. Knittel, Eds., Proc. of 1970 Symposium at PIB, Artech House, 1972, pp. 292–300.

Rahim, T. and Davies, D. E. N., "Effect of Directional Elements on the Directional Response of Circular Arrays," *Proc. IEE*, Vol. 128, Part 11, Feb. 1982, pp. 18–22.

Redlich, R. W., "Sampling Synthesis of Ring Arrays," *Trans. IEEE*, Vol. AP-18, Jan. 1970, pp. 116–118.

Rengarajan, S. R., "Mutual Coupling between Slots Cut in Rectangular Cylindrical Structures: Spectral Domain Technique," *Radio Sci.*, Vol. 31, Nov.-Dec. 1996, pp. 1651–1661.

Royer, G. M., "Directive Gain and Impedence of a Ring Array of Antennas," *Trans. IEEE*, Vol. AP-14, Sept. 1966, pp. 566–573.

Schrank, H. E., "Basic Theoretical Aspects of Spherical Phased Arrays," in *Phased Array Antennas*, A. A. Oliner and G. H. Knittel, Eds., Proc. of 1970 Symposium at PIB, Artech House, 1972, pp. 323–327.

Schwartzman, L. and Kahn, W. K., "Maximum Efficiency for Cylindrically Disposed Multiple-Beam Antenna Arrays," *Trans. IEEE*, Vol. AP-12, Nov. 1964, pp. 795–797.

Sengupta, D. L., Smith, T. M., and Larson, R. W., "Radiation Characteristics of a Spherical Array of Circularly Polarized Elements," *Trans. IEEE*, Vol. AP-16, Jan. 1968a, pp. 2–7.

Sengupta, D. L. et al., "Experimental Study of a Spherical Array of Circulatory Polarized Elements," *Proc. IEEE*, Vol. 56, Nov. 1968b, pp. 2048–2051.

Shapira, J., Felsen, L. B., and Hessel, A., "Ray Analysis of Conformal Antenna Arrays," *Trans. IEEE*, Vol. AP-22, Jan. 1974a, pp. 49–63.

Shapira, J., Felsen, L. B., and Hessel, A., "Surface Ray Analysis of Mutually Coupled Arrays on Variable Curvature Cylindrical Surfaces," *Proc. IEEE*, Vol. 62, Mar. 1974b, pp. 1482–1492.

Sheleg, B., "A Matrix-Fed Circular Array for Continuous Scanning," *Proc. IEEE*, Vol. 56, July 1968, pp. 2016–2027.

Sheleg, B., "Butler Submatrix Feed Systems for Antenna Arrays," *Trans. IEEE*, Vol. AP-21, Mar. 1973, pp. 228–229.

Sheleg, B., "Circular and Cylindrical Arrays," *Workshop on Conformal Antennas*, Naval Air Sys. Command, Apr. 1975, pp. 107–138, AD-A015 630.

Shestag, L. N., "A Cylindrical Array for the TACAN System," *Trans. IEEE*, Vol. AP-22, Jan. 1974, pp. 17–25.

Sinnott, D. H. and Harrington, R. F., "Analysis and Design of Circular Antenna Arrays by Matrix Methods," *Trans. IEEE*, Vol. AP-21, Sept. 1973, pp. 610–614.

Skahil, G. and White, W. D., "A New Technique for Feeding of Cylindrical Array," *Trans. IEEE*, Vol. AP-23, Mar. 1975, pp. 253–256.

Stewart, G. E. and Golden, K. E., "Mutual Admittance for Axial Rectangular Slots in a Large Conducting Cylinder," *Trans. IEEE*, Vol. AP-19, 1971, pp. 120–133.

Stockton, R. and Hockensmith, R. P., "Microprocessor Provides Multi-Mode Versatility for the ESSA Antenna System," *IEEE APS Symposium Record*, Vol. 2, 1979, pp. 469–472.

Sureau, J. C. and Hessel, A., "Element Pattern for Circular Arrays of Axial Slits on Large Conducting Cylinders," *Trans. IEEE*, Vol. AP-17, Nov. 1969, pp. 799–803.

Sureau, J. C. and Hessel, A., "Element Pattern for Circular Arrays of Waveguide-Fed Axial Slits on Large Conducting Cylinders," *Trans. IEEE*, Vol. AP-19, Jan. 1971, pp. 64–74.

Sureau, J. C. and Hessel, A., "Realized Gain Function for a Cylindrical Array of Open-Ended Waveguides," in *Phased Array Antennas*, A. A. Oliner and G. H. Knittel, Eds., Proc. of 1970 Symposium at PIB, Artech House, 1972, pp. 315–322.

Thiele, G. A. and Donn, C. "Design of a Small Conformal Array," *Trans. IEEE*, Vol. AP-22, Jan. 1974, pp. 64–70.

Tillman, J. D., Jr., *The Theory and Design of Circular Antenna Arrays*, University of Tennessee Eng. Exp. Station, 1968.

Tillman, J. D., Patton, W. T., Blakely, C. E., and Schultz, F. V., "The Use of a Ring Array as a Skip Range Antenna," *Proc. IEEE*, Vol. 43, June 1955, pp. 1655–1660.

Tillman, J. D., Hickman, C. E., and Neff, H. P., "The Theory of a Single Ring Circular Array," *Trans. AIEE*, Vol. 80, Pt. 1, 1961, p. 110.

Tomasic, B. and Hessel, A., "Periodic Structure Ray Method for Analysis of Coupling Coefficients in Large Concave Arrays — Part I: Theory,"; "Part II: Application," *Trans. IEEE*, Vol. AP-37, Nov. 1989, pp. 1377–1385, 1386–1397.

Tomasic, B., Hessel, A., and Ahn, H., "Asymptotic Analysis of Mutual Coupling in Concave Circular Cylindrical Arrays — Part I: Far-Zone,"; "Part II: Near-Zone," *Trans. IEEE*, Vol. AP-41, Feb. 1993, pp. 121–136, 137–145.

Villeneuve, A. T., "Conical Phased Array Antenna Investigations," Hughes Aircraft Co., Culver City, Calif., Final Report HAC-FR-79-27-857, Apr. 1979, AD-B039 121.

Vinneneuve, A. T., Behnke, M. C., and Kummer, W. H., "Wide-Angle Scanning of Linear Arrays Located on Cones," *Trans. IEEE*, Vol. AP-22, Jan. 1974, pp. 97–103.

Villeneuve, A. T., Alexopoulos, N. G., and Kummer, W. H., "Hemispherically Scanned Antennas," Hughes Aircraft Co., Culver City, Calif., Final Report, May 1975, AD-A015 766.

Villeneuve, A. T., Behnke, M. C., and Kummer, W. H., "Radiating Elements for Hemispherically Scanned Arrays," *Trans. IEEE*, Vol. AP-30, May 1982, pp. 457–462.

Wait, J. R., *Electromagnetic Radiation from Cylindrical Structures*, Pergamon Press, 1959, pp. 30, 45.

Walsh, J. E., "Radiation Patterns of Arrays on a Reflecting Cylinder," *Proc. IRE*, Vol. 29, Sept. 1951, pp. 1074–1081.

Watson, G. N., "The Diffraction of Electric Waves by the Earth," *Proc. Roy. Soc. Lon.*, Vol. 93, Ser. A, 1918, pp. 83–89.

CHAPTER TWELVE

Measurements and Tolerances

There are excellent books on measurements, listed below. Thus only special topics pertinent to arrays will be discussed herein. These are: measurement of low sidelobe antennas; array diagnostics; and *scan impedance* simulators. Because array performance is affected by random errors, tolerances are also discussed in this chapter.

A. E. Bailey, *Microwave Measurement*, IEE/Peregrinus, 1985.

G. H. Bryant, *Principles of Microwave Measurement*, IEE/Peregrinus, 1988.

G. E. Evans, *Antenna Measurement Techniques*, Artech House, 1990.

J. E. Hansen, *Spherical Near-field Antenna Measurements*, IEE/Peregrinus, 1988.

J. S. Hollis, T. J. Lyon, and L. Clayton, *Microwave Antenna Measurements*, Scientific Atlanta, 1970.

IEEE Standard Test Procedures for Antennas, Std 149, 1979.

12.1 MEASUREMENT OF LOW-SIDELOBE PATTERNS

As mentioned in Chapter 3, the conventional far-field distance of $2D^2/\lambda$ is usually inadequate for accurate measurement of patterns with low sidelobes; antenna diameter is D. Consider patterns such as the Taylor $\bar{n}$. As the observation distance R moves in from infinity, the first sidelobe rises and the null starts filling. Then the sidelobe becomes a shoulder on the now wider main beam, and the second null rises; see Fig. 12.1. This process continues as the distance decreases. To first order the results are dependent only on design sidelobe level. Figure 12.2 shows the distance at which the first sidelobe is raised a given amount, for values of SLR from 13 (uniform) to 60 b dB (Hansen, 1984). This graph also shows distances at which the first sidelobe is shouldered.

Thus the accuracy of measurement may be traded against the measurement distance.

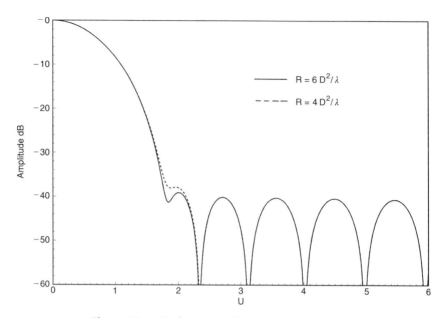

Figure 12.1 Taylor pattern for SLR= 40 dB, $\bar{n} = 11$.

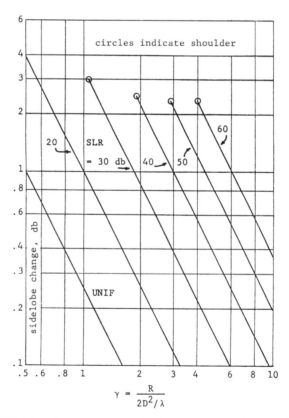

Figure 12.2 Sidelobe change vs. normalized measurement distance for Taylor $\bar{n}$ line source.

Bayliss and other difference patterns possess the same type of extended far-field distances as the low-sidelobe designs of Chapter 3. Again the first sidelobe height and null depth deteriorate as the distance decreases from $2D^2/\lambda$. The progressive raising and shouldering of sidelobes continues as R decreases. Figure 12.3 shows a case where the first sidelobe has already been absorbed by the main beam, and the second sidelobe and null are raised. A universal chart (Hansen, 1992), Fig. 12.4, gives first sidelobe increase, and shoulder value, versus normalized distance, for SLR from 13 (uniform) to 50 dB.

It may be noticed that the difference patterns require a slightly larger measurement distance than the corresponding sum patterns.

12.2 ARRAY DIAGNOSTICS

Element errors in arrays are seldom due to mechanical tolerances, but instead are due to less tangible factors such as mutual coupling affecting both phase and amplitude of excitation, uncertain loss in phasers and feed networks, and phase errors in phasers. The most direct way of determining these is through probing, where a movable probe is coupled closely to one element at a time. An obvious problem is the effect of the element–probe coupling. Modulated small scatterers have been used instead of a probe to ease this difficulty. A review of

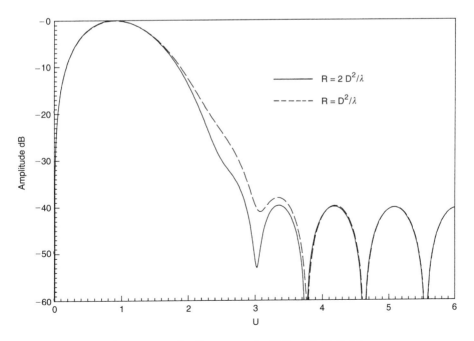

Figure 12.3 Bayliss pattern for SLR= 40 dB, $\bar{n}=11$.

456 MEASUREMENTS AND TOLERANCES

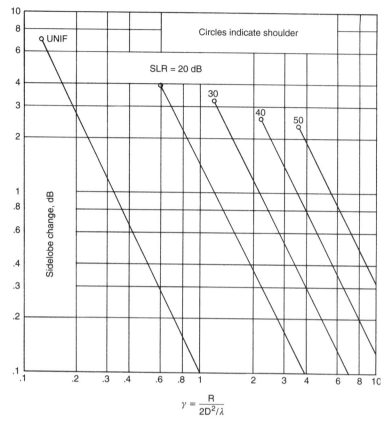

Figure 12.4 Sidelobe change vs. normalized measurement distance for Bayliss line source.

near-field measurements with an extensive bibliography has been given by Dyson (1973). See also Aumann and Willwerth (1987) and Gillespie (1988).

In all of these procedures, the measured near-field is transformed to a complex far-field pattern. This pattern is then back-transformed to the array aperture. When the near-field measurements are over a planar surface, the FFT can be used for both transformations. It is important to include a correction for the probe two-dimensional pattern. To this end the probe field is expressed as a plane wave expansion. The probe effects can then be handled (Paris et al., 1978; Joy et al., 1978). Useful references on near-field measurements as a diagnostic tool include Lee et al. (1988), Garneski (1990), Repjar et al. (1991), MacReynolds et al. (1992), Friedel et al. (1993), and Guerriere et al. 1(1995).

Similar transformations can be applied to cylindrical near-field measurements (Langsford et al., 1989) and to bipolar near-field measurements (Williams et al., 1994; Yaccarino et al., 1994). Probe correction and transformations are more complicated with spherical near-field measurements, but are readily manageable with modern computers (J. E. Hansen, 1988; Rahmat-Samii and Lemanczyk, 1988). Of course, far-field measurements can substitute for the near-field measurements (Ekelman, 1987; Johansson and Svensson, 1990; Guler et al., 1992); however, it is necessary to measure over a plane extending well beyond the array, or to measure over close to a hemisphere. In addition, both phase and amplitude are needed; many far-field ranges or chambers have marginal coverage for this.

Examples of planar near-field diagnostics are shown in Figs. 12.5 and 12.6; these are calculated array face errors for an 8 × 16 element microstrip patch subarray. In Fig. 12.5 element excitation amplitude is shown, and two defective elements are noted. Excitation phase is given in Fig. 12.6, where one element indicates improper amplitude and phase (Repjar et al., 1991). Near-field diagnostics has become a powerful tool for verifying the realization of the desired aperture distribution. Another way of validating, and in this case monitoring, the aperture distribution utilizes a second array located near to, but weakly coupled to, the primary array (Acoraci, 1990). Measured or calculated mutual couplings are then used in an algorithm to determine the primary array distribution. Finally, with a digitally beamformed receiving array, element failure can be partially corrected for one or two signals by calculating the fields at

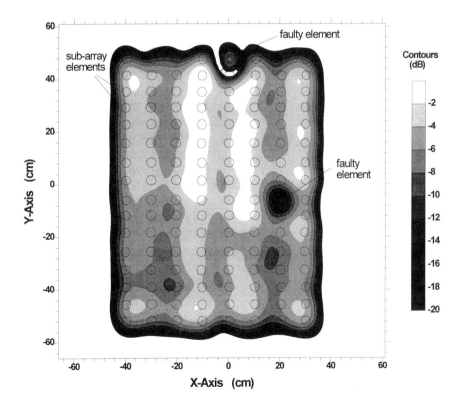

Figure 12.5 Array face amplitude. (Courtesy Repjar, A. G. et al., "Determining Faults on a Flat Phased Array Antenna Using Planar Near-Field Techniques," *Proceedings AMTA Symposium*, Boulder Co., Oct. 1991, pp. 8-11–8-19.)

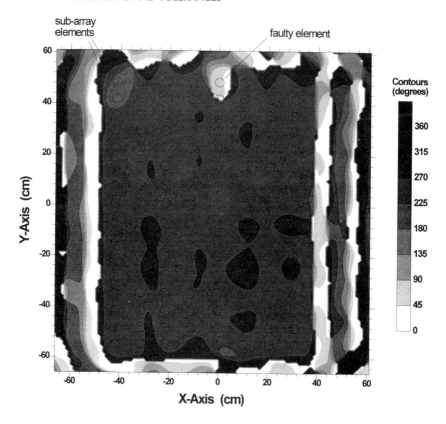

Figure 12.6 Array face phase. (Courtesy Repjar, A. G. et al., "Determining Faults on a Flat Phased Array Antenna Using Planar Near-Field Techniques," *Proceedings AMTA Symposium*, Boulder, Co., Oct. 1991, pp. 8-11–8-19.)

adjacent elements, and then interpolating to supply the missing element input (Mailloux, 1996).

12.3 WAVEGUIDE SIMULATORS

A simple microwave tool that allows the measurement of *scan impedance* in one scan direction and for one frequency is the waveguide simulator, developed by Brown and Carberry (1963) and Hannan, Meier, and Balfour (1963). Although the unit cell virtual walls are, in general, impedance walls, there are some symmetry planes and scan angles that allow perfectly conducting walls. These walls can then be a metallic waveguide, terminated in a section of the array containing a small number of elements. Another way of visualizing the process utilizes the decomposition of a TE_{01} waveguide mode into two plane waves that reflect off the side walls, as sketched in Fig. 12.7. The angle of the

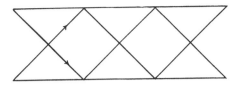

Figure 12.7 Plane waves comprising TE_{10} mode.

plane wave with the guide axis corresponds to the θ_0 scan angle. When neither mode index is zero, an additional pair of plane waves reflect off the broad walls (Collin, 1966). For the less often used triangular waveguide simulator, there are three sets of plane waves involved. Allowable scan planes are sketched in Fig. 12.8. In a rectangular waveguide simulator the dominant modes allow principal plane scans, while diagonal plane scans require higher modes. For the triangular waveguide simulator the lowest TE mode provides scan planes along the symmetry planes. Again, higher modes allow scan planes at other angles. Figure 12.9 shows a square lattice array, with two of the many possible simulators using rectangular guide. Note the use of half and quarter elements. A symmetric hexagonal lattice array is depicted in Fig. 12.10, along with possible simulators. Note that rectangular waveguide simulators can be used for triangular lattice arrays. Also note that simulators may include only partial elements.

A typical simulator consists of a transition section from standard guide to the simulator guide, the simulator waveguide, and the element port. Across the end of the simulator waveguide is placed a thin metallic sheet with the slot

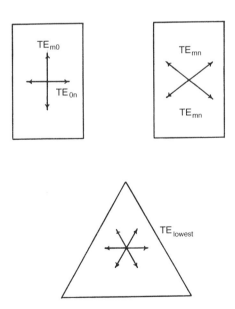

Figure 12.8 Simulator scan planes.

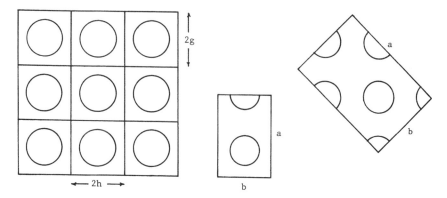

Figure 12.9 Square lattice array with simulators.

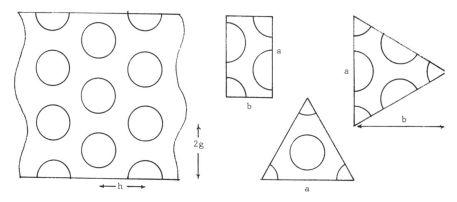

Figure 12.10 Hexagonal lattice array with simulators.

elements appearing as holes. Behind this sheet is a waveguide that includes all the complete elements, and that is terminated in a matched load; this is called the element port. For other elements than slots, or for complex element structures, the element plate is modified accordingly. The simulator waveguide must be sufficiently long to allow decay of evanescent modes. All diagonal plane simulators contain some fractional elements, as the simulator waveguide is always larger than the array unit cell. Triangular waveguide simulators also contain fractional elements, for any scan plane. Since regular arrays have rectangular or triangular lattices, the simulator waveguide is limited to polygonal cross sections with 3 or 4 sides, and internal angles of 30, 45, 60, or 90 deg. Of course most simulators use rectangular or square waveguide because the modes are simple and the connection to standard guide is easy. Because the waveguide mode constituent plane waves reflect in the H-plane, simulation of H-plane scan, either in a principal plane or in a diagonal plane, is direct. Conversely, since there is no situation where there are only two plane wave constituents that reflect in the E-plane, there are no linearly polarized E-plane simulators. Approximate simulation in the E-plane will be discussed below.

From the previous discussion it is clear that each simulator provides limited *scan impedance* data, although frequency variation may provide limited additional data. Thus the choice of the fewest and most economical simulators for a given application is like a chess game, as observed by Wheeler (1972). Consider a rectangular simulator waveguide of dimensions a and b, and call the spacing between symmetry planes in the array g and h; see Fig. 12.9 where the array lattice is square, and the unit cell is $2g \times 2h$. The cutoff wavelength in the simulator waveguide is

$$\lambda_c = \frac{2}{\sqrt{(m/a)^2 + (n/b)^2}} \tag{12.1}$$

and the scan angles are

$$\sin\theta = \frac{\lambda}{\lambda_c}, \quad \tan\phi = \frac{na}{mb}. \tag{12.2}$$

Then

$$\frac{m}{a} = \frac{2}{\lambda}\sin\theta\cos\phi, \quad \frac{n}{b} = \frac{2}{\lambda}\sin\theta\sin\phi. \tag{12.3}$$

Simulator waveguide dimensions are now chosen as integral multiples of symmetry plane spacings:

$$a = pg, \quad b = qh, \tag{12.4}$$

where p and q are integers. Finally,

$$\frac{m}{p} = \frac{2g}{h}\sin\theta\cos\theta, \quad \frac{n}{q} = \frac{2h}{\lambda}\sin\theta\sin\phi. \tag{12.5}$$

Figure 12.11 shows simulators for a square lattice array (Hannan et al., 1963). The H-plane tools use the TE_{10} mode, and are principal plane (C) or diagonal plane (IC). The TM_{11} mode is used in the E-plane tools. All are for an array with element spacing 0.575λ. Calculation of the beam angle is simple: For the near-broadside simulator, $p = 8$ and $1/8 = 0.575 \sin 12.56$. For the principal E-plane simulator, $p = 3$ and $1/3 = 0.575 \sin 35.43$. Finally for the diagonal E-plane simulator $p = 4$ and $1/4 = (0.575/\sqrt{2})\sin 37.94$. Equilateral triangular lattice array simulators are shown in Fig. 12.12 (Balfour, 1965). The TE_{10} mode is used in the 14, 34, and 57 deg H-plane simulators, while the TE_{20} mode is used in the 40 deg H-plane model. The rectangular E-plane simulators use the TM_{11} mode; the triangular E-plane simulators use the lowest TM mode (see Schelkunoff, 1943). Scan angle calculation is again simple. Element spacing is 0.6λ, and for the near-broadside tool, $p = 4$ and $1/4 = 0.6\sqrt{3}\sin 13.92$. The adjacent tool uses $p = 3$ and $1/3 = 0.6\sqrt{3}\sin 18.71$. For the TE_{20} mode

462 MEASUREMENTS AND TOLERANCES

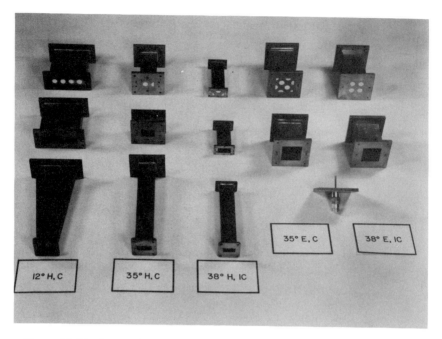

Figure 12.11 Square lattice waveguide simulators. (Courtesy Hazeltine Corp.)

Figure 12.12 Triangular lattice waveguide simulators. (Courtesy Hazeltine Corp.)

simulator, $p = 3$, and $2/3 = 0.6\sqrt{3}\sin 39.90$. It may be noted in the photographs that some of the partial elements are not loaded.

To use the simulator the desired mode(s) is excited in the simulator waveguide, with the element port(s) match loaded. Partial elements need not be connected to an element port, or loaded, if they do not couple to the waveguide mode. However, all elements must include all structure that affects stored energy, such as dielectrics, iris, etc. The reflection coefficient is now measured in the simulator waveguide; this gives the impedance seen by the incident plane wave. For multielement simulators a transformation is necessary to get *scan impedance* from the wave impedance. This is based on the junction of two waveguides and involves line lengths and a transformer (Hannan and Balfour, 1965). Later, one-port simulators that give *scan impedance* directly will be discussed. Surface waves have been observed in simulators of dielectric loaded elements (Hannan, 1967).

Simulation of a broadside beam is not possible; small angles require large simulators. Also E polarization produced by a linear element cannot be simulated exactly. The E-plane simulators previously shown use circularly polarized or cross-linear-polarized elements. E-plane simulation in a diagonal plane can utilize a pair of modes, TE and TM, with identical nonzero indices. A detailed review of waveguide simulator design and practice has been given by Wheeler (1972). See also Hannan and Balfour (1965) and Gregorwich et al. (1971). Byron and Frank (1968) show excellent agreement with direct measurements.

"One-port" simulators have only one whole element port, along with partial ports. Thus only the single port needs to be terminated. These simple simulators are useful for H-plane, wide-angle, principal plane work, where no grating lobes appear (Balfour, 1967). Figure 12.13 shows three simulators for an array with a square lattice. All use the TE_{01} mode. Figure 12.13a uses $p = 4$, $q = 2$, with a principal plane beam at 30 deg. Figure 12.13b uses $p = 3$, $q = 2$ with a principal plane beam at 41.8 deg. Figure 12.13c uses $p = q = 2$ at 45 deg. Element spacing is $\lambda/2$. Equilateral triangular (symmetric hexagonal) array simulators are shown in Fig. 12.14. Element spacing to the nearest neighbor is $1/\sqrt{3}$. Figure 12.14b is for an H-plane principal plane beam at 60 deg; $p = q = 2$. In Fig. 12.14a is shown a diagonal plane simulator, again with

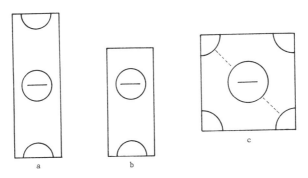

Figure 12.13 Square lattice single port simulators.

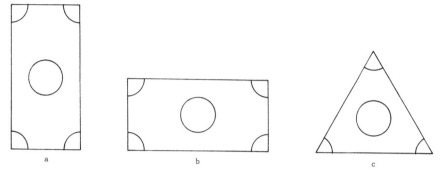

Figure 12.14 Hexagonal lattice single port simulators.

$p = q = 2$, and with beam at 30 deg. A triangular simulator is shown in Fig. 12.14c, where the index is 3, and a diagonal plane beam is at 41.81 deg.

Microstrip patch arrays have also been simulated. Solbach (1986) used two H-plane waveguide simulators, corresponding to Fig. 12.13a and b. Here it was important to carefully reproduce the patch feeds and load resistances. These simulators gave angles of 11.92 and 43.53 deg. This work was continued by Aberle and Pozar (1990) and Aberle et al. (1994). Arrays of probe fed patches, and patches with parasitic patches were simulated using a single centered patch with four corner quarter patches, with TE_{10}; see Fig. 12.14. The array lattice was symmetrical hexagonal, and the element spacing gave a beam angle of 22.48 deg.

A more general-purpose simulator was developed by Gustincic (1972) and Derneryd and Gustincic (1979), which allows *scan admittance* at various scan angles to be determined. Grating lobes can also be encompassed. The simulator uses many elements, with all elements and partial elements but two terminated. Transmission measurements are then made between these two elements. In turn, the two active elements occupy all positions in the simulator, which allows *scan admittance* data to be calculated for N scan angles if there are N simulator elements.

A waveguide simulator can also be modeled in a computer, where the waveguide walls and elements are represented by moment method patches, or by grids as used by finite element methods and FDTD procedures.

12.4 ARRAY TOLERANCES

Errors in arrays include those in element position and orientation, and element excitation phase and amplitude. All errors are of two types: systematic and random. Many errors, of course, are combinations of the two. Systematic errors (correlated errors and bias errors) usually require an array calculation including the specific errors; there are no general rules for these effects because there are many types and combinations of these errors. Random errors, however, are subject to statistical manipulation, and this allows general, and often simple, results to be obtained.

In well-designed and well-manufactured arrays element position and orientation errors are small, and can be replaced by equivalent phase (and, if needed, amplitude) errors. Thus this section is concerned with the effects of phase and amplitude errors only. Quantization of phase, and the associated phenomena, are treated in detail in Chapter 2. Array optimization subject to constraints on tolerances is discussed in Chapter 4. Array errors are observed in three areas: directivity (gain) reduction; raising of average sidelobe level and of some individual sidelobes; and main beam pointing and perhaps shape change. Each of these is analyzed separately. All random errors will be assumed to be normally distributed (Gaussian) with zero mean and variance of σ^2. Arrays will be assumed sufficiently large that the central limit theorem applies. This means for a linear array $N \geq 10$, and desirably $N \geq 20$. Linear arrays are analyzed; extensions to planar arrays are shown later.

12.4.1 Directivity Reduction and Average Sidelobe Level

Consider a linear array of isotropic elements, with space factor

$$f_0(u) = \sum A_n \exp jn\pi u. \qquad (12.6)$$

With errors, the coefficient becomes

$$A_n(1 + \delta_a) \exp j\delta_\phi \qquad (12.7)$$

where the amplitude error is δ_a and the phase error is δ_ϕ in radians. The array power pattern becomes

$$f^2 = \sum\sum A_n A_m^*(1 + \delta_{an})(1 + \delta_{am}) \exp\{j[\delta_{\phi n} - \delta_{\phi m} + (n-m)\pi u]\}. \qquad (12.8)$$

Under the assumptions above, the mean of the sum of independent variables equals the sum of the means, giving the average pattern

$$f^2 = f_0^2 + \frac{\sigma^2}{G_0}, \qquad (12.9)$$

where G_0 is the error-free directivity, $\sigma^2 = \sigma_a^2 + \sigma_\phi^2$ where σ_a is the amplitude standard deviation, and σ_ϕ is the phase standard deviation, in radians. Directivity, normalized to G_0 is

$$\frac{G}{G_0} \simeq \exp(-\sigma^2) \simeq \frac{1}{1+\sigma^2}. \qquad (12.10)$$

For example, 1σ errors of 10 deg phase and 1 dB amplitude reduce directivity by 0.193 dB. Ruze (1952, 1966) developed this theory, which also shows that the pattern is raised slightly, although the increase is apparent only in the sidelobe region. This increase is less for higher directivity arrays. Clearly the sidelobe increase affects low-sidelobe designs more than uniform amplitude −13 dB sidelobes. From the formula above, the mean sidelobe level (SLL) is related to the design sidelobe level (SLL$_0$) by

$$\text{SLL} = \text{SLL}_0\sqrt{1 + \sigma^2/(G_0 \cdot \text{SLL}_0^2)}. \tag{12.11}$$

This is shown in Fig. 12.15 versus the universal parameter $\sigma/\sqrt{G_0}$. Note that the sidelobe levels are in voltage in this expression. Clearly larger (higher directivity) arrays are less sensitive. An approximate result for average sidelobe level with respect to isotropic is based on a filled aperture of area A, where directivity is $4\pi A/\lambda^2$ (Mailloux, 1994):

$$\text{SLL} \simeq \pi(\sigma_a^2 + \sigma_\phi^2). \tag{12.12}$$

Here SLL is in power. Because the variances of phase and amplitude add, the plot of SLL versus σ_a and σ_ϕ is a circle, as seen in Fig. 12.16. From this plot it is easy to find combinations of errors that produce a specified average sidelobe level.

12.4.2 BEAM POINTING ERROR

Assuming that element position and orientation errors are subsumed into phase and amplitude errors, the beam pointing variance for linear array becomes (Rondinelli, 1959; Carver et al., 1973; Steinberg, 1976)

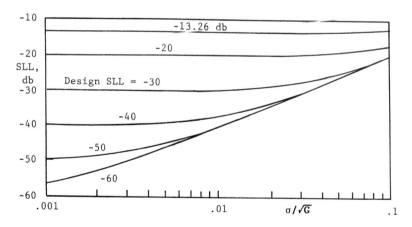

Figure 12.15 Mean sidelobe level vs. universal factor.

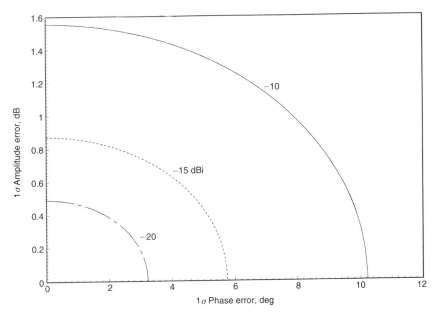

Figure 12.16 Average sidelobe level due to random errors.

$$\sigma_\theta^2 = \frac{\sigma^2 \sum\limits_{-N/2}^{N/2} A_n^2 k^2 d^2 n^2}{\left[\sum\limits_{-N/2}^{N/2} A_n k^2 d^2 n^2\right]^2}. \tag{12.13}$$

This is for a symmetric equally spaced array at broadside. An array with uniform excitation has $\sigma_\theta^2 = 3\sigma^2/(\pi^2 N^3 d^2/\lambda^2)$; larger arrays have smaller beam pointing error, but the beamwidth is correspondingly smaller. Since the uniform linear array beamwidth is $0.886\lambda/Nd$ with d the element spacing, the standard deviation of the beam pointing error normalized to the 3 dB beamwidth is

$$\frac{\sigma_\theta}{\theta_3} = \frac{\sqrt{3}\sigma}{0.886\pi\sqrt{N}} = \frac{0.622\sigma}{\sqrt{N}}, \tag{12.14}$$

where σ^2 is the variance of phase and amplitude errors. Typically the beam pointing errors are small, and as a fraction of beamwidth they decrease slowly with number of elements, for a uniform linear array. For a uniform square planar array, of N elements per side, the standard deviation of the beam pointing error, in terms of the 3 dB beamwidth is

$$\frac{\sigma_\theta}{\theta_3} = \frac{\sqrt{3}\sigma}{0.886\pi N} = \frac{0.662\sigma}{N}. \tag{12.15}$$

12.4.3 Peak Sidelobes

The probability that peak sidelobes will exceed a specified SLL is found from the probability density function (PDF); this is a Ricean distribution, as the field at any angle is the vector sum of the error-free field E_0 and the error components (Bennett, 1956; Skolnik, 1969). This PDF is (Rondinelli, 1959; Hsaio, 1984)

$$\text{PDF} = \frac{\text{SLL} \cdot \text{SLL}_0}{\Delta^2} \exp\left(-\frac{\text{SLL}^2 + \text{SLL}_0^2}{2\Delta^2}\right) I_0\left(\frac{\text{SLL} \cdot \text{SLL}_0}{\sigma^2}\right), \quad (12.16)$$

where the variance $\sigma^2 = \Delta^2/2\,G_0$ and I_0 is a modified Bessel Function. Now the probability of individual sidelobes equal to or less than a specified sidelobe level is given by the integral of the PDF from zero to the specified SLL:

$$P(\text{SLL} \leq \text{SLL}_0) = \int_0^{\text{SLL}} \text{PDF}\, d\text{SLL}. \quad (12.17)$$

Through a change of variable this can be written as a Marcum Q function (Marcum, 1960):

$$P = 1 - Q(A \cdot B). \quad (12.18)$$

Here the parameters are $A = \text{SLL}_0/\Delta$ and $B = \text{SLL}/\Delta$. The Q function definition is

$$Q(A, B) = \int_B^\infty t \exp\left[\tfrac{1}{2}(A^2 + t^2)\right] I_0(At)\, dt. \quad (12.19)$$

When the relative errors are small, the Ricean distribution reduces to a Rayleigh distribution; this allows the cumulative probability to be simply integrated. However, when the sidelobe region of interest is low, or when the errors are not small, the more accurate results above should be used.

An iterative algorithm for Q function calculation of Brennan and Reed (1965) has been improved by McGee (1970). With double-precision arithmetic this routine is suitable for all but small Q values and large A, B values. Asymptotic formulas (Helstrom, 1968) have proved disappointing.

Figure 12.17 shows the probability that sidelobes will be below a given increase over the design sidelobe level SLL_0, as a function of the increase in dB, for values of Δ/SLL_0 of 0.1, 0.5, and 1. Since probability is more likely to be specified, in Fig. 12.18, curves of constant probability are plotted against the

ARRAY TOLERANCES 469

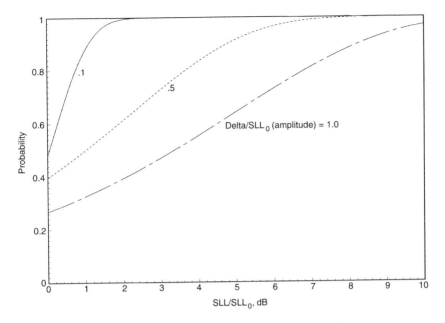

Figure 12.17 Probability of increase in sidelobe level.

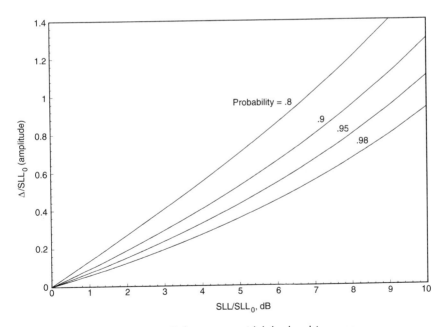

Figure 12.18 Tolerance vs. sidelobe level increase.

sidelobe increase, versus Δ/SLL_0. To give an example, let the probability be 0.9 that all sidelobes will be no larger than 3 dB above SLL_0. Then the Q function solutions are $A = 3.40731$ and $B = 4.81295$, giving $B/A = 3.0\,\mathrm{dB}$. From this, $\Delta/\mathrm{SLL}_0 = 1/A = 0.2935$, a value that can be obtained directly from Fig. 12.18. Further assume an array gain of 20 dB and a design sidelobe level of $-30\,\mathrm{dB}$. Then $\Delta = 0.009281$, and since $\sigma^2 = \sigma_a^2 + \sigma_\phi^2 = \Delta^2/2G_0$, the total error variance $\sigma^2 = 0.01723$. If now the amplitude and phase variances are equal, $\sigma_a = 0.0928$ and $\sigma_\phi = 5.32\,\mathrm{deg}$. Thus error standard deviations of 9% in amplitude and 5 deg in phase should have a 90% probability of raising the sidelobe level 3 dB, for this example. In general, as mentioned above, longer arrays are more tolerant, as are higher design sidelobes. Figure 12.18 allows a rapid determination of acceptable tolerances.

ACKNOWLEDGMENT

Photographs courtesy of Peter Hannan.

REFERENCES

Aberle, J. T. and Pozar, D. M., "Analysis of Infinite Arrays of One- and Two-Probe-Fed Circular Patches," *Trans. IEEE*, Vol. AP-38, Apr. 1990, pp. 421–432.

Aberle, J. T., Pozar, D. M., and Manges, J., "Phased Arrays of Probe-Fed Stacked Microstrip Patches," *Trans. IEEE*, Vol. AP-42, July 1994, pp. 920–927.

Acoraci, J. H., "On-Line Bite to Accurately Monitor Beam Position, Beam Shape and System Performance of Electronically Scanned Phased Array Antennas," *Proceedings AMTA Symposium*, Philadelphia, Pa., 1990, pp. 9-25–9-28.

Aumann, H. M. and Willwerth, F. G., "Near-Field Testing of a Low-Sidelobe Phased Array Antenna," *Proceedings AMTA Symposium*, Seattle, Wash., 1987, pp. 3–7.

Balfour, M. A., "Phased Array Simulators in Waveguide for a Triangular Arrangement of Elements," *Trans. IEEE*, Vol. AP-13, May 1965, pp. 475–476.

Balfour, M. A., "Active Impedance of a Phased-Array Antenna Element Simulated by a Single Element in a Waveguide," *Trans. IEEE*, Vol. AP-15, Mar. 1967, pp. 313–314.

Bennett, W. R., "Methods of Solving Noise Problems," *Proc. IRE*, Vol. 44, May 1956, pp. 609–638.

Brennan, L. E. and Reed, I. S., "A Recursive Method of Computing the Q Function," *Trans. IEEE*, Vol. IT-11, Apr. 1965, pp. 312–313.

Brown, C. R. and Carberry, T. F., "A Technique to Simulate the Self and Mutual Impedance of an Array," *Trans. IEEE*, Vol. AP-11, May 1963, pp. 377–378.

Byron, E. V. and Frank, J., "On the Correlation between Wide Band Arrays and Array Simulators," *Trans. IEEE*, Vol. AP-16, Sept. 1968, pp. 601–603.

Carver, K. R., Cooper, W. K., and Stutzman, W. L., "Beam-Pointing Errors of Planar-Phased Arrays," *Trans. IEEE*, Vol. AP-21, Mar. 1973, pp. 199–202.

Collin, R. E., *Foundations for Microwave Engineering*, McGraw-Hill, 1966.

Dernryd, A. G. and Gustincic, J. J., "The Interpolation of General Active Array Impedance from Multielement Simulators," *Trans. IEEE*, Vol. AP-27, Jan. 1979, pp. 68–71.

Dyson, J. D., "Measurement of Near Fields of Antennas and Scatterers," *Trans. IEEE*, Vol. AP-21, July 1973, pp. 446–460.

Ekelman, E. P., "Antenna Diagnosis Using Microwave Holographic Techniques on a Far-Field Range," *Proceedings AMTA Symposium*, Seattle, Wash., 1987, pp. 180–186.

Friedel, J. E., Keyser, R. B., Johnson, R. E., "Interpretation of Near-Field Data for a Phased Array Antenna," *Proceedings AMTA Symposium*, Dallas, Tex., Oct. 1993, pp. 42–47.

Garneski, D., "A New Implementation of the Planar Near-Field Back Projection Technique for Phased Array Testing and Aperture Imaging," *Proceedings AMTA Symposium*, Philadelphia, Pa., 1990, pp. 9-9–9-14.

Gillespie, E. S., "Special Issue on Near-Field Scanning Techniques," *Trans. IEEE*, Vol. AP-36, June 1988.

Gregorwich, W. S., Hessel, A., and Knittel, G. H., "A Waveguide Simulator Study of a Blindness Effect in a Phased Array," *Microwave J.*, Vol. 14, Sept., 1971, pp. 37–41.

Guerrieri, J. et al., "Planar Near-Field Measurements and Microwave Holography for Measuring Aperture Distribution on a 60 GHz Active Array Antenna," *Proceedings AMTA Symposium*, Wiliamsburg, Va., Nov. 1995, pp. 295–299.

Guler, M. G. et al., "Far-Field Spherical Microwave Holography," *Proceedings AMTA Symposium*, Columbus, Oh., Oct. 1992, pp. 8-3–8-7.

Gustincic, J. J., "The Determination of Active Array Impedance with Multielement Waveguide Simulators," *Trans. IEEE*, Vol. AP-20, Sept. 1972, pp. 589–595.

Hannan, P. W., "Discovery of an Array Surface Wave in a Simulator," *Trans. IEEE*, Vol. AP-15, July 1967, pp. 574–576.

Hannan, P. W. and Balfour, M. A., "Simulation of a Phased-Array Antenna in Waveguide," *Trans. IEEE*, Vol. AP-13, May 1965, pp. 342–353.

Hannan, P. W., Meier, P. J., and Balfour, M. A., "Simulation of Phased Array Antenna Impedance in Waveguide," *Trans. IEEE*, Vol. AP-11, Nov. 1963, pp. 715–716.

Hansen, J. E., Ed., *Spherical Near-Field Antenna Measurements*, IEE/Peregrinus, 1988.

Hansen, R. C., "Measurement Distance Effects on Low Sidelobe Patterns," *Trans. IEEE*, Vol. AP-32, June 1984, pp. 591–594.

Hansen, R. C., "Measurement Distance Effects on Bayliss Difference Patterns," *Trans. IEEE*, Vol. AP-40, Oct. 1992, pp. 1211–1214.

Helstrom, C. W., *Statistical Theory of Signal Detection*, Pergamon, 1968, Appendix F.

Hsiao, J. K., "Normalized Relationship among Errors and Sidelobe Levels," *Radio Sci.*, Vol. 19, Jan.-Feb. 1984, pp. 292–302.

Johansson, M. and Svensson, B., "Array Antenna Diagnosis and Calibration," *Proceedings AMTA Symposium*, Philadelphia, Pa., 1990, pp. 10-27–10-32.

Joy, E. B., Leach, W. M., and Paris, D. T., "Applications of Probe-Compensated Near-Field Measurements," *Trans. IEEE*, Vol. AP-26, May 1978, pp. 379–389.

Langsford, P. A., Hayes, M. J. C., and Henderson, R. I., "Holographic Diagnostics of a Phased Array Antenna from Near-Field Measurements," *Proceedings AMTA Symposium*, Monterey, Calif., Oct. 1989, pp. 10-32–10-36.

Lee, J. J. et al., "Near-Field Probe Used as a Diagnostic Tool to Locate Defective Elements in an Array Antenna," *Trans. IEEE*, Vol. AP-36, June 1988, pp. 884–889.

MacReynolds, K. et al., "Phased-Array Testing and Diagnostics Using Planar Near-Field Scanning," *Proceedings AMTA Symposium*, Columbus, Oh., Oct. 1992, pp. 8-8–8-13.

Mailloux, R. J., *Phased Array Antenna Handbook*, Artech House, 1994.

Mailloux, R. J., "Array Failure Correction with a Digitally Beamformed Array," *Trans. IEEE*, Vol. AP-44, Dec. 1996, pp. 1543–1550.

Marcum, J. L., "Studies of Target Detection by a Pulsed Radar—Mathematical Appendix," *Trans. IEEE*, Vol. IT-6, 1960, pp. 159–160, 227–228.

McGee, W. F., "Another Recursive Method of Computing the Q Function," *Trans. IEEE*, Vol. IT-16, July 1970, pp. 500–501.

Paris, D. T., Leach, W. M., and Joy, E. B., "Basic Theory of Probe-Compensated Near-Field Measurements," *Trans. IEEE*, Vol. AP-26, May 1978, pp. 373–379.

Rahmat-Samii, Y. and Lemanczyk, J., "Application of Spherical Near-Field Measurements to Microwave Holographic Diagnosis of Antennas," *Trans. IEEE*, Vol. AP-36, June 1988, pp. 869–878.

Repjar, A. G. et al., "Determining Faults on a Flat Phased Array Antenna Using Planar Near-Field Techniques," *Proceedings AMTA Symposium*, Boulder, Co., Oct. 1991, pp. 8-11–8-19.

Rondinelli, L. A., "Effects of Random Errors on the Performance of Antenna Arrays of Many Elements," *IRE National Convention Record.*, Pt. 1, 1959, pp. 174–189.

Ruze, J., "The Effect of Aperture Errors on the Antenna Radiation Pattern," *Nuovo Cimento*, Vol. 9 Suppl., 1952, pp. 364–380.

Ruze, J., "Antenna Tolerance Theory—A Review," *Proc. IEEE*, Vol. 54, Apr. 1966, pp. 633–640.

Schelkunoff, S. A., *Electromagnetic Waves*, Van Nostrand, 1943.

Skolnik, M. I., "Nonuniform Arrays," in *Antenna Theory*, Part 1, R. E. Collin and F. J. Zucker, Eds., McGraw-Hill, 1969, Chapter 6.

Solbach, K., "Phased Array Simulation Using Circular Patch Radiators," *Trans. IEEE*, Vol. AP-34, Aug. 1986, pp. 1053–1058.

Steinberg, B. D., *Principles of Aperture and Array System Design*, Wiley-Interscience, 1976, Chapter 13.

Wheeler, H. A., "A Survey of the Simulator Technique for Designing a Radiating Element," in *Phased Array Antennas*, A. Oliner and G. H. Knittel, Eds., Proc. 1970 Phased Array Antenna Symp., Artech House, 1972, pp. 132–148.

Williams, L. I., Rahmat-Samii, Y., and Yaccarino, R. G., "The Bi-Polar Planar Near-Field Measurement Technique, Part I: Implementation and Measurement Comparisons," *Trans. IEEE*, Vol. AP-42, Feb. 1994, pp. 184–195.

Yaccarino, R. G., Rahmat-Samii, Y., and Williams, L. I., "The Bi-Polar Planar Near-Field Measurement Technique Part II: Near-Field to Far-Field Transformation and Holographic Imaging Methods," *Trans. IEEE*, Vol. AP-42, Feb. 1994, pp. 196–204.

Author Index

Aberle, J. T., 249, 265, 464, 470
Abramowitz, M., 101, 102, 226, 265
Acoraci, J. H., 457, 470
Agrawal, A. K., 190, 212
Agrawal, V. D., 96, 98, 102
Ahn, H., 420, 452
Ajioka, J. S., 184, 185, 186, 211, 358, 380
Alexopoulos, N. G., 153, 162, 443, 452
Alford, A., 146, 158
Ali, S. M., 156, 162
Allen, J. L., 29, 45, 221, 265, 369, 372, 381
Altman, J. L., 166, 211, 265
Amitay, N., 4, 219, 235, 255, 265, 272, 301
Andersen, J. B., 223, 260, 265, 373, 381
Andreasen, 97, 102
Antonucci, J., 395, 446
Archer, D., 337, 341, 355, 381
Ares, F., 90, 96, 102, 103
Auman, H. M., 456, 470

Bach, H., 309, 310, 311, 327
Bahl, I. J., 150, 158
Bailey, A. E., 453
Bailey, M. C., 229, 265, 273, 301
Bailin, L. L., 182, 211, 443, 445, 446
Baklanov, Y. V., 123, 125
Balanis, C. A., 132, 158, 244, 266

Balfour, M. A., 458, 461, 463, 470, 471
Balzano, Q., 413, 420, 446, 447
Bamford, L. D., 223, 266
Bandler, J. W., 92, 103, 123, 125
Baracco, J. M., 193, 211
Barber, E. E., 358, 383
Barbiere, D., 53, 103
Bargeliotes, P. C., 413, 440, 443, 446
Barkeshli, S., 266, 269
Barlatey, L., 156, 158
Bateman, D. G., 229, 266
Bates, R. H. T., 255, 266
Bayard, J-P R., 250, 251, 252, 266
Bayliss, E. T., 78, 103, 120, 125
Bedrosian, S. D., 102, 103
Begovich, N. A, 182, 211, 228, 266
Behnke, M. C., 271, 443, 446, 449, 452
Bennett, W. R., 468, 470
Bertoni, H. L., 162, 205, 214
Bhartia, P., 150, 158
Bhat, B., 188, 213
Bickmore, R. W., 63, 103
Bird T. S., 229, 266, 273, 301, 420, 446
Biswell, D. E., 389, 446
Blackman, R. B., 71, 103
Blakely, C. E., 384, 451
Blank, S. J., 384, 330, 381
Blass, J., 335, 381, 385, 446
Bloch, A., 39, 45, 304, 328
Bock, E. L., 133, 158

473

Bodeep, G. E., 205, 214
Bogner, B. F., 394, 446
Bolljahn, J. T., 133, 158
Booker, H. G., 227, 266
Borgiotti, G. V., 233, 235, 266. 420, 446, 447
Bostian, C. W., 273, 301
Bouwkamp, C. J., 304, 328
Bowling, D. R., 327, 328
Bowman, J. J., 420, 424, 447
Boyns, J. E., 424, 426, 447
Brachat, P., 193, 211
Breithaupt, R. W., 145, 158
Bremmer, H., 424, 447
Brennan, L. E., 468, 470
Brown, C. R., 458, 470
Brown, G. H., 1, 131, 158
Brown, J., 341, 381
Brown, J. L., 58, 103, 318, 328
Brown, L. B., 54, 67, 103
Brown, R. C., 330, 381
Brown, R. M., 341, 382
Bruce, E., 1
Brunner, G., 40, 46, 169, 212, 225, 268, 378, 381
Bryant, G. H., 453
Burrell, G. A., 246, 269
Butler, C. M., 133, 136, 161
Butler, J. K., 325, 329
Butler, J. L., 331, 333, 335, 381
Byron, E. V., 255, 266, 463, 470

Carberry, T. F., 458, 470
Carin, L., 162, 300, 301
Caron, P. R., 372, 382
Carslaw, H. S., 299, 301
Carson, C. T., 139, 160
Carter, P. S., 1, 224, 266, 385, 397, 447
Carver, K. R., 155, 158, 228, 268, 466, 470
Cátedra, M. F., 235, 266, 273, 301
Cha, A. G., 273, 301
Chadwick, G. G., 338, 381
Chan, A. K., 445, 447
Chan, K. K., 443, 447
Chang, Y., 207, 211

Chang, D. C., 133, 161
Charalambous, C., 92, 103, 123, 125
Chattoraj, T. K., 281, 303
Chen, W., 150, 160
Cheng, D. K., 123, 126, 182, 214, 324, 328
Cheston, T. C., 180, 211
Chew, W. C., 229, 266
Chignell, R. J., 144, 158
Christopher, E. J., 394, 447
Chu, R.-S., 250, 267
Clarricoats, P. J. B., 159, 273, 301, 330, 381
Clavin, A., 262, 264, 267
Clayton, L., 453
Cohen, E. D., 190, 211
Collin, R. E., 263, 265, 267, 397, 447, 459, 470, 472
Compton, R. T., 5, 397, 447
Cook, J. S., 210, 211
Cooley, M. E., 150, 158, 251, 252, 266, 267
Cooper, W. K., 466, 470
Corey, L., 426, 447
Cornbleet, S., 356, 381
Croney, J., 184, 211
Cummings, W. C., 336, 381, 426, 447

Dahele, J. S., 193, 212
Damiano, J.-P., 443, 447
Darcie, T. E., 205, 214
Davidson, D. B., 142, 160
Davies, D. E. N., 330, 382, 385, 388, 393, 394, 447, 449, 450
Davies, D. K., 202, 203, 212
de Bruijn, N. G., 304, 328
Debski, T. R., 219, 267
Dell-Imagine, R. A., 424, 449
Denison, D. R., 276, 301
Derneryd, A. G., 464, 471
Deschamps, G. A., 150, 158
Deshpande, M. D., 273, 301
Detrick, D. L., 337, 339, 381
Diamond, B. L., 231, 235, 255, 267, 281, 282, 284, 285, 301
Dinger, R. J., 327, 328
Dion, A. R., 363, 381

Dolfi, D., 207, 211
Dolph, C. L., 51, 103, 304, 328
Donn, C., 443, 451
Dorne. A., 133, 158
Dowling, T. B., 413, 446
Drane, C. J., 58, 103, 318, 328
Dudley, D. G., 229, 268
DuFort, E. C., 83, 104, 258, 267, 363, 376, 381, 449
DuHamel, R. H., 58, 103, 305, 318, 323, 328, 385, 386, 447
Duncan, R. H., 415, 447
Dunne, C. L., 372, 382
Dyson, J. D., 456, 471

Edelberg, S., 235, 257, 259, 267
Eichmann, G., 397, 449
Ekelman, E. P., 457, 471
Elachi, C., 193, 211
Eleftheriades, G. V., 148, 158
Elliott, R. S., 20, 45, 88, 89, 93, 94, 95, 96, 102, 103, 104, 123, 124, 126, 139, 142, 145, 158, 161, 169, 179, 211, 375, 381
Erlinger, J. J., 106, 125
Esman, R. D., 206, 211
Evans, G. E., 453
Evans, S., 147, 158

Fano, R. M., 156, 158
Farrell, G. F., Jr., 256, 267
Felsen, L. B., 300, 301, 424, 447, 451
Fenby, R. G., 385, 394, 447
Ferris, J. E., 146, 147, 162
Fikioris, J. G., 153, 162
Fitzsimmons, G. W., 197, 211
Fletcher, R., 91, 92, 103, 122, 123, 125
Fock, V. A., 424, 447
Fong, A. K. S., 355, 383
Forbes, G. R., 131, 160
Forooraghi, K., 258, 267
Foster, D., 184, 211
Fourikis, N., 5, 147, 159
Franchi, P., 395, 446
Frank, J., 181, 211, 255, 266, 463, 470
Frankel, M. Y., 206, 211
Franz, K., 304, 328

Fray, A., 223, 266
Friedel, J. E., 456, 471
Friedman, B., 324, 328
Friis, H. T., 78, 104, 224, 270
Fry, D. W., 166, 212
Fukao, S., 273, 301
Fukuoka, Y., 157, 163
Fulton, R. D., 265, 269

Gabriel, W. F., 426, 447
Galindo, V., 4, 219, 255, 262, 265, 267, 272, 273, 301
Gallegro, A. D., 286, 287, 301
Gammon, D., 288, 294, 299, 302
Gardiol, F. E., 150, 159, 163
Garneski, D., 456, 471
Gately, A. C. Jr., 222, 267
Gazit, E., 147, 159
Gee, W., 338, 381
Gent, H., 341, 381
Gerlin, A., 385, 448
Gething, P. J. D., 393, 448
Gibson, P. J., 147, 159
Gilbert, E. N., 323, 328
Gill, P. E., 92, 104, 123, 125
Gillespie, E. S., 456, 471
Gladman, B. R., 387, 450
Gobert, J. F., 385, 448
Goldberg, J. J., 231, 267
Golden, K. E., 413, 448
Goutzoulis, A. P., 202, 203, 212
Goward, F. K., 166, 212
Granholm, J., 192, 212
Gregorwich, W. S., 255, 267, 394, 448, 463, 471
Griffel, G., 205, 214
Griffin, D. W., 229, 269
Guerriere, J., 456, 471
Guler, M. G., 457, 471
Guo, Y. C., 33, 46, 150, 159
Gustafson, L. A., 184, 212
Gustincic, J. J., 464, 471

Hadge, E., 189, 212
Hall, C. M., 192, 193, 212
Hall, P. S., 150, 159, 192, 193, 212, 213, 223, 266, 267

Hamid, M. A. K., 229, 267
Hammerstad, E., 153, 154, 159
Hamming, R. W., 299, 301
Haneishi, 228, 267
Hannan, P. W., 210, 212, 219, 222, 255, 258, 267, 268, 426, 448, 458, 461, 463, 471
Hansell, C. W., 1
Hansen, J. E., 453, 456, 471
Hansen, R. C., 3, 4, 5, 6, 37, 40, 43, 45, 46, 67, 76, 104, 107, 112, 116, 125, 128, 132, 152, 159, 161, 169, 189, 211, 212, 213, 225, 226, 262, 268, 270, 278, 288, 294, 296, 299, 300, 301, 302, 305, 309, 310, 312, 313, 326, 328, 345, 378, 381, 386, 389, 395, 404, 412, 414, 417, 426, 430, 443, 448, 450, 453, 455, 471
Hansen, W. W., 304, 308, 328
Harmuth, H. F., 373, 381
Harrington, R. F., 97, 104, 229, 269, 273, 302, 385, 397, 448, 451
Hayden, E. C., 393, 448
Hayes, M. J. C., 456, 471
Helstrom. C. W., 468, 471
Henderson, R. I., 456, 471
Herczfeld, P. R., 204, 212
Herd, J. S., 260, 271
Herper, J. C., 420, 421, 448
Hersey, H. S., 123, 125
Hessel, A., 246, 247, 248, 249, 255, 256, 257, 268, 269, 420, 421, 422, 423, 445, 448, 451, 452, 471
Heyman, E., 156, 160
Hickman, C. E., 385, 393, 448, 451
Hilburn, 385, 448
Hill, D. R., 144, 159
Hockensmith, R. P., 445, 448
Hockham, G. A., 184, 212
Hoffman, M., 445, 448
Holley, A. E., 424, 449
Hollis, J. S., 453
Holzman, E. L., 190, 212
Howard, J. E., 439, 443, 449
Hsiao, J. K., 273, 301, 468, 471
Huang, J., 193, 212
Hudson, J. E., 5

Huebner, D. A., 264, 267
Hunt, C. J., 150, 159

Idselis, M. H., 91, 104, 122, 126
Imbriale, W. A., 330, 381
Ince, W. J., 188, 212
Irzinski, E. P., 394, 449
Ishimaru, A., 97, 104, 273, 302, 445, 447
Itoh, T., 157, 163
Ittipiboon, A., 157, 159
Iwasaki, H., 193, 212

Jaggard, D. L., 100, 104
James, J. R., 150, 159, 193, 212, 267
James, P. W., 385, 449
Jan, C.-G., 144, 159
Janaswamy, R., 147, 148, 159
Jansen, R. H., 154, 160
Jasik, H., 5
Jedlicka, R. P., 228, 268
Jemison, W. D., 204, 212
Jensen, O, 153, 159
Johansson, M., 457, 471
Johns, S. T., 206, 213
Johnson, R. C., 5, 185, 186, 211
Johnson, R. E., 456, 471
Johnson, W. A., 229, 268
Jones, E. M. T., 156, 160, 189, 213
Jones, W. R., 83, 104, 262, 268
Josefsson, L., 121, 125
Joubert, J., 142, 159
Joy, E. B., 456, 471, 472

Kahn, W. K., 222, 223, 259, 260, 268, 271, 372, 373, 382, 401, 449, 451
Kales, M. L., 341, 382
Kaminow, I. P., 139, 159
Kasilingam, D. P., 150, 161
Kastner, R., 156, 160
Kelleher, K. S., 333, 383
Kelly, A. J., 264, 268
Kerr, J. L., 147, 152, 153, 160
Keyser, R. B., 456, 471
Khac, T. V., 139, 160, 169, 213
Khamas, S. K., 326, 328
Kilburg, F. J., 264, 267

Kildal, P. S., 258, 267
Kim, Y., 100, 104
Kim, Y. S., 149, 160
Kim, Y. U., 123, 124, 125, 126
King, D. D., 97, 104
King. H. E., 133, 137, 160, 162, 224, 268
King, M. H., 44, 46
King, R. W. P., 128, 160
Kinsey, R. R., 83, 84, 104
Kirkpatrick, G. M., 78, 104
Kirschning, M., 154, 160
Knight, P., 5
Knittel, G. H., 4, 210, 212, 255, 256, 257, 261, 262, 263, 264, 268, 270, 271, 301, 421, 423, 426, 446, 448, 449, 450, 451, 471, 472
Knudesn, H. L., 485, 449
Koepf, G. A., 208, 213
Kong, F. N., 147, 158
Kong, J. A., 156, 162
Kornbau, T. W., 265, 269
Koster, N. H. L., 154, 160
Kotthaus, U., 150, 160
Koul, S. K., 188, 213
Kouyoumjian, R. G., 422, 450
Kritikos, H. N., 330, 382
Kuhn, D. H., 256, 267
Kumar, A., 200, 213
Kummer, W. H., 168, 176, 210, 213, 271, 435, 438, 440, 443, 446, 449, 452
Kurss, H., 222, 268, 372, 382
Kurtz, L. A., 169, 211
Kwan. B. W., 228, 270

La Paz. L., 304, 329
Ladner, A. W., 1
Lam, P. T., 338, 381
Lancaster, M. J., 326, 328
Langdon, H. S., 156, 160
Lange, J., 189, 213
Langsford, P. A., 456, 471
Larson, C. J., 246, 268
Larson, R. W., 445, 451
LaRussa, F. J., 372, 382
Lawson, J. D., 90, 91, 92, 105

Leach, W. M., 471, 472
Lechtreck, I. W., 255, 268
Lee, J. J., 190, 202, 204, 207, 213, 258, 268, 358, 363, 382, 456, 472
Lee, K. F., 150, 160
Lee, K.-M., 250, 267
Lee, K. S., 397, 449
Lee, S. W., 5, 25, 46, 133, 161, 262, 268, 380, 382, 385, 420, 449
Lemanczyk, J., 456, 472
Lepage, W. R., 397, 448
Lerner, D. W., 210, 212
Lessow, H. A., 223, 265
Levin, D., 250, 268, 280, 302
Levy, R., 84, 104, 189, 213
Lewin, L., 373, 382
Lewis, J. T., 125, 126
Lewis, W. D., 78, 104
Libelo, L. F., 3, 6
Liebman, P. M., 361, 382
Lim, J. C., 385, 449
Lindenblad, N. E., 1
Lioutas, N., 147, 159
Liu, C. C., 246, 247, 248, 249, 268, 269
Lo, Y. T., 5, 25, 46, 96, 97, 98, 102, 104, 127, 132, 133, 153, 155, 160, 161, 228, 269, 318, 324, 329, 380, 382, 385, 445, 449, 450
Long. S. A., 169, 213
Longstaff, I. D., 385, 394, 449
Lopez, A. R., 83, 104
Lowe, R., 331, 381
Lubin, Y., 249, 269
Luebbers, R. J., 156, 160, 246, 269
Luzwick, J., 229, 269, 273, 302
Lyon, J. A. M., 229, 230, 269
Lyon, T. J., 453

Ma, M. T., 5, 385, 449
MacPhie, R. H., 445, 450
MacReynolds, K., 456, 472
Magill, E. G., 263, 264, 269
Mahapatra, S., 147, 161
Mailloux, R. J., 5, 34, 46, 257, 265, 269, 372, 382, 458, 466, 472

Malech, R. G., 144, 161, 222, 235, 238, 244, 270, 273, 278, 302, 420, 450
Malherbe, J. A. G., 142, 160, 384, 450
Maloney, J. G., 133, 160
Manges, J., 249, 265
Marcum, J. L., 468, 472
Marcuvitz, N., 424, 447
Marin, M., 229, 266, 269
Martin, A. M., 327, 328
Martin, N. M., 229, 269
Matthaei, G. L., 156, 160
Matthew, P. J., 206, 211
Mayhan, J. T., 330, 382
McCormick, A. H. I., 143, 161
McFarland, J. L., 337, 338, 358, 380, 381, 382
McGee, W. F., 468, 472
McNamara, D. A., 84, 85, 86, 87, 104, 384, 450
Medhurst, R. G., 39, 45
Meier, P. J., 458, 461, 471
Metzen, P., 331, 382
Miller, C. J., 29, 46
Miller, G. A., 304, 329
Miller, T. W., 5
Milne, K., 5
Mink, J. W., 155, 158
Moeschlin, L., 121, 125
Moffet, A. T., 102, 104
Mohammadian, A. H., 228, 229, 269
Monser, G., 150, 160
Montgomery, C. G., 214, 222, 269
Monzingo, R. A., 5
Moody, H. J., 333, 382
Moreno, E., 102, 103
Moreno, T., 326, 329
Morgan, S. P., 323, 328
Mosig, J. R., 156, 158, 246, 269, 273, 303
Mucci, R. A., 123, 126
Munger, A. D., 387, 391, 420, 443, 450
Munk, B. A., 246, 263, 265, 268, 269, 270, 273, 280, 281, 302, 303
Munson, R. E., 150, 160

Murray, W., 92, 104, 123, 125
Musa, L., 354, 355, 382

Nakajima, T., 193, 212
Neff, H. P., 393, 448, 451
Nehra, C. P., 249, 269
Neikirk, D. P., 151, 161
Nelson, E. A., 30, 46
Nelson, J. A., 129, 133, 158, 160
Newman, E. H., 155, 161, 228, 270, 323, 329
Ng, K. T., 263, 265, 270
Ng, W. W., 202, 213
Nocedal, J., 123, 126
Nolen, J. N., 335, 382

Odlum, W. J., 209, 213
Oliner, A. A., 4, 139, 143, 144, 161, 222, 235, 238, 244, 255, 256, 257, 259, 267, 268, 270, 273, 278, 301, 302, 420, 421, 423, 446, 448, 450, 451, 472
Oltman, H. G., 157, 161
Olver, A. D., 5
Orchard, H. J., 96, 104
Orlow, J. R., 106, 125
Oseen, C. W., 304, 329
Ostroff, E. D., 190, 213

Packard, R. F., 97, 104
Page, H., 385, 450
Panicali, A. R., 445, 450
Parini, C. G., 273, 301
Paris, D. T., 456, 471, 472
Park, P. K., 139, 145, 161, 184, 213
Patel, D. P., 363, 382
Pathak, D. C., 422, 450
Pathak, P. H., 266, 269
Patton, W. T., 384, 451
Pearson, L. W., 228, 270
Pecina, R. G., 210, 211
Perini, J., 91, 104, 122, 126
Perrott, R. A., 273, 302
Pistolkors, A. A., 1
Pistorius, C. W. I., 384, 450
Poe. M. T., 228, 268
Pool, S. E., 39, 45

Povinelli, M. J., 149, 161
Powell, M. J. D., 91, 103, 123, 125
Powers, E. J., 78, 105
Pozar, D. M., 147, 153, 157, 159, 161, 228, 247, 248, 265, 270, 273, 302, 464, 470
Prasad, S. N., 147, 161
Pridmore-Brown, D. C., 413, 443, 448, 450
Pritchard, R. L., 305, 329
Provencher, J. H., 426, 450
Pues, H., 153, 162

Rahim, T., 388, 450
Rahmat-Samii, Y., 456, 472
Rao, J. B. L., 84, 105, 356, 357, 362, 363, 382
Rao, K. V. S., 150, 158
Rasmussen, H. H., 260, 265, 373, 381
Raudenbush, E., 296, 302
Reale, J. D., 250, 270
Rebeiz, G. M., 150, 161
Redlich, R. W., 97, 105, 385, 450
Ree, I. S., 468, 470
Reeves, C. M., 91, 103, 123, 125
Reich, H. J., 130, 158, 160
Reid, D. G., 304, 329
Rengarajan, S. R., 103, 143, 161, 258, 270, 417, 450
Repjar, A. G., 457, 458, 471, 472
Rhodes, D. R., 232, 234, 235, 270, 311, 321, 329
Ribero, J. M., 443, 447
Riblet, H. J., 57, 105, 189, 213, 304, 317, 329
Ricardi, L. J., 363, 381
Richards, W. F., 153, 160
Richie, J. E., 330, 382
Richmond, J. H., 226, 270
Rispin, L. W., 133, 161
Roberts, J., 144, 158
Roederer, A., 273, 302
Rondinelli, L. A., 466, 468, 472
Roscoe, A. J., 273, 302
Rosenberg, T. J., 337, 339, 381
Rothenberg, C., 362, 383
Rothman, M., 62, 105

Rotman, W., 341, 343, 382
Royer, G. M., 385, 450
Rudduck, R. C., 115, 126
Rudge, A. W., 5, 330, 382
Rutledge, D. B., 150, 161
Ruze, J., 342, 363, 382, 465, 472

Sabban, A., 156, 160
Safavi-Nairi, S., 420, 449
Salzer, H. E., 58, 105, 318, 329
Sandler, S. S., 97, 105
Sangster, A. J., 143, 145, 161
Scharp, G. A., 54, 67, 103
Scharstein, R. W., 276, 301
Schaubert, D. H., 147, 148, 150, 157, 159, 162, 247, 248, 251, 252, 253, 254, 266, 270, 271
Schelkunoff, S. A., 1, 48, 105, 146, 162, 224, 270, 304, 329, 461, 472
Schiffman, B. M., 189, 213
Schjaer-Jacobsen, H., 223, 265
Schrank, H. E., 445, 450
Schroeder, K. G., 337, 383
Schultz, F. V., 384, 451
Schuss, J. J., 337, 382
Schwartzman, L., 361, 382, 401, 451
Scott, W. R., Jr., 133, 160
Scotto, M., 443, 447
Seaton, A. F., 435, 443, 449
Secrest, D., 281, 303
Sengupta, D. L., 146, 147, 153, 162, 445, 451
Senior, T. B. A., 420, 424, 447
Shafai, L., 157, 162
Shanks, D., 250, 271, 280, 302
Shapira, J., 420, 451
Sharp, E. D., 25, 46
Sheleg, B., 394, 429, 451
Shelton, J. P., 333, 343, 369, 383
Shestag, L. N., 394, 451
Shimizu, J. K., 189, 213
Shin, J., 252, 253, 254, 270, 271
Shmoys, J., 246, 247, 248, 249, 268, 269
Shubert, K. A., 280, 302
Shuley, N. V., 147, 159
Sichak, W., 150, 158

Sigelmann, R. A., 445, 447
Silver, S., 443, 445, 446
Silvestro, J. W., 273, 302
Simons, R. N., 148, 149, 162
Singh, R., 250, 271, 280, 302
Singh, S., 250, 271, 280, 302
Sinnott, D. H., 385, 451
Skahill, G., 394, 451
Skinner, J. P., 281, 302, 303
Skolnik, M. I., 468, 472
Skiverik, A. K., 273, 303
Smith, G. S., 133, 160
Smith, M. S., 33, 46, 193, 213, 337, 350, 354, 355, 382, 383
Smith, P., 145, 161
Smith, T. M., 445, 451
Smith, W. T., 330, 383
Sneddon, N., 299, 303
Sohtell, V., 121, 125
Sokolnikoff, E. S., 305, 329
Sokolnikoff, I. S., 305, 329
Solbach, K., 464, 472
Sole, G. C., 337, 383
Solomon, D., 153, 160
Solymar, L., 316, 323, 324, 329
Soref, R. A., 203, 205, 207, 213
Southworth, G. C., 1, 8, 46
Spellmire, R. J., 63, 103
Sphicopoulos, T., 156, 158
Spradley, J. L., 209, 213
Stangel, J., 362, 382
Stark, P. A., 89, 93, 105, 170, 213
Stavis, G., 129, 160
Stearns, S. D., 5
Stegen, I. A., 226, 265
Stegen, L., 101, 102
Stegen, R. J., 53, 105, 139, 141, 159, 305, 329
Stein, S., 369, 376, 383
Steinberg, B. D., 97, 100, 105, 466, 472
Stellitano, P., 144, 162
Sterba, E. J., 1
Stern, G. J., 93, 94, 95, 96, 103, 104, 139, 161
Stevenson, A. F., 138, 162
Stewart, G. E., 413, 443, 448, 450

Steyskal, H., 5, 219, 260, 271, 273, 303
Stockton, R., 445, 448
Stroud, A. H., 281, 303
Stutzman, W. L., 330, 383, 466, 470
Sullivan, P. L., 157, 162
Sureau, J. C., 420, 421, 448, 451
Suzuki, Y., 193, 212, 228, 267
Svensson, B., 457, 471

Taflove, A., 148, 162
Tai, C. T., 169, 213, 307, 329
Tamir, T., 205, 214
Tang, R., 261, 271, 358, 383
Taylor, T. T., 50, 59, 105, 112, 126
Temes, G. S., 92, 105, 123, 126
Temme, D. H., 188, 212
Thiele, E., 148, 162
Thiele, G. A., 443, 451
Thomas, D. T., 369, 370, 383
Thomas, R. K., 44, 46, 97, 104
Tillman, J. D., 4, 385, 393, 448, 451
Tomar, R. S., 150, 158
Tomasic, B., 420, 421, 448, 452
Tong, D. T. K., 207, 214
Toughlian, E. M., 200, 208, 214
Trunk, G. V., 363, 382
Tsandoulas, G. N., 261, 262, 263, 271
Tseng, F.-I., 123, 126, 182, 214
Tufts, D. W., 125, 126
Tukey, J. W., 71, 103
Tulintseff, A. N., 156, 162
Tulyathan, P., 155, 161
Tun, S. M., 273, 301
Turner R. F., 341, 343, 382
Tyrell, W. A., 189, 214

Unz, H., 97, 105
Uslenghi, P. L. E., 420, 424, 447
Usoff, J. M., 273, 303
Uzkov, A.I., 307, 329
Uzsoky, M., 316, 323, 324, 329
Uzunoglu, N. K., 153, 162

Valentino, P. A., 362, 383
Van de Capelle, A., 153, 162
Van Der Maas, G. J., 53, 105
Vandesande, J., 153, 162

Varon, D., 259, 271
Ventresca, P., 150, 159
Villeneuve, A. T., 74, 105, 264, 271, 440, 443, 446, 449, 452
Voges, R. C., 325, 329
Von Aulock, W. H., 20, 46
Vowinkel, B., 150, 160

Wait, J. R., 397, 398, 399, 452
Walmsley, T., 1
Walsh, J. E., 389, 452
Walters, L. C., 385, 449
Wasylkiwskyj, W., 222, 223, 260, 271, 273, 303, 373, 383
Waterhouse, R. B., 249, 271
Watson, G. N., 36, 46, 424, 452
Watson, W. H., 137, 143, 144, 162, 165, 214
Wentworth, R. H., 205, 214
Westlake, J. R., 231, 271
Whaley, C. C., 281, 303
Wheeler, H. A., 215, 216, 217, 219, 232, 234, 235, 263, 264, 269, 271, 461, 463, 472
Whicker, L. R., 188, 214
White, W. D., 367, 383, 394, 451
Whittaker, E. T., 36, 46
Widrow, B., 5
Wilkinson, E. J., 189, 214
Willey, R. E., 101, 105
Williams, L. I., 456, 472
Williams, W. F., 337, 383
Wilwerth, F. G., 456, 470
Wimp, J., 226, 271
Winter, C. F., 369, 383
Wolff, E. A., 132, 162
Wolfson, R. I., 184, 212
Wong, J. L., 133, 137, 160, 162
Wong, N. S., 261, 271, 272, 358, 383
Woodward, P. M. 90, 91, 92, 105
Woodyard, J. R., 304, 308, 328
Wu, C., 156, 162
Wu, C. P., 4, 219, 255, 261, 262, 265, 267, 272
Wu, M. C., 207, 214
Wunsch, A. D., 133, 162
Wunsch, G., 253, 270

Yaccarino, R. G., 456, 472
Yang, R. F. H., 385, 448
Yang, X. H., 157, 162
Yaru, N., 323, 329
Yee, H. Y., 139, 144, 162, 169, 214
Yngvesson, K. S., 147, 149, 160, 163
Yost, T., 204, 212
Young, L., 156, 160, 189, 214
Young, L. B., 189, 214

Zai, D. Y. F., 92, 105, 123, 126
Zhang, Q., 157, 163
Zmuda, H., 200, 207, 208, 214
Zucker, F. J., 447, 472
Zurcher, J.-F., 150, 163
Zysman, G. I., 259, 271

Subject Index

Active element pattern, 2
Active impedance, 219
Admittance, *see* Mutual admittance
Aperture distribution, *see* Pattern synthesis
Aperture efficiency, *see* Directivity
Array diagnostics, 455
Array distribution, *see* Pattern synthesis
Array efficiency, *see* Directivity
Array feeds, 164
  corporate, 188, 196
  distributed, 189
  frequency scan, 181
  phasers, 184
  photonic, 200
  resonant, 164
  series, 164, 195
  shunt, 188, 196
  standing wave, *see* Resonant
  systematic errors, 208
  travelling wave, 170
  two-dimensional, 195
Array geometry
  arc, 389, 422
  conformal, 384
  conical, 430
  cylindrical, 395
  distributed, 189
  finite, 273
  finite-by-infinite, 276
  linear, 7, 47
  minimum redundancy, 102
  planar, 41, 106
  ring, 385
  semi-infinite, 273
  spherical, 445
  small, 281
  thinned, 96
Array pattern, *see* Pattern synthesis
Array topology
  Blass matrix, 335
  bootlace lens, 356
  Butler matrix, 331, 394
  digital, 364
  dome lens, 361
  endfire, 10, 308, 315, 438
  McFarland matrix, 337
  multiple beam 330
  Rotman lens, 338, 341
  smart skin, 384
  see also Array feeds
Average sidelobe level, 73, 465

Baffles, 257
Balun, 129, 149
Bayliss difference pattern, 78, 116, 455
Beam cophasal excitation, 393
Beam efficiency, 61, 70, 111, 115
Beam interpolation, 370
Beam pointing error, 25, 466
Beam port, 342
Beam squint, 16, 171

SUBJECT INDEX

Beamforming network, 330
Bickmore-Spellmire pattern, 63
Blass matrix, 335
Blind angles, 250, 254
Bootlace lens, 356
Bragg cell, 204
Butler matrix, 331, 394

Canonical minimum scattering antennas, 222
Carter mutual impedance, 224, 244
Chebyshev polynomial, 51
Clavin slot, 264
Conformal array, 384
Conical array, 430
Constrained optimization, 317
Convergence acceleration, 280
Corporate feed, 188, 196
Current sheet, 215
Cylindrical array, 395
   beam co-phasal excitation, 393
   comparison with planar, 424
   grating lobes, 401
   sector, 422
   slot element, 397
   slot mutual admittance, 411

Decollimation, 33
Depolarization, 407, 435
Difference pattern, 77, 78, 83, 84, 106, 121
Digital beamforming, 364
Dilation factor, 65, 75
Dipole, 127, 224, 235, 250
Dipole arrays, 40
   collinear, 41, 379
   parallel, 40, 313, 380
Direction cosine, 18
Directivity
   Bayliss, 82
   Circular Bayliss, 120
   Dolph-Chebyshev, 55, 319
   endfire, 44, 308
   Hansen One-Parameter, 110
   Hansen-Woodyard, 308
   linear array, 34, 38, 62, 305
   planar array, 41, 45
   sector or arc array, 422
   superdirective, 305, 317, 325
   Taylor circular $\bar{n}$, 115
   Taylor $\bar{n}$, 68
   Taylor One-Parameter, 59
   Villeneuve, 76
Dispersion, 3
Dispersive fiber, 205
Distributed feed, 180
Dolph-Chebyshev pattern, 51, 57, 317, 406
Dome lens, 361
Dynamic range, see Spur free dynamic range

Edge effects, 276, 286
Edge slot, 143
Effective aperture, 45
Efficiency, see Array efficiency
Electronic scan, 193, 195
Element port, 342
Elements, 127
   bow-tie dipole, 133
   Clavin, 264
   dipole, 39, 127, 224, 235, 250
   folded dipole, 132
   horn, 228
   isotropic, 36
   microstrip patch, 150, 156, 228, 246
   open sleeve dipole, 133
   open waveguide 145, 262
   short dipoles, 38, 234
   stripline slot, 145
   TEM horn, 146, 252
   waveguide slot, 137, 143, 145, 235
Elliott synthesis, 93
Errors, 208, 316, 350, 357, 464
External wave filter, 264

Fano matching limits, 156
Ferroelectric lens, 362
Fiber optics, see Photonic feed
Finite array, 273
Finite-by-infinite array, 276
Flat plane slot array, 106, 191
Floquet's theorem, 235
Forced excitation, 217

Fractal array, 100
Free excitation, 217
Frequency scan, 181

Generalized pattern function, 246
Gibbsian model, 297
Grating lobe series, 232
Grating lobes, 11, 20, 23, 296, 391, 401

Hamming pattern, 71
Hansen circular pattern, 107
Hansen-Woodyard, 308
High temperature superconductors, see superconductors
Horn, 229, see TEM horn

Ideal distribution, 65
Ideal element pattern, 217
Impedance, see Mutual impedance
Impedance crater, 233
Impedance matching, see Matching
IP3, see Spur free dynamic range

Large array method, 274
Leaky wave, 256
Least P^{th} optimization, 92, 123
Levin transform, 280
Linear array, 7
  bandwidth, 15, 164, 309
  beam efficiency, 61, 70
  beam pointing, 16, 171, 182, 466
  beamwidth, 9, 54, 60, 65
  directivity, 34, 56, 62, 68, 305, 308
  grating lobes, 11
  patterns, 7, 47, 53, 60, 66, 79, 81, 86
  quantization lobes, 26, 31
  sidelobes, 11, 59, 64, 467
Longitudinal slot, 138
Lossy dipole, 132, 316
Low sidelobes, 71, 180, 365, 453

Matching, 156, 325
McFarland 2-D matrix, 337
Measurements, 453, 455
Microstrip patch, 150, 156, 228, 246, 249

Minimum redundancy array, 102
Minimum scattering antennas, 222, 259
Modal series, 413, 445
Module, 190, 194, 197, 200
Multimode elements, 261
Multiple beam array, 330
  beam interpolation, 370
  Blass matrix, 335
  bootlace lens, 356
  Butler matrix, 331, 394
  McFarland matrix, 337
  Nolen matrix, 335
  orthogonality, 372
  power divider, 331
  Rotman lens, 330, 341
Mutual admittance, 139, 411, 417
Mutual coupling, see Mutual impedance
Mutual impedance, 215, 223, 228

Newton-Raphson iteration, 89, 93, 170, 180
Nolen matrix, 335

Open waveguide element, 145
Optimization, 317
  directivity, 305, 308
  gradient, 91, 97, 122
  sum and difference pattern, 83, 106, 121, 191
Orthogonal beams, 367, 372
Overlapped sub-array, 35, 363

Parasitic patch, 156, 249
Patch, see Microstrip patch
Pattern linear array, 7, 47, 53, 60, 66, 74, 79
Pattern synthesis, 47
  Bayliss, 78, 116
  Bickmore-Spellmire, 63
  circular Bayliss, 116
  difference, 77, 84
  Dolph-Chebyshev, 51, 317
  Elliott, 93
  Hamming, 71
  Hansen One-Parameter, 107

SUBJECT INDEX **485**

Hansen-Woodyard, 308
low sidelobe, 71
Orchard, 96
ring sidelobe, 123
shaped beams, 90
sidelobe shaping, 85
superdirective, 305
Taylor circular n̄, 112
Taylor n̄, 64
Taylor One-Parameter, 59
Taylor principles, 50
unit circle, 48
Villeneuve n̄, 74
Woodward-Lawson, 90
Zolotarev, 84
Periodic cell, *see* Unit cell
Phase shifter, *see* Phaser
Phased array, 1
Phaser, 184
    quantization lobes, 26
    scan, 184
Photonic Feed, 200
    acousto-optical switched, 204
    binary switched delay, 202
    dispersive fiber, 205
    optical FT, 207
Planar array, 17
    beam efficiency, 110
    beam steering, 25, 466
    beamwidth, 18, 109, 113
    comparison with sector array, 424
    design, 177
    directivity, 41, 45, 110, 115, 120
    grating lobes, 20, 23, 296
    hexagonal lattice, 23
    pattern synthesis, 106, 112, 116, 123
    patterns, 17
    quantization lobes, 29
Poisson sum formula, 245, 389
Printed circuit antennas, *see*
    Microstrip patch, TEM horn
Printed horn, *see* TEM horn

Q, 256, 310, 319
Quantization lobes, 26
    decollimation, 33
    phaser, 26

sub-array, 31

Range equation, 3
Resistive tapering, 366
Resonant array, 164, 168
Ring array, 385
Ring array grating lobe, 391
Rotman lens, 338, 341

Scan compensation, 257
    baffles, 257
    external filters, 264
    multimodes, 261
    networks, 258
    slot with monopoles, 264
    WAIM, 263
Scan element pattern, 2, 218, 219, 274, 420
Scan impedance, 2, 217, 219, 238, 274
Scan reactance, 237
Scan reflection coefficient, 218
Scan resistance, 221, 237
Scattering scan element pattern, 2
Sector array, 42
Semi-infinite array, 273
Sequential excitation, 193
Series feed, 164, 195
Serpentine feed, *see* Frequency scan
Shanks transform, 280
Shaped beam, 90
Shunt feed, 188, 196
Shunt slot, 138
Sidelobe ratio, 50
Sidelobe shaping, 85
Sidelobes, 28, 71, 365, 453, 465, 468
Signal dispersion, 3
Simulator, *see* Waveguide simulator
Slope difference pattern, 82
Slot, 137, 234, 397, 413
Slot admittance, 139, 144, 411, 417
Slot impedance, 142
Smart skin array, 384
Snake feed, *see* Frequency scan
Space factor, 8; *also see* Pattern synthesis
Space tapered array, 101
Spatial domain, 224, 278

486 SUBJECT INDEX

Spectral domain, 235, 280
Spherical array, 445
Spur free dynamic range, 203
Staggering elements, 402, 434
Stripline slot, 145
Sub-array, 190
 difference pattern, 83
 overlapped, 35
 quantization lobes, 31
Superconductors, 316, 326
Superdirective array, 305, 317, 325
Surface wave, 255
Systematic errors, 208
Systems factors, 2

Tandem feed, 83
Tapered slot, *see* TEM horn
Taylor circular n̄ pattern, 112
Taylor n̄ pattern, 64, 454
Taylor One-Parameter pattern, 59
Taylor principles, 50
TEM horn, 146, 252
Thinned array, 96, 101

Tolerances, *see* Errors
Travelling wave array, 170
Travelling wave conductance, 175, 177

Unit cell, 222, 235, 254, 261, 458
Unit circle, 48, 89, 95, 235

Villeneuve n̄ patterns, 74
Visible space, 13
Vivaldi, *see* TEM horn

Waveguide lens, 364
Waveguide, open, 145
Waveguide simulator, 254, 458
 many element, 464
 one port, 463
Waveguide slot, *see* Slot
Wide-angle impedance matching, 263
Woodward-Lawson pattern, 90

Zolotarev difference pattern, 84

WILEY SERIES IN MICROWAVE AND OPTICAL ENGINEERING

KAI CHANG, Editor
Texas A&M University

PHASED ARRAY ANTENNAS • *R. C. Hansen*
FIBER-OPTIC COMMUNICATION SYSTEMS, Second Edition • *Govind P. Agrawal*
COHERENT OPTICAL COMMUNICATIONS SYSTEMS • *Silvello Betti, Giancarlo De Marchis and Eugenio Iannone*
HIGH-FREQUENCY ELECTROMAGNETIC TECHNIQUES: RECENT ADVANCES AND APPLICATIONS • *Asoke K. Bhattacharyya*
COMPUTATIONAL METHODS FOR ELECTROMAGNETICS AND MICROWAVES • *Richard C. Booton, Jr.*
MICROWAVE RING CIRCUITS AND ANTENNAS • *Kai Chang*
MICROWAVE SOLID-STATE CIRCUITS AND APPLICATIONS • *Kai Chang*
DIODE LASERS AND PHOTONIC INTEGRATED CIRCUITS • *Larry Coldren and Scott Corzine*
MULTICONDUCTOR TRANSMISSION-LINE STRUCTURES: MODAL ANALYSIS TECHNIQUES • *J. A. Brandão Faria*
PHASED ARRAY-BASED SYSTEMS AND APPLICATIONS • *Nick Fourikis*
FUNDAMENTALS OF MICROWAVE TRANSMISSION LINES • *Jon C. Freeman*
MICROSTRIP CIRCUITS • *Fred Gardiol*
HIGH-SPEED VLSI INTERCONNECTIONS: MODELING, ANALYSIS, AND SIMULATION • *A. K. Goel*
HIGH-FREQUENCY ANALOG INTEGRATED CIRCUIT DESIGN • *Ravender Goyal (ed.)*
MICROWAVE APPROACH TO HIGHLY IRREGULAR FIBER OPTICS • *Huang Hung-chia*
NONLINEAR OPTICAL COMMUNICATION NETWORKS • *Eugenio Iannone, Antonio Mecozzi, Francesco Matera, and Marina Settembre*
FINITE ELEMENT SOFTWARE FOR MICROWAVE ENGINEERING • *Tatsuo Itoh, Giuseppe Pelosi and Peter P. Silvester (eds.)*
OPTICAL COMPUTING: AN INTRODUCTION • *M. A. Karim and A. S. S. Awwal*
INTRODUCTION TO ELECTROMAGNETIC AND MICROWAVE ENGINEERING • *Paul R. Karmel, Gabriel D. Colef, and Raymond L. Camisa*
MILLIMETER WAVE OPTICAL DIELECTRIC INTEGRATED GUIDES AND CIRCUITS • *Shiban K. Koul*
MICROWAVE DEVICES, CIRCUITS AND THEIR INTERACTION • *Charles A. Lee and G. Conrad Dalman*
ADVANCES IN MICROSTRIP AND PRINTED ANTENNAS • *Kai-Fong Lee and Wei Chen (eds.)*
OPTOELECTRONIC PACKAGING • *A. R. Mickelson, N. R. Basavanhally, and Y. C. Lee (eds.)*
ANTENNAS FOR RADAR AND COMMUNICATIONS: A POLARIMETRIC APPROACH • *Harold Mott*
INTEGRATED ACTIVE ANTENNAS AND SPATIAL POWER COMBINING • *Julio A. Navarro and Kai Chang*
FREQUENCY CONTROL OF SEMICONDUCTOR LASERS • *Motoichi Ohtsu (ed.)*
SOLAR CELLS AND THEIR APPLICATIONS • *Larry D. Partain (ed.)*
ANALYSIS OF MULTICONDUCTOR TRANSMISSION LINES • *Clayton R. Paul*
INTRODUCTION TO ELECTROMAGNETIC COMPATIBILITY • *Clayton R. Paul*

INTRODUCTION TO HIGH-SPEED ELECTRONICS AND OPTOELECTRONICS •
Leonard M. Riaziat
NEW FRONTIERS IN MEDICAL DEVICE TECHNOLOGY • *Arye Rosen and Harel Rosen (eds.)*
NONLINEAR OPTICS • *E. G. Sauter*
FREQUENCY SELECTIVE SURFACE AND GRID ARRAY • *T. K. Wu (ed.)*
ACTIVE AND QUASI-OPTICAL ARRAYS FOR SOLID-STATE POWER COMBINING •
Robert A. York and Zoya B. Popović (eds.)
OPTICAL SIGNAL PROCESSING, COMPUTING AND NEURAL NETWORKS • *Francis T. S. Yu and Suganda Jutamulia*